Special Relativity

Special Relativity

A Heuristic Approach

Sadri Hassani

University of Illinois at Urbana-Champaign, Urbana, IL, USA

Illinois State University, Normal, IL, USA

Elsevier
Radarweg 29, PO Box 211, 1000 AE Amsterdam, Netherlands
The Boulevard, Langford Lane, Kidlington, Oxford OX5 1GB, United Kingdom
50 Hampshire Street, 5th Floor, Cambridge, MA 02139, United States

Library of Congress Cataloging-in-Publication Data
A catalog record for this book is available from the Library of Congress

British Library Cataloguing-in-Publication Data
A catalogue record for this book is available from the British Library

ISBN: 978-0-12-810411-8

For information on all Elsevier publications
visit our website at https://www.elsevier.com/books-and-journals

Publisher: John Fedor
Acquisition Editor: Anita Koch
Editorial Project Manager: Amy Clark
Production Project Manager: Anitha Sivaraj
Designer: Christian J. Bilbow

Typeset by VTeX

To Sarah, Dane, and Daisy

Contents

Preface

This is an intermediate textbook on special relativity written with mainly physics juniors and seniors in mind. However, a sophomore with a strong background in calculus and calculus-based introductory physics, or a graduate student who feels that (s)he lacks the necessary knowledge in special relativity, can also benefit from the book. Instructors will find a multitude of carefully chosen illustrative examples and close to three hundred end-of-chapter problems—whose detailed solutions are available upon purchase or adoption—extremely helpful pedagogical tools for teaching relativity.

Why write another special relativity textbook when there are so many (some excellent) books already available? The answer is (a) the approach and (b) the detailed coverage of topics that are only glossed over in the existing books. The approach is the emphasis and abundant use of the Lorentz transformation. The topics, emphasized in this book but mentioned only in passing in many others, include relativistic photography, interstellar travel (encompassing a detailed discussion of ground-based laser propulsion), spacetime triangle inequality and the proof of the fact that acceleration reduces aging, connection between special relativity, spin, and antimatter, etc.

At the heart of spacial relativity lies the Lorentz transformation. While length contraction and time dilation are less abstract, they can lead to "obvious" results which turn out to be incorrect. Lorentz transformation, on the other hand, once mastered and applied correctly, will always yield the right results, even though it is more abstract. This book relies almost exclusively on Lorentz transformations, and applies them to a large number of situations, some appearing for the first time in this book. To truly appreciate the relevance and power of Lorentz transformations, the novice needs to see not only their numerous applications, but also their origin and be shown that they are firmly based on experimental results.

Students learn new abstract concepts best when these concepts are connected—through a well-designed analogy—to familiar ideas. I believe that the most heuristic derivation of Lorentz transformation is to connect it to the concept of the relativistic spacetime distance, because the idea of "distance" is intuitively obvious to the beginner. So, if the familiar Euclidean distance is presented in such a way as to make a transparent contact with the spacetime distance, the latter will not be as mysterious as it first appears to be, and its byproduct, the Lorentz transformation, can then be appreciated.

Starting with some intuitive and obvious assumptions concerning distance in the two dimensions in which *flatlanders* live—and for whom "height" is not as directly accessible as length and width—the book derives the three-dimensional Euclidean distance between two points in terms of their coordinates. Then, assuming the invariance of this distance, it deduces the familiar orthogonal coordinate transformation. I have presented the derivation in such a way that the transition to spacetime becomes self-evident. Thus, following exactly the same procedure, I derive the Minkowskian distance and the corresponding transformation that respects the invariance of that distance: the Lorentz transformation.

The first chapter is conceptual and requires no math beyond high school algebra. It is a good intuitive introduction to relativity of simultaneity and length contraction. Chapter 2 derives the time dilation and length contraction formulas and applies them to various examples, including some unique and insightful applications to relativistic photography in anticipation of Chapter 7. Chapter 3 derives the invariant spacetime distance and Lorentz transformation for a one-dimensional relative motion after showing students how *flatlanders* can derive Euclidean distance even though they can't "feel" the third dimension of height.

Pictures are worth a thousand words. That's why geometry can be extremely useful when algebra becomes tedious. The geometry of two-dimensional spacetime is introduced in Chapter 4 and applied to many examples such as Doppler effect, re-derivation of Lorentz transformations, and the well known "train and tunnel" paradox. The spacetime triangle inequality—the fact that the sum of the lengths of two sides of a triangle is *less* than the length of the third side—is derived in this chapter and used to show that a straight line in spacetime geometry is the *longest* distance between two events. A direct consequence of this property is that inertial observers age faster than accelerated observers, thus explaining the twin paradox.

The natural relativistic substitute for the Newtonian universal time in kinematics is identified as the proper time. In Chapter 5, I generalize the ordinary concept of velocity in a plane (the derivative of displacement with respect to the universal time) to the velocity in a spacetime plane as the derivative of the spacetime displacement with respect to proper time. Spacetime momentum is the immediate consequence of spacetime velocity, and relativistic momentum and energy are identified as the components of spacetime momentum.

Chapter 6 defines the general Lorentz transformation using matrices and, with the help of Appendix B, extends the two-dimensional Lorentz transformations to the full four-dimensional spacetime. The concept of 4-vectors and their dot product are introduced as are the notions of 4-velocity, 4-acceleration, 4-momentum, 4-force, and the second law of motion. Various examples illustrate the power of 4-vectors and their dot products.

Chapter 7 is, as far as I know, unique to this book. It is the arena in which the power of Lorentz transformation and the limitation, even the incorrectness, of the use of the concepts of time dilation and length contraction is illustrated (see Problem 7.1 for a popular mistake). Conditions under which length contraction can be *photographed* are investigated, and the reason why ordinary photography of relativistically moving objects does not reveal this length contraction is elaborated. One of the outstanding results of this chapter is the *rigorous proof* that the photograph of a relativistically moving sphere in a pinhole camera is an ellipse *elongated* along the direction of motion, and that the image of such a sphere is a circle if (and only if) it moves directly toward or away from the camera. This does not contradict the results of Penrose and Terrell, who demonstrate that the *cone* converging on a *point observer* is circular. By actually constructing these circular cones, the chapter shows that, if the pinhole camera points at the center of the sphere, the image formed on its photographic plate cannot be circular. In fact, the chapter proves in three different ways that the image is an ellipse stretched in the direction of motion.

Chapter 8 sees an important application of 4-vectors and four-dimensional dot product to particle interactions. The notions of center of mass and lab frames are introduced and used in various examples to investigate particle collisions and decays. The chapter ends with a simple introduction to the Dirac's discovery of antimatter and the connection between special relativity and spin.

Chapter 9, another chapter unique to this book, applies four-momentum conservation of Chapter 8 to rocket propulsion. The motion of fuel carrying spacecraft is analyzed for massive as well as photon exhaust cases and their impracticality demonstrated. The alternative ground-based laser propulsion is covered in lengthy detail in light of the recent interest in such projects. Certain peculiarities of relativistic motion in one dimension, such as the fact that force is an invariant quantity, are pointed out and the importance of Lorentz covariance, a crucial prerequisite of the application of special relativity, is emphasized. It is then shown that the requirement of Lorentz covariance invalidates many existing approaches to the problem.

Chapter 10 is an elementary introduction to tensor algebra as applied to special relativity and used mainly in Chapter 11 on relativistic electrodynamics. The latter derives the Lorentz transformation for electric and magnetic fields and calculates the electric and magnetic fields of a uniformly moving charge starting from the Coulomb force of a static charge. Electromagnetic field tensor leads naturally to the Lorentz force law and an elegant way of writing Maxwell's equations.

Although somewhat outside its main thrust, the book ends with a very accessible introduction to the standard cosmological model. A few of the topics developed in the book are used to illustrate the importance of relativity in the physics of the early universe. Along the way, a number of ideas, not related to relativity, are discussed and the relevant equations derived in detail. So, Chapter 12 is also an accessible exposure to such topics as EM radiation in cavities, Planck formula of the black body radiation, Stefan-Boltzmann law, Friedmann equation, and the evolution of the early universe.

Discussions (sometimes heated!) with some of my colleagues clarified many subtle ideas of relativity and consolidated my belief that Lorentz transformation is the bedrock of special relativity, and should be at the core of teaching the subject to the novice. I would like to thank Robert Schrock for reading an early version of Chapter 7 and giving constructive comments on it. I also thank Sören Holst and John Mallinckrodt for many exchanges of email, which helped me decipher the intricacies of applying Lorentz transformations to various interesting examples, some of which appear throughout the book. Needless to say, I am solely responsible for the accuracy of the content of the book.

Sadri Hassani

Urbana, IL, USA
December 2016

List of Symbols, Phrases, and Acronyms

$\hat{a}$ unit vector in the direction of $\vec{a}$.

antiparticle a particle whose mass and spin are exactly the same as its corresponding particle, but the sign of all its "charges" are opposite. If a particle is represented by the letter p, then it is customary to denote its antiparticle by $\bar{p}$. If a particle is represented by the letter q^- (or q^+), then it is customary to denote its antiparticle by q^+ (or q^-).

as arcsecond; an arcsecond is an angle 1/3600 of a degree.

baryon a hadron whose spin is an odd multiple of $\hbar/2$. Baryons are composed of three quarks. Examples of baryons are protons and neutrons.

$\vec{\beta}$ fractional velocity of one observer relative to another, $\vec{\beta} = \vec{v}/c$.

boson a particle whose spin is an integer multiple of $\hbar$. All gauge particles are bosons as are all mesons, as well as the Higgs particle.

causally connected referring to two events. If an observer or a light signal can be present at two events, those events are said to be causally connected.

causally disconnected referring to two events. If an observer or a light signal cannot be present at two events, those events are said to be causally disconnected.

CBR Cosmic Background Radiation.

CM center of mass.

CS coordinate system.

$\hat{\mathbf{e}}_x, \hat{\mathbf{e}}_y, \hat{\mathbf{e}}_z$ unit vectors along the three Cartesian axes.

EM electromagnetic or electromagnetism, one of the four fundamental forces of nature.

equilibrium temperature temperature of the universe at which matter and radiation densities are equal.

eV electron volt, unit of energy equal to 1.6×10^{-19} J.

fermion a particle whose spin is an odd multiple of $\hbar/2$. Fermions obey Pauli's exclusion principle: no two identical fermions can occupy a single quantum state. Electrons, protons, and neutrons are fermions, so are all leptons and quarks, as well as all baryons.

γ the Lorentz factor, $\gamma = 1/\sqrt{1-\beta^2} = 1/\sqrt{1-(v/c)^2}$.

gauge bosons According to the modern theory of forces, fundamental particles interact via the exchange of gauge bosons. Excluding gravity, whose microscopic behavior is not well understood, there are 12 gauge bosons whose exchange explains all the interactions: Z^0, $W^{\pm}$ and γ (photon) are responsible for electroweak interaction, while 8 gluons are responsible for strong interaction.

gluons the particles responsible for strong interactions: two or more quarks participate in strong interaction by exchanging gluons. There are four gluons, which with their antiparticles comprise the eight gluons whose exchange binds quarks together.

GTR general theory of relativity; the relativistic theory of gravity.

Gyr gigayear, equal to 10^9 years.
hadron a particle capable of participating in strong nuclear interactions. Examples of hadrons are protons, neutrons and pions. All hadrons are made up of quarks and/or antiquarks.
half-life the time interval in which one half of the initial decaying particles survive.
LAV Law of Addition of Velocities.
lepton a particle that participates only in electromagnetic and weak nuclear interactions, but not in strong nuclear interactions. Leptons are elementary particles in the sense that they are not made up of anything more elementary. There are three electrically charged leptons: electron, muon, and tauon. Each charged lepton has its own neutrino. So, altogether there are six leptons.
LHC Large Hadron Collider.
light cone (at an event E) The set of all events that are causally connected to E.
light hour the *distance* that light travels in one hour, $\approx 1.08 \times 10^{12}$ m.
light minute the *distance* that light travels in one minute, $\approx 1.8 \times 10^{10}$ m.
light second the *distance* that light travels in one second, $\approx 3 \times 10^8$ m.
lightlike referring to two events, when $c\Delta t = \Delta x$ or $(\Delta s)^2 = 0$.
luminally connected referring to two events. If a light signal can be present at two events, those events are said to be luminally connected.
ly light year; one light year is 9.467×10^{15} m.
mean time the time interval in which $1/e$ of the initial decaying particles survive.
meson a hadron whose spin is an integer multiple of $\hbar$. Mesons are composed of one quark and one antiquark. Examples of mesons are pions.
MeV million electron volt, unit of energy equal to 1.6×10^{-13} J.
μm micrometer $= 10^{-6}$ m.
Minkowskian distance also called "spacetime distance,"

$$(\Delta s)^2 = (c\Delta t)^2 - (\Delta x)^2$$

is an expression involving the coordinates of two events which is independent of the coordinates used to describe those events.
Mly million light years.
MM clock sometimes called "light clock" is described on page 17.
Mpc Megaparsec.
muon an elementary particle belonging to the group of particles named "leptons," to which electron belongs as well. Muon is called a "fat electron" because it behaves very much like an electron except that it is heavier.
neutrino a neutral lepton with very small mass. Neutrinos participate only in weak nuclear force. That's why they are very weakly interacting.
ns nanosecond or 10^{-9} s.
Parsec a distance of about 3.26 light years. One parsec corresponds to the distance at which the mean radius of the Earth's orbit subtends an angle of one second of arc.
positron the antiparticle of the electron.
quarks elementary particles which make up all hadrons. There are six quarks: up, down, strange, charm, bottom, top. Quarks participate in all interactions, in particular, the strong interaction.
RF reference frame.
spacelike referring to two events, when $c\Delta t < \Delta x$ or $(\Delta s)^2 < 0$.
spacetime distance see Minkowskian distance.
STR special theory of relativity.
tauon an elementary particle belonging to the group of particles named "leptons," to which electron belongs as well. It is the heaviest lepton discovered so far.
timelike referring to two events, when $c\Delta t > \Delta x$ or $(\Delta s)^2 > 0$.

Note to the Reader

If you want to master spacial relativity, you'll have to know how to apply Lorentz transformations. They are relativistic coordinate transformations that connect two different observers. Sort of generalizations of the orthogonal transformations used in Euclidean geometry.

To master Lorentz transformations, you'll have to really understand the concept of an "event." Once you realize what an event is, the next (and hardest step) is to be able to visualize how that event looks in a "moving" reference frame. Consider two events: the launch of a spacecraft and its landing on an exoplanet. For an Earth observer, the launch occurs here and now, the spacecraft starts moving away at launch, the landing occurs there and then, and at landing, the craft approaches the exoplanet.

Now suppose that you are on the spacecraft. If you can imagine that *you are not moving*, that the launch occurs here and now, with the *Earth* starting to move away, that the landing occurs *here* and then, and that in landing the *exoplanet approaches* you, then you are halfway to understanding Lorentz transformations.

There are plenty of examples to train your mind to picture such Lorentz transformations. Go through them and try to reproduce the solutions *with the book closed*. Yes, with the book closed, because simply "following the solutions" in the book will not teach you Lorentz transformation (or physics, or math). Once you have gone through all the examples, do as many of the end-of-chapter problems as you can ... the more the merrier!

A Solutions Manual providing solutions to all of the end-of-chapter problems is available for download at https://www.elsevier.com/books-and-journals/book-companion/9780128104118.

CHAPTER 1

Qualitative Relativity

The seeds of relativity theory were planted on a spring day in a lecture hall on the campus of the University of Copenhagen in 1820. As Professor Hans Christian Örsted was demonstrating the power of large electric currents, he noted the deflection of a nearby compass needle as soon as a large current was established in the circuit. Thus he stumbled on one of nature's best-kept secrets, namely that electric currents can produce magnetic fields.

1.1 IT BEGAN WITH MAXWELL

You might think that because currents are produced by moving charges and because motion is a relative concept, there is already a connection between electromagnetism and relativity. After all, if you move with the charges producing the electric current, they will appear motionless to you and you should not detect any magnetic field.[1] While this is certainly true, it contains little more than the fact that the magnetic field disappears for an observer moving alongside the electric charges. To make contact with *the* theory of relativity, you need the full machinery of electromagnetism as described by Maxwell's equations.

If you have some familiarity with vector analysis and have not seen how Maxwell derived his famous equations, you *have to* read Appendix A. It truly affirms the "power of the mind," not in the way that mystic health gurus want you to believe, but the power instigated by mathematics and vindicated by countless practical uses of its implications. It also lays the foundation of relativity in a statement repeated here:

Second Postulate of Relativity

Note 1.1.1. *Electromagnetic waves travel at* $c = 299\,792\,458$ *m/s, the speed of light in vacuum, regardless of the motion of their source.*

[1] In a real wire, the motion of the negative electrons creates a current. If you move with the electrons, then the positive background charges move in the opposite direction, and the magnetic field will not disappear. I am not considering real wires, but the case in which only charges of one sign are present.

Special Relativity. DOI:10.1016/B978-0-12-810411-8.00001-8

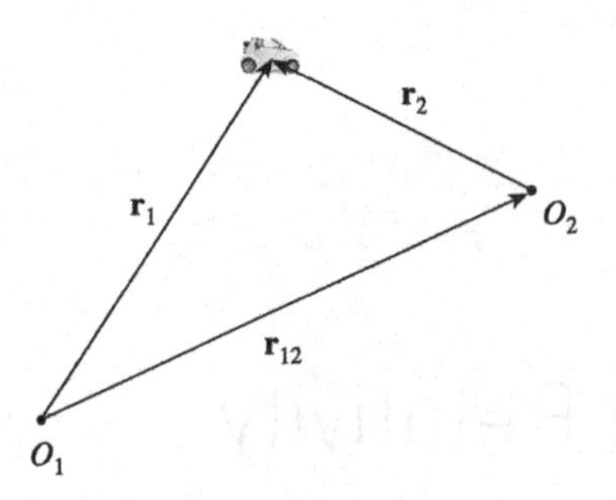

FIGURE 1.1 Instantaneous location of a moving object relative to two observers.

The value of the speed of light given here is exact. In 1983, the speed of light was given this exact value, while the unit of length, meter, was defined as the distance light travels in 1/299 732 458 second. Second itself is defined as 9 192 631 770 periods of the oscillation of certain transition of cesium 133 atom.

The statement in Note 1.1.1 is sometimes called the **second postulate of relativity**, and as you'll see, the entire theory of special relativity is based on it. For completeness, let's also state the **first postulate of relativity**:

First Postulate of Relativity

Note 1.1.2. *The laws of physics are the same for all inertial observers.*

A consequence of this postulate is that it is impossible to detect the (absolute) uniform motion of a reference frame (RF) through measurements done entirely in that RF. Here is why: If you *could* measure the absolute motion and you got a number as the speed of a particular RF, then by Note 1.1.2 this number should be the same for *all* RFs. But if all RFs have the same absolute speed, they would not be moving relative to one another.

The second postulate of relativity is in conflict with the intuitive classical notion of the **law of addition of velocities** (LAV). Figure 1.1 shows the instantaneous location of a moving object relative to observers O_1 and O_2. From your elementary mechanics course, you know that these position vectors satisfy $\mathbf{r}_1 = \mathbf{r}_{12} + \mathbf{r}_2$. Differentiating this equation with respect to time, you obtain

Law of Addition of Velocities

$$\mathbf{v}_1 = \mathbf{v}_{12} + \mathbf{v}_2, \tag{1.1}$$

which states that the velocity of the object relative to O_1 is the sum of the velocity of the object relative to O_2 and the velocity of O_2 relative to O_1. This is the LAV. As an example, suppose that Sonya is riding a train moving at 200 km/hr. She fires a bullet from her gun toward the front of the train. If the speed of the bullet relative to the gun is 300 km/hr, then Sam, standing on the ground outside, sees the bullet fly by at 500 km/hr.

Now suppose that Sonya, riding on a super train and holding her laser pointer outside the train, sends a beam of light toward the front of the train. Sam, who is standing outside, hundreds of kilometers away from the train (therefore, not even seeing the source of the beam), detects the laser light and measures its speed to be 299 792 458 m/s, regardless of how fast the train is moving! That's the essence of the second postulate of relativity in Note 1.1.1. On the other hand, LAV instructs our intuition to add the speed of the train to the speed of the light relative to the laser gun. This is the conflict between LAV and the second postulate.

The LAV was so firmly established in the nineteenth century that the notion of light traveling with the same speed relative to two different RFs—and therefore, the second postulate of relativity—was considered a logical fallacy! So, what was the alternative? One possibility was to consider the Earth as the special RF in which Maxwell's equations hold. But this was immediately dismissed because Copernicus, three centuries earlier, had already cautioned about the dangers of bestowing upon the Earth a special position in the universe. Moreover, the orbital motion of the Earth, with its semi-annual 180-degree directional flip, would have to have distorting effects on the light of stars and planets, which would be easily detectable. Lack of such distortions immediately rules out the Earth as a privileged frame. If the Earth is not the privileged RF, then, for the same reason, no other planet, nor the Sun nor any other star or galaxy can be privileged.

The rise and fall of ether theory.

The only remaining alternative was the "medium" surrounding the celestial bodies and filling the space between them. Just as water wave is the oscillation of water and sound is the oscillation of air, the undulation of this medium, called **ether**, manifested itself as light. The idea of ether was invented and used by Huygens as early as 1690, but it became the subject of intense scrutiny in the nineteenth century after Maxwell's prediction of the electromagnetic (EM) waves. Ether theory demanded so many strange and complicated notions that by the last quarter of the nineteenth century, physicists started to doubt its existence. The final blow to ether came through the works of Albert Michelson (1852–1931) and Edward Morley (1838–1923), one of whose crucial experiments showed that there was no detectable motion of Earth relative to ether, thereby undermining the theory and the concept of ether.

1.1.1 Einstein and Signal Velocity

When he was only 16 years old in 1895, Einstein wondered how the world would appear to him if he could catch up with light; in particular, what would light itself look like. You can appreciate the intriguing nature of this question by an analogy. Imagine that you are quietly moving over circular water waves at exactly the same speed as their expansion rate. How will the waves appear to you? Concentrate only on the local waves[2] and assume that you ride just

How would water waves look if you caught up with them?

[2] Clearly you cannot move with all the waves as they go in different directions.

above a crest. You see the crest right underneath you, and since you move with it at exactly the same speed, it will remain underneath. The other crests and troughs in your immediate vicinity also move at the same speed, and will also remain static. Thus, locally at least, the wave ceases to exist! It will appear as a static and permanent deformation of water surface with no motion and no oscillation.

How would light look if you caught up with it?

If you could catch up with light, and if ether really existed, then the ether wave (i.e., light) would also cease to exist! What is even more intriguing is that, since ether was supposedly invisible (transparent), there would be no trace of light left! Although in the case of water waves, a static deformation of water, the medium of transmission, was left to remind you of the once traveling wave, no such reminder will remain in the case of light. As soon as you catch up with it, light will be completely gone!

In the intervening years between 1895 and 1905, the question of the behavior of light and motion was constantly on Einstein's mind. As his scientific and mathematical abilities matured, the question took on the more definite form of how light behaved in moving frames. In the conflict between Maxwell's constancy of the speed of light and Newton's LAV, Einstein took sides with Maxwell and resolved the conflict by the analysis of the concept of time. Here is Einstein's account of his discovery of STR:

Einstein explains how he created STR.

> Why do these two concepts [LAV and Maxwell's prediction] contradict each other? I realized that this difficulty was really hard to resolve. I spent almost a year in vain trying to modify the idea of Lorentz in the hope of resolving this problem.
>
> By chance a friend of mine in Bern (Michele Besso) helped me out. It was a beautiful day when I visited him with this problem. I started the conversation with him in the following way: "Recently I have been working on a difficult problem. Today I come here to battle against that problem with you." We discussed every aspect of this problem. Then suddenly I understood where the key to this problem lay. Next day I came back to him again and said to him, without even saying hello, "Thank you. I've completely solved the problem." An analysis of the concept of time was my solution. Time cannot be absolutely defined, and there is an inseparable relation between time and signal velocity. With this new concept, I could resolve all the difficulties completely for the first time.
>
> Within five weeks the special theory of relativity was completed. I did not doubt that the new theory was reasonable from a philosophical point of view. ...This is the way the special theory of relativity was created.

There is an inseparable relation between time and the speed of light.

The second postulate is most intriguing. You already saw the example of the bullet and the laser gun. To tickle your common sense a little more, assume that you see a light beam passing you at 299 792 458 m/s. Immediately, you get on your super fast spaceship and chase the beam with a speed of 299 792 450 m/s. Newtonian physics (and common sense) tells you that the speed of that light beam relative to you must be a mere 8 m/s. The second postulate says that even though you are chasing the light beam at 299 792 450 m/s, as you measure the speed of the light beam just ahead of you, you find it to be 299 792 458 m/s! Incredible, strange, impossible, you say. Incredible? Yes. Strange and impossible? No. There is nothing less strange and impossible than a proven law of nature. It is we that are strange, judging a perfectly normal nature[3] by our crude senses and declaring the imperfections we perceive through them as "normal" and "possible."

Strange consequences of second principle.

Nature is not strange. We are!

1.2 RELATIVITY OF SIMULTANEITY

When we say "two events are simultaneous," we are using two common words whose meanings need to be made precise. **Events** are the most essential building blocks of relativity. An event is any abrupt physical phenomenon whose spatial position and time of occurrence can be recorded. An event is relativity's equivalent of a point in geometry, and like the latter, relativity always deals with ideal events, i.e., those that do not have any extension either in time or in space. All observers agree on a single ideal event. In other words, if a physical phenomenon is stamped as an event by one observer, then it is stamped as an event by all observers.

Events as building blocks of STR.

Event defined.

Note 1.2.1. *An event is described by a single point in space and a single instant in time. The occurrence of an event is a universal phenomenon: If it occurs for one observer, it occurs for all.*

As a prototype of events, we consider the explosion of a firecracker in the sequel. It is not, however, an ideal event, just as a point drawn on a piece of paper is not an ideal geometric point. If we consider the explosion of a firecracker microscopically, the chemical processes leading up to the explosion do not take place instantaneously at a point in space, but require a finite time interval and take place in a volume. We ignore such microscopic details, just as we ignore the finite radius and the thickness of the disk left by a pencil on paper as we draw figures when we study geometry.

1.2.1 Simultaneity in a Single RF

Now that you know the precise notion of an event, you need to understand the precise meaning of **simultaneity**. How can you determine whether two events are simultaneous or not? Firecrackers A and B explode somewhere in space and you receive the signals of their explosions. Question: Did the explosions

Simultaneity of two events occurring at the same point.

[3] After all, what is more normal or natural than nature itself?

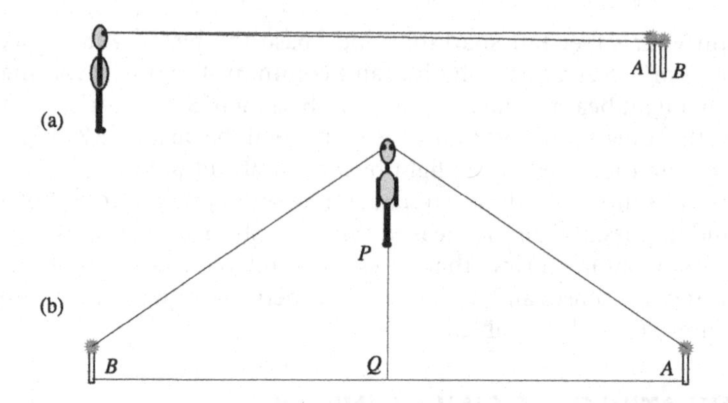

FIGURE 1.2 (a) Two adjacent firecrackers are said to have exploded at the same time if the observer receives their signals at the same time. (b) Two widely separated firecrackers are said to have exploded at the same time if the observer receives their signals at the same time while being located on the perpendicular bisector of the line segment joining the firecrackers.

occur at the same time? If A and B were at the same point when they exploded, it is easy to answer this question: If you receive their light signals at the same time, they must have occurred at the same time [Figure 1.2(a)]. Because of the special property of light stated in the second postulate, this conclusion is independent of the state of motion of the firecrackers: A and B could be moving at very high speeds in arbitrary directions at the time they reach the common point in the figure.

You can appreciate the significance of light signals used in communicating simultaneity by looking at events A and B which occur in *moving* reference frames. Suppose bullets are used instead of light signals. Sam and Sonya are standing on two trains parked next to each other at point C a distance of 300 m from James. They fire bullets simultaneously from their rifles moving at 100 m/s *toward* James.[4] Three seconds later, James sees the bullets pass by him together. He concludes that the two events of firing the bullets must have occurred simultaneously because he knows that the two rifles were at the same location and the speed of the bullets were equal.

Now suppose that Sam is moving away from James while Sonya is moving toward James both at 50 m/s. When they reach C at the same time, they fire bullets toward James. Sonya's bullet has a speed of 150 m/s while Sam's bullet speed is 50 m/s, both relative to James. So, although fired simultaneously,

[4] . . . who is (hopefully) not directly in front of the rifles!

Sonya's bullet reaches James in 2 seconds while Sam's reaches him in 6 seconds, and he cannot conclude that the two events occurred simultaneously. The motion of the rifles has affected the speed of the bullets relative to James. Light does not have this problem. If Sam and Sonya were moving at half the speed of light in opposite directions, the light signals they emit at C will reach James *at exactly the same time*. That is why, as Einstein alluded to it in the quote above,

EM waves are the only communicators of simultaneity.

Note 1.2.2. *Light (or any other electromagnetic wave) is the only communicator of simultaneity.*

What if A and B are not located at the same point? Figure 1.2(b) shows the line segment $\overline{AB}$ and its perpendicular bisector $\overline{PQ}$. If the observer happens to be on $\overline{PQ}$ and receives the two light signals at the same time, he can conclude that they must have occurred at the same time. Again, due to the invariance of light speed, this definition of simultaneity is independent of the state of motion of A, B, and the observer. Thus, to conclude whether two events separated in space are simultaneous, you must locate yourself on the perpendicular bisector of the line segment separating them and see if you receive the light signals at the same time. If you are off the perpendicular bisector, you must allow for the difference in the travel time of light signals.

Simultaneity of two events occurring at different points.

To facilitate the understanding of simultaneity, assume that, for each pair of events, every reference frame has a designated observer O_{des}, who happens to be equidistant from that given pair. Then, we can define simultaneity as follows:

Designated observer of a reference frame.

Note 1.2.3. *A pair of events A and B are **simultaneous** in a reference frame if the designated observer of that RF receives the light signal from A and B at the same time.*

Having defined the notion of simultaneity, I can now ask: If the designated observer O_{des} decides that the events A and B are simultaneous, can another observer O call them simultaneous? If O_{des} and O do not move relative to one another, i.e., if both are in the same RF, the answer is yes, and here is how simultaneity can be communicated to the other members of the RF.

Suppose O_{des} has a (very accurate) watch that emits a light signal every second. Any other observer, including O, *at a fixed distance from* O_{des}, will receive those signals every second, because, for any given observer, each signal has exactly the same distance to travel. Therefore, O's clock can be compared and made to run at exactly the same rate as the clock held by O_{des}, and vice versa. In particular, O_{des} can send a coded message to O at the exact "second" that she receives the two simultaneous signals from events A and B, informing O that those two events occurred at the same time.

All clocks of an RF can be made to run at the same rate and be synchronized.

Furthermore, all clocks of the RF can be completely synchronized to a particular reference clock C. This is the most accurate clock available. In fact, it is so accurate that you can call it an *ideal clock*.[5] You can then synchronize any other clock with C even if it is located far, very far—light years—away from C. To synchronize any clock C′, an operator at C determines the distance between C and C′ by measuring the time interval between sending a light signal to C′ and receiving the reflected echo at C. Denoting this time interval by T, she calculates the distance between C and C′ to be $Tc/2$.

Once the distance between C′ and C is determined, the operator can send a code at 12:00 containing the sentence "As soon as you receive this signal, set your time to 12:00 plus $T/2$!" Then the operator at C′ can set his time to be exactly the same as C. This is what takes place at the Earth's RF.[6] All clocks are set to Greenwich Mean Time. However, since it would be silly for the residents of Tokyo to be forced to call it 12:00 noon when everywhere is dark, the synchronization of the Earth clocks follows a more democratic rule by incorporating the geographic location in the process.

It is crucial to note that the synchronization procedure is applicable because all clocks are stationary relative to one another. As you'll see shortly, relative motion between two clocks affects their operation. Theoretically, you can synchronize all the ideal clocks of a single RF to tick in perfect harmony:

Note 1.2.4. *All observers in a single reference frame keep the same time and agree on the notion of simultaneity.*

1.2.2 Simultaneity in Different RFs

What happens to simultaneity when two RFs move relative to one another? To investigate this question, consider two firecrackers located at the two ends of a train car. Sam is on the ground and Sonya stands at the midpoint of the train car. Suppose that the firecrackers produce smoke that can leave permanent marks on the ground. Suppose also that Sam finds himself right in the middle of these two marks.[7] Finally, assume that he receives the signals from

---

[5] An ideal clock never runs fast or slow. Such a clock does not exist, of course, but *atomic clocks* approximate this ideal clock to a very high degree of accuracy.

[6] Strictly speaking, not all residents of Earth have the same RF because of the Earth's spin: People in China and the US move in opposite directions, and very accurate global synchronizations take this motion into account.

[7] The fact that I am placing Sam and Sonya *in the middle* of the two events does not restrict the conclusions I reach concerning simultaneity (or lack thereof) in their RF. Think of them as designated observers who can transmit their observation to all observers in their respective RF.

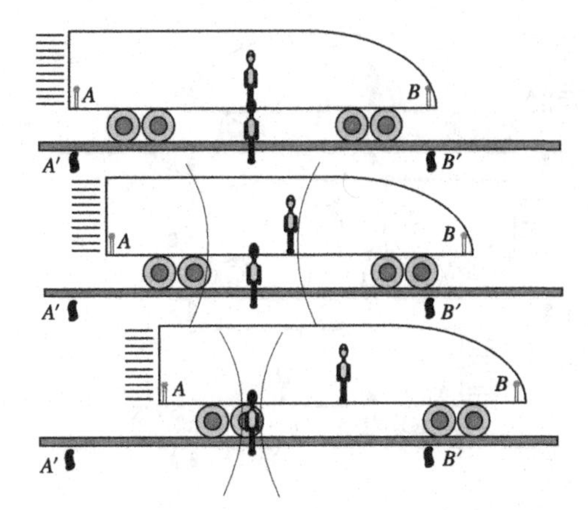

FIGURE 1.3 Different observers disagree on the simultaneity of two events. The ground observer (Sam) sees the two events as simultaneous, while the train observer (Sonya) does not. The marks labeled A' and B' are left on the ground by the explosion of the firecrackers A and B, respectively. These marks are as seen by Sam.

A and B at the same time. Figure 1.3 shows the situation as seen by Sam. The top figure is the moment the explosions occur, as *calculated* by Sam after he receives the signals (allowing for the signals' travel time). The middle figure shows the wave fronts moving away from their sources. Note that Sonya has received the signal from B, and that the two wave fronts are equidistant from Sam and from the positions—according to Sam—of the firecrackers *at the time of explosions* (i.e., positions of A and B in the top figure, or equivalently, positions of A' and B'). Finally, in the bottom figure the two wave fronts have reached Sam while Sonya is still waiting for the signal from A.

After receiving the two signals simultaneously, Sam walks to A' counting his steps, walks to B' counting his steps, notices that the two distances are equal, and concludes that events A and B were indeed simultaneous. What about Sonya? After receiving the two signals, she measures her distance from A and B and verifies that she was halfway between them. She therefore concludes that B must definitely have occurred before A.

Although simultaneity of two events is not universal, the fact that Sam sees A and B as simultaneous, is. Not only does he say "A and B are simultaneous for me," but also Sonya and all other observers in the universe say "A and B are simultaneous for Sam," despite the fact that they themselves do not measure the occurrence of A and B to be simultaneous. We can actually demonstrate this by equipping Sam with a special firecracker C that he can trigger when the two signals reach him at the same time. Thus, the explosion of C heralds the simultaneity of A and B as seen by Sam, and all observers keeping an eye on

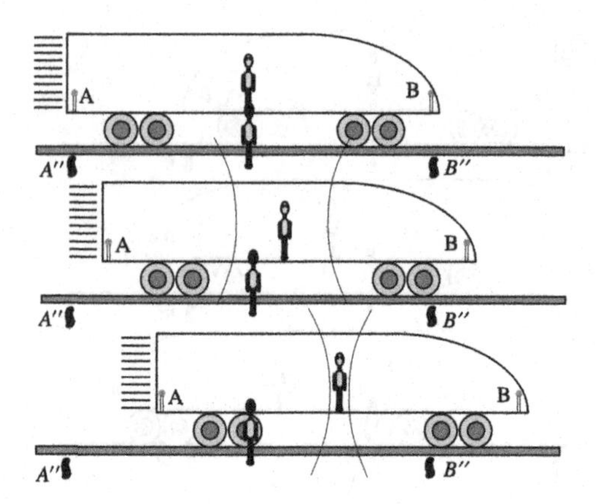

FIGURE 1.4 Here Sonya sees the two events as simultaneous. Sam sees A before B. The marks on the platform are as seen by Sonya.

C can conclude that Sam saw A and B at the same time (if they see the single explosion of C) or not (if they don't see the explosion of C).

The preceding discussion may have given you the impression that Sam is getting special treatment because he is the one who sees the two events simultaneously, or that he is the one who "is not moving." To erase any impression of such special treatment, give Sonya the "honor" of observing the two events at the same time. Figure 1.4 shows the situation as seen by Sonya. The top figure is the moment the explosions occur, as *calculated* by Sonya after she receives the signals, allowing for the signals' travel time. The middle figure shows the wave fronts moving away from their sources. Note that Sam has (almost) received the signal from A, and that the two wave fronts are equidistant from Sonya and from the positions of the firecrackers. In the bottom figure the two wave fronts have reached Sonya while Sam is still waiting for the signal from B. As before, after receiving the two signals, Sam and Sonya measure their distances from A and B (or their marks on the platform) and verify that they were both halfway between them. Sonya concludes that the two events occurred simultaneously while Sam decides that A must have occurred before B.

Note 1.2.5. *Simultaneity is a relative phenomenon. If two events occur simultaneously in one RF, in general, they do not occur simultaneously in other RFs moving relative to the first. However, the fact that a given RF sees two events as simultaneous (or not) is universal.*

1.3 RELATIVITY OF LENGTH

How to measure the length of an object in motion.

Before scrutinizing the effect of motion on length, you need to know how to measure the length of an object when it moves. You could jump on the object and use a tape measure. But this would be the same as measuring the length of the object when it *does not move,* because motion is relative and when you are on the object, the object is not moving relative to you. You may try to use a tape measure while the object is moving—put the head of the tape measure at the front of the object when it reaches you—but by the time you get to the end of the object, it has moved from where it was when you started the measurement. This should tell you that, somehow you must put the two ends of the tape measure *simultaneously* at the two ends of the object in motion. One way to do this is to have two events occur simultaneously (according to you, of course) at the two ends of the object, and make sure that those events leave permanent marks in your reference frame. The distance between these two marks, which you can measure at your leisure, is the length of the object.

Note 1.3.1. *For observer O to be able to measure the length of a moving object, two events at the two ends of the object must occur at the same time according to O and must leave permanent marks in the reference frame of O. The distance between those marks is the length of the object as measured by O.*

Let's look at the events of Figure 1.3, assuming that we are *repeating* the exact same experiment. So, there are already *black* marks on the ground from the previous experiment. From Sam's perspective, $\overline{A'B'}$ is the length of the train. Sonya is standing in the middle of the train, anticipating the occurrence of B. She eventually receives the signal from B and concludes that (a fraction of a second earlier) the front end of the train must have coincided with the black mark at B'. Immediately she looks back, and since the signal from A has not arrived yet, she concludes that the rear end of the train has not reached the mark on the ground yet. She, therefore, concludes that the *train is longer* than the distance between the ground marks!

It seems that Sam and Sonya are in disagreement concerning the length of the train. Sam's measurement of the length of the train yields a shorter length than Sonya's. Could it be that their contradicting conclusions are due to the fact that Sam saw the two events simultaneously? That if Sonya sees the two events at the same time the conclusions will be reversed? To see that that is not the case, let's conduct the (double) experiment in which Sonya sees the explosion of the firecrackers simultaneously, as shown in Figure 1.4. Sam is standing in the middle of the two marks on the platform, anticipating the occurrence of A. He eventually receives the signal from A and concludes that (a fraction of a second earlier) the rear end of the train must have coincided with the black mark at A''. Immediately he looks to his right, and since the signal from B has

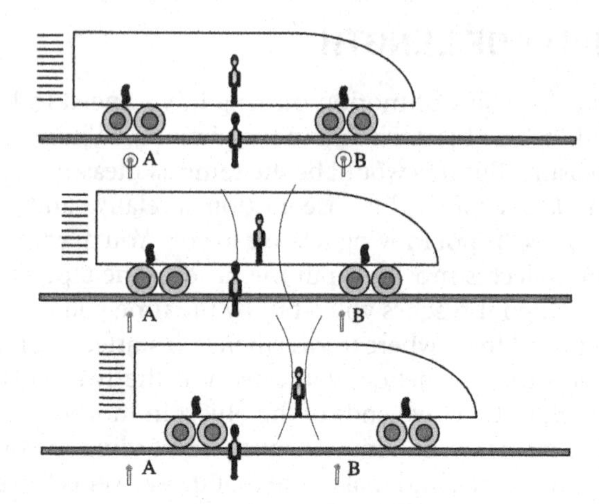

FIGURE 1.5 Sonya sees the two events as simultaneous and decides that the length $\overline{AB}$ is the distance between the two marks on the train. Sam sees A before B and decides that the distance between the marks is shorter than $\overline{AB}$.

not arrived yet, he concludes that the front end of the train has not reached the mark on the ground yet. He, therefore, decides that the *train is shorter* than the distance between the ground marks! It doesn't matter who sees the two events at the same time; *the train is shorter for Sam.*

Is there anything special about the train? Is it because it is moving and the platform is not? It can't be, because motion is relative. To Sonya it is the platform that is moving, and if relativity is correct, lengths on the platform should appear shorter to Sonya. Is this indeed the case? Place the firecrackers on the platform instead of the train (leaving marks on the train when they explode) and let Sonya and Sam measure the distance between them (or their marks). Since the preceding discussions have shown that the distance measurement (whether it is shorter or longer) is independent of who sees the explosions simultaneously, assume that Sonya, riding on the train, sees the firecrackers explode simultaneously. Sam, on the other hand, receives the signal from A before that from B. Figure 1.5 explains this situation.

As usual, assume that an identical experiment has been done before, and it is now being repeated. So, there are already black marks on the train from the previous explosions indicating the distance between the firecrackers *as measured by Sonya.* Sam is standing in the middle of the two firecrackers on the platform, anticipating the occurrence of A. He eventually receives the signal from A and concludes that a fraction of a second earlier the black mark at the rear end of the train must have coincided with A. Immediately he looks to the front, and since the signal from B has not arrived yet, he concludes that the mark in front of the train has not reached B yet. He, therefore, decides that the distance between the two firecrackers on the platform is larger than the marks on the train. Therefore, *Sonya measures the distance between A and B to be less than what Sam measures.*

Note the symmetry between the two observers. Neither is any more special than the other. Both are claiming that lengths measured in their RF are longer than the moving lengths.[8] In other words, they both conclude that *moving lengths shrink*. And the shrinkage is not due to some kind of a mechanical compression of the train or the platform (or cars, trucks, planes, or meter sticks). The two ends of a stick merely indicate two points in space. For instance, these two points could be the locations of two stars. As seen from Earth, these two stars, say the Sun and Alpha Centauri, which are almost fixed relative to the Earth's RF,[9] are seen to be about 4 light years away from one another. The crew of a spaceship, traveling at very high speed toward Alpha Centauri, see this distance contracted because, to them, the length between the Sun and Alpha Centauri is in motion. Thus, *motion affects the space itself*.

Moving lengths shrink.

All the contractions treated so far occur for lengths that are along the path of motion. Would the same effect occur if the length moved along a path that was perpendicular to itself? Since the motion of Sonya's train has helped us so much, I'll use it one more time. Assume that perpendicular lengths also shrink when in motion.[10] When Sonya's train is stationary, its wheels rest on the tracks, so that the distance between the wheels is equal to the distance between the tracks. When the train moves, Sonya sees the tracks moving relative to her. So she concludes that the tracks get closer together, and therefore, the train will fall *outside* of the tracks. Sam, on the other hand, sees the train moving and concludes that the distance between the wheels must shrink and, therefore, that the train must fall *inside* of the tracks. But derailing of a train is an actual physical process (an event) and thus must be independent of Sonya and Sam. The only sound conclusion is to say that neither shrinks. Summarizing, we conclude that

Only lengths parallel to motion shrink.

Note 1.3.2. *All moving lengths parallel to the direction of motion shrink. Lengths perpendicular to the direction of motion are not affected.*

1.4 PROBLEMS

1.1. A rod of length L emits light from all of its points simultaneously (in its rest frame) when a remote switch is turned on. Its center is on the x-axis and is moving on the axis in a plane parallel to a very large photographic plate and

[8] Remember that it is *the platform* that is moving relative to Sonya!

[9] Although the Sun and Alpha Centauri are moving relative to Earth, their speed is so small compared to light speed that their motion can be ignored.

[10] It doesn't matter whether they shrink or expand. The conclusion will be the same.

infinitesimally close to it. When it reaches the middle of the plate, the switch is turned on.

(a) Compare the length $L_\parallel$ of the image on the photographic plate with L when the rod is along the x-axis: $L_\parallel > L$, $L_\parallel = L$, or $L_\parallel < L$? Give a reason for your answer.

(b) Compare the length $L_\perp$ of the image on the photographic plate with L when the rod is perpendicular to the x-axis: $L_\perp > L$, $L_\perp = L$, or $L_\perp < L$? Give a reason for your answer.

1.2. A rod is placed along the x-axis with its center at the origin. A pinhole camera C_1 is located on the z-axis and takes a picture of the stationary rod. Now the rod starts moving along the x-axis parallel to itself from $-\infty$. Camera C_1 is removed and another pinhole camera C_2 replaces it on the z-axis. As soon as the center of the rod reaches the origin (call it $t = 0$), C_2 takes a picture.

(a) Is the pinhole of C_2 collecting the light rays from the two ends of the rod that were emitted at $t = 0$?

(b) Is the pinhole collecting the light rays from the two ends of the rod that were emitted simultaneously, but not at $t = 0$?

(c) If the answer to (b) is no, which end emitted its light first, the trailing end or the leading end?

(d) Is it possible for the image of the rod in C_2 to be *longer* than its image in C_1? Hint: Consider the location of each end as it emits the light ray captured by C_2.

1.3. A rod is placed along the y-axis with its center at the origin. A pinhole camera C_1 is located on the z-axis and takes a picture of the stationary rod. Now the rod starts moving along the x-axis parallel to itself from $-\infty$. Camera C_1 is removed and another pinhole camera C_2 replaces it on the z-axis. As soon as the center of the rod reaches the origin (call it $t = 0$), C_2 takes a picture.

(a) Is the pinhole of C_2 collecting the light rays from the two ends of the rod that were emitted at $t = 0$?

(b) Is the pinhole collecting the light rays from the two ends of the rod that were emitted simultaneously, but not at $t = 0$?

(c) If the answer to (b) is no, which end emitted its light first, the top or the bottom?

(d) Is it possible for the image of the rod in C_2 to be longer than its image in C_1? Hint: Consider the locations of the ends as they emit their light rays captured by C_2, the distance between those locations and the pinhole, and the angle they subtend at the pinhole.

1.4. A circular ring emits light from all of its points simultaneously (in its rest frame) when a remote switch is turned on. It is moving in a plane parallel to a photographic plate and infinitesimally close to it. When it reaches the plate, the switch is turned on. What is the shape of the image of the photograph? Hint: See Problem 1.1.

1.5. A conveyor belt moving at relativistic speed carries cookie dough. A circular stamp cuts out cookies as the dough rushes by beneath it. What is the shape of these cookies? Are they flattened in the direction of the belt, stretched in that direction, or circular?

1.6. A conveyor belt moving at relativistic speed carries cookie dough. A laser gun one meter above the belt emits a beam in the shape of the surface of a circular cone that cuts the dough perpendicularly. Are these cookies flattened in the direction of the belt, stretched in that direction, or circular? Hint: Concentrate on the two ends of the diameter of the beam along the dough, and note that their light beams arrive simultaneously at the stationary bed on which the dough is moving. Now consider how the two events appear in the RF of the moving dough and what implication it has on the length of the diameter. The discussion surrounding Figure 1.3 may be helpful.

1.7. A circular ring is centered at the origin in the xy-plane. A pinhole camera C_1 is located on the z-axis and takes a picture of the stationary ring. Now the ring starts moving along the x-axis from $-\infty$. Camera C_1 is removed and another pinhole camera C_2 replaces it on the z-axis. As soon as the center of the rod reaches the origin (at $t=0$), C_2 takes a picture.

(a) Is the pinhole of C_2 collecting the light rays from the two ends of the horizontal diameter (along the x-axis) of the ring that were emitted at $t=0$? Hint: Look at Problem 1.2.
(b) Is the pinhole of C_2 collecting the light rays from the two ends of the horizontal diameter of the ring that were emitted simultaneously, but not at $t=0$?
(c) If the answer to (b) is no, which end emitted its light first, the trailing end or the leading end?
(d) Is it possible for the image of the horizontal diameter in C_2 to be *longer* than its image in C_1?
(e) Is the image of the vertical diameter (along the y-axis) in C_2 equal to, longer than, or shorter than its image in C_1? Hint: Look at Problem 1.3.
(f) Can you guess what the shape of the image of the ring is in C_2?

CHAPTER 2

Relativity of Time and Space

The qualitative discussion of the last chapter made it clear that motion affects time and space. In this chapter, you'll learn about the dilation of time for a moving clock and contraction of length for a moving ruler. You'll see that the result derived in this chapter [Equation (2.3)] is at the heart of relativity theory, and that the rest of the book is essentially based on that formula.

2.1 TIME DILATION

As Einstein said "Time is what is measured by clocks." So, let us look at the effect of motion on a clock. The "effect of motion" does not mean the effects of the bumps and puddles of a rough road which may actually damage a clock. You should picture the smoothest possible ride such as the sailing of a spaceship in outer space. Any effect on the clock is therefore an effect on time itself, although a particular clock is used to detect the effect.

The clock best suited for this investigation is one of the arms of the apparatus that Michelson and Morley used in their experiments that showed that Earth was not moving relative to ether. This clock, shown in Figure 2.1, consists of a tube with a source S of light (or any other electromagnetic wave) at one end and a mirror M at the other. The distance between S and M is L which we can conveniently take to be, say 1.5 m. Then it takes light 5 ns (nanoseconds) to go from S to M, and 10 ns to go to M and come back to S. Place a light sensitive "ticker" at S and the clock ticks for us every 10 ns. Call such a clock a Michelson-Morley clock, or **MM clock** for short.

The Michelson-Morley clock.

Now place an MM clock perpendicular to the direction of motion on a train moving to the right and let two observers O (on the ground) and O' (on the train) watch it. Each tick consists of three events: E_1, the emission of a light beam at S; E_2, its reflection at M; and E_3, its reception at S. How does O' see the ticking of the clock? The clock is sitting right beside her, and to her, the process of ticking is the light going straight up and coming straight down. She concludes that the time interval, denoted by $\Delta\tau$, is simply twice the time it

Calculation of the tick of a moving MM clock.

Special Relativity. DOI:10.1016/B978-0-12-810411-8.00002-X

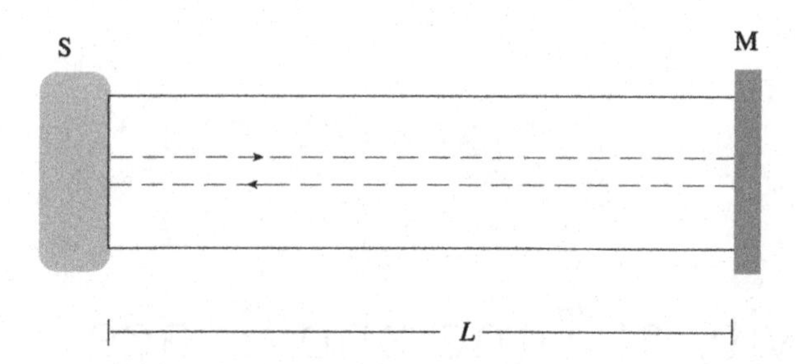

FIGURE 2.1 A Michelson-Morley clock. A "tick" of this clock occurs when the light signal makes a round trip along the length L.

takes light to travel the distance $\overline{SM} = L$:

$$\Delta\tau = 2\frac{L}{c}. \tag{2.1}$$

How does O perceive the succession of these three events? Since the clock is moving to the right, the light signal that leaves S will reach M only after M has moved to the right. Thus, to O, the events E_1 and E_2 are separated not only by a vertical distance, but also by a horizontal distance. Let me clarify this a little more. Suppose that the light signal sent by S and reflected by M is represented by a black dot. Figure 2.2(a) shows five snapshots of the clock. In the first snapshot the dot is produced at E_1. A little later (therefore a little to the right) the dot is at the midpoint of the clock tube. Still a little later (and a little further to the right) the dot reaches the mirror at the top. Finally, the dot passes the midpoint again on its return and reaches S at event E_3.

Section 1.3 showed that the vertical distance is unaffected by motion. So, the length of the MM clock is still L, and because of the addition of the horizontal distance—the line segment E_1A in Figure 2.2(a)—O decides that $\overline{E_1E_2} > \overline{AM} = L$. The *speed of light being the same for all observers*, he concludes that it takes light more than L/c to travel $\overline{E_1E_2}$. The other leg, $\overline{E_2E_3}$, has exactly the same length as $\overline{E_1E_2}$. Thus, to O, the total travel time from S to M and back takes *longer* than $2L/c$. Therefore, he concludes that the *clock on the train must tick slower!*

Moving clocks slow down.

Note 2.1.1. *A clock that is moving* **relative** *to an observer slows down for* **that** *observer.*

It is crucial that you keep the relativity of motion in mind. Although the train is "moving," the clock is *not moving* relative to the observer on the train!

To quantify the above statement, let's refer to the triangle E_1AE_2 of Figure 2.2. Pythagorean theorem implies that

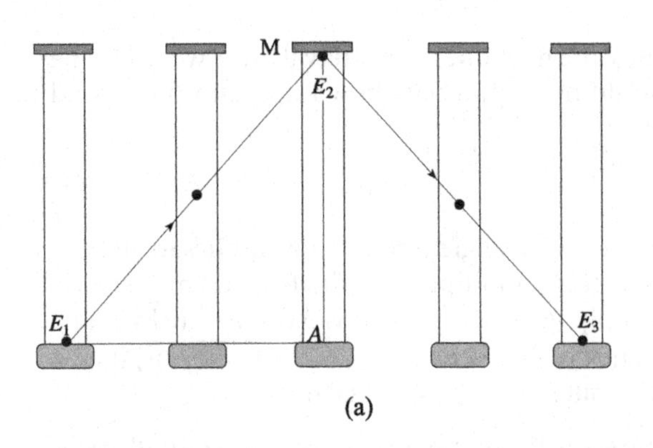

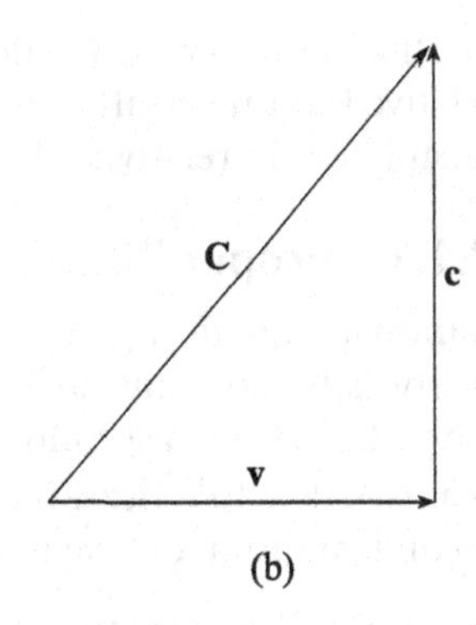

FIGURE 2.2 (a) A moving Michelson-Morley clock. The path of light (represented by a black dot) is not a vertical line but a slanted one due to the motion of M. (b) Law of addition of velocities applied to the light signal in the MM clock.

$$\left(\overline{E_1E_2}\right)^2 = \left(\overline{E_1A}\right)^2 + \left(\overline{AE_2}\right)^2.$$

Let the speed of the train be v and the light beam's travel time from S to M be δt according to O. Then $\overline{E_1A} = v\delta t$ and $\overline{E_1E_2} = c\delta t$ with c the (universal) speed of light. Putting all of this in the above equation gives

$$(c\delta t)^2 = (v\delta t)^2 + L^2. \tag{2.2}$$

Solving for δt yields

$$(\delta t)^2 = \frac{L^2/c^2}{1 - v^2/c^2} \quad \text{or} \quad \delta t = \frac{L/c}{\sqrt{1 - v^2/c^2}}.$$

Denote by Δt the duration of the light's round trip as seen by O. Then

Time dilation formula.

$$\Delta t = 2\delta t = \frac{2L/c}{\sqrt{1 - v^2/c^2}} = \frac{\Delta\tau}{\sqrt{1 - (v/c)^2}}, \tag{2.3}$$

after using Equation (2.1). This is the **time dilation** formula.

It is instructive to investigate the consequence of abandoning the second postulate of relativity and restoring the law of addition of velocities. If this law were true, then O would measure the speed of light—as it goes from E_1 to E_2—to be C as shown in Figure 2.2(b). Then Equation (2.2) would become

The crucial role of the second postulate!

$$(C\delta t)^2 = (v\delta t)^2 + L^2.$$

But the Pythagorean theorem, as applied to the right triangle in Figure 2.2(b), tells us that $C^2 = v^2 + c^2$. Therefore,

The second postulate is at the heart of relativity theory.

$$(v^2 + c^2)(\delta t)^2 = v^2(\delta t)^2 + L^2 \quad \text{or} \quad c^2(\delta t)^2 = L^2$$

or $c\delta t = L$ or $\delta t = L/c$, leading to

$$\Delta t = 2\delta t = 2L/c = \Delta\tau,$$

i.e., the two observers would measure the same time interval: time would *not* be relative but universal! You should now appreciate how important the second postulate is in relativity theory.

2.1.1 Proper Time

Although Equation (2.3) is derived for a single tick, it really applies to all time intervals because any such interval is a multiple of a single tick. For instance, when $L = 1.5$ m, a second is simply 10^8 ticks, an hour is 3.6×10^{11} ticks, and a year is 3.16×10^{15} ticks. If each tick is altered by a factor of $\sqrt{1-(v/c)^2}$, then a second, an hour, or a year is also altered by the same factor.

What is the best way to describe the difference between Δt and $\Delta\tau$? Both measure the time interval between two events, E_1 and E_3, but $\Delta\tau$ measures the time interval in a RF in which E_1 and E_3 occur at the *same spatial point*: The emission and reception of light occur at the same point S. For this reason $\Delta\tau$ is called **proper time**.

Ticks are just prototypes of events. Therefore, in Equation (2.3), $\Delta\tau$ can be the proper time between *any two events*. If the clock, and therefore the observer, is present at both events, the events must occur at the same spatial point for that observer. For example, to the captain of a spaceship who looks at the take-off from Earth and landing on another planet from the window of the spaceship, these two events are taking place at the same spatial point: just outside the window. We can thus define proper time in two different ways:

Proper time defined.

Note 2.1.2. *The* **proper time** *interval between two events is the time measured by an observer who is present at both events or, equivalently, for whom the two events occur at the same* spatial *point.*

Example 2.1.3. Time Dilation in Muon Decay

Muon is an elementary particle that decays into an electron and a couple of neutrinos with a mean life of $T = 2.197 \times 10^{-6}$ s. This means that if a sample of muons consists of N_0 particles at time $t = 0$, the number of surviving muons at time t is given by

$$N(t) = N_0 e^{-t/T}.$$

Muons are created naturally through the interaction of cosmic rays with the atmosphere. At 20 km above the surface of the Earth the flux of muons is about 100 per cm^2 per second. These muons rush toward Earth with a speed of $0.99c$. It will therefore take them

$$\frac{20000}{0.99 \times 3 \times 10^8} = 6.73 \times 10^{-5}\ \text{s} = 30.65\,T$$

to cover 20 km. Hence, the expected flux at sea level should be

$$N(t) = 100\,e^{-30.65} = 4.88 \times 10^{-12}\ \text{muons per cm}^2\ \text{per second.}$$

So, one has to wait $1/(4.88 \times 10^{-12})$ seconds or about 6500 years before detecting a single muon on a cm^2 of the Earth's surface! In contrast, one observes about 1 muon per cm^2 per second on Earth. What's going on?

The mean life quoted above is as measured in the *rest frame* of muons, i.e., it is a proper time interval (between the event at which $N = N_0$ and the event at which $N = N_0 e^{-1}$). As muons pour down toward Earth, we measure their mean lifetime to be

$$\Delta t = \frac{2.197 \times 10^{-6}}{\sqrt{1 - 0.99^2}} = 1.56 \times 10^{-5} \text{ s} = 4.32T.$$

Therefore, instead of $30.96T$, muons cover the distance of 20 km in $4.32T$, and the flux at sea level becomes

$$N(t) = 100\, e^{-4.32} = 1.33 \text{ muons per cm}^2 \text{ per second},$$

in agreement with observation! ■

The time interval Δt, called the **coordinate time**,[1] is the time measured by an observer O'—moving relative to O (who measures the proper time) with speed v—for whom the two events occur at two different spatial points.

Equation (2.3) can be written more succinctly by introducing two frequently used symbols:

$$\Delta t = \frac{\Delta \tau}{\sqrt{1 - \beta^2}} \equiv \gamma \Delta \tau, \qquad \beta \equiv v/c, \quad \gamma = \frac{1}{\sqrt{1 - \beta^2}}. \tag{2.4}$$

The symbol β is the **fractional speed** of one observer relative to the other. The symbol γ is called the **Lorentz factor** or **gamma factor**. It determines whether relativistic considerations are essential or not. If the speed v of an object is very small compared to c, i.e., when $\beta << 1$, then γ is very nearly 1 and relativistic effects can be ignored. On the other hand, when γ is much larger than 1, i.e., when β approaches 1, classical mechanics fails and relativity theory must be invoked.

Fractional speed and gamma (Lorentz) factor introduced.

Equation (2.4) is at the heart of the relativity theory, and as you'll see in the sequel, almost all relativistic results are obtained from it. The equation applies to situations in which the relative velocity of the two observers remains constant. If there is an abrupt change in the speed or the direction of motion, we have to apply (2.4) *separately* to those portions of the motion in which the velocity does not change. Before discussing the implications of Equation (2.4), let's look at some examples.

Example 2.1.4. The captain of the Spaceship Enterprise cruising at 1.8×10^8 m/s, or $0.6c$ (or $\beta = 0.6$), wakes up at time zero (event E_1), goes about doing his chores for 16 hours—according to his clock, of course—and goes to bed (event E_2). How long does this appear to the Earth observers?

---

[1] This strange naming of Δt will make more sense when t becomes a coordinate in Chapter 4.

The proper time $\Delta\tau$ is 16 hours and is measured by the spaceship clock, because the two events occur at the same point in the spaceship, the bed. Stated equivalently, this time interval is measured by the captain's clock, which is present at both events. To calculate Δt, the time interval between E_1 and E_2 according to the Earth clock, note simply that $\Delta t = 16/\sqrt{1-(0.6)^2} = 20$ hrs. The Earth observers, therefore, conclude that the captain has a long 20-hour work day! ■

Space travel beyond the closest stars may seem physically impossible. The ambition of exploring stars hundreds of light years away appears to be futile.[2] After all, it takes light, the fastest object in the universe, 100 years to reach Earth from a star that is 100 light years away. How then can the crew of a space probe, even moving infinitesimally close to the speed of light, hope to survive the journey? All members of the crew will be dead by the time the probe reaches the star. The following example shows that, due to the relativity of time, such space travels are possible if we can achieve speeds close to light speed.[3]

Possibility of space travel.

Example 2.1.5. The spaceship Viking is a super fast cosmic explorer that can attain a speed of $0.999c$. This spaceship is charged with exploring the star system Zeta Leporis located at a distance of 70 light years from Earth. How long does it take Viking to reach (one of the planets of) Zeta Leporis

(a) according to the Earth clock, and
(b) according to the spaceship clock?

For (a), simply use

$$\text{time} = \frac{\text{distance}}{\text{speed}} = \frac{70 \text{ light years}}{0.999c} = \frac{70 \text{ years} \cdot c}{0.999c} = \frac{70}{0.999} \text{ years}$$

or 70.07 years. So, as expected, it takes a little over 70 years for Viking to get to Zeta Leporis, as seen by ground observers.

To obtain (b), note that the spaceship clock is present at the two events E_1 (spaceship leaves the Earth), and E_2 (spaceship lands on Zeta Leporis). Hence, the spaceship clock measures the proper time between the two events. Thus

$$70.07 \text{ years} = \frac{\Delta\tau}{\sqrt{1-(0.999)^2}}$$

or $\Delta\tau = 3.13$ years. So, it takes only a little over 3 years for the crew of Viking to reach Zeta Leporis! ■

The captain of the Spaceship Enterprise is 30 years old when he starts on a mission that takes him and his crew to (one of the planets of) a distant star, leaving

---

[2] A light year is a *distance* obtained by multiplying a year by the speed of light. Multiplying 3.15569×10^7—the number of seconds in a year—by 3×10^8 m/s, the speed of light, you obtain the result that 1 light year $= 9.467 \times 10^{15}$ m. In some calculations it is more convenient to write light year as *year* $\cdot\, c$, as is done in Example 2.1.5.

[3] I am ignoring the tremendous amount of energy needed to achieve and maintain speeds close to light speed. For the practical impossibility of such trips, see Chapter 9.

Daughters older than their fathers!

his wife and his newborn daughter behind. After spending a few months on the planet, he and the crew of the Enterprise head home and reach Earth 15 years after they took off. When he asks for his wife, he is told that she died of old age decades ago, and as he looks in the crowd welcoming the crew, he finds a 60-year-old lady who resembles his wife. He is shocked to find out that the lady is his daughter! If you don't believe the story, look at the following example: Numbers don't lie!

Example 2.1.6. The spaceship Enterprise is charged with an exploratory mission that takes it to a planet in a star system far away. The captain of the ship, who has just had a baby, is 30 years old when Enterprise takes off with a speed of $0.98c$. It takes the crew of the spaceship 5 years to get to their destination. They spend 6 months on the planet and then head back home with a speed of 285,000 km/s. How does the age of the father compare with that of his daughter?

The journey can be naturally divided into three parts: moving at $0.98c$ toward the planet; landing and staying on the planet on which their speed is zero (or very small); heading back home with a speed of 285,000 km/s. When using Equation (2.4), you have to apply it *separately* to the three portions of the journey.

The proper time interval—measured by the crew—between take off from Earth and landing on the planet is 5 years. The time interval between the same two events as measured by Earth people is

$$\Delta t = \gamma \Delta \tau = \frac{\Delta \tau}{\sqrt{1-\beta^2}} = \frac{5}{\sqrt{1-(0.98)^2}} = 25.13 \text{ years.}$$

This is the age of the "baby" when the spaceship lands on the planet.

How far is the planet from Earth? It takes 25.13 years for the ship to get there while moving at 98% light speed. Therefore, the distance is

$$\text{speed} \times \text{time} = 0.98c \times 25.13 \text{ years} = 24.63\, c \cdot \text{yr} = 24.63 \text{ ly.}$$

How long does it take the Earth inhabitants for the captain to travel from the planet back to Earth? Now that we have the distance and the speed (285,000 km/s is $0.95c$), we can find the time interval:

$$\text{time} = \frac{\text{distance}}{\text{speed}} = \frac{24.63\, \cancel{l}\text{y}}{0.95\cancel{c}} = 25.93 \text{ years}$$

How long does it take the captain to travel from the planet back to Earth? The crew measures the proper time again. The coordinate time Δt is 25.93 years. Therefore,

$$\Delta \tau = \Delta t \sqrt{1-\beta^2} = 25.93\sqrt{1-(0.95)^2} = 8.1 \text{ years.}$$

How old is the "baby" when her father arrives back on Earth? Simply add the time intervals, keeping in mind that the 6 months that the crew spends on the planet is the same for Earth observers. Thus,

$$\text{age of the "baby"} = 25.13 + 0.5 + 25.93 = 51.56 \text{ years.}$$

And how old is the captain? Again, just add the time intervals to his initial age:

$$\text{age of the captain} = 30 + 5 + 0.5 + 8.1 = 43.6 \text{ years.}$$

The father is almost 8 years younger than his daughter! ■

Never faster than light!

The factor γ puts severe restrictions on physical processes. A physical quantity cannot be the square root of a negative number. Therefore, $1 - \beta^2 \geq 0$, i.e., $\beta \leq 1$ or $v/c \leq 1$, leading to the conclusion that nothing can travel faster than light. Objects can carry information, and information can be interpreted as signals. Therefore,

> **Note 2.1.7.** *No signal can be transmitted with a speed greater than the speed of light.*

No massive object can move at the speed of light!

What about c itself? Can we move *at* the speed of light? As you'll discover in Section 5.2.2, relativity predicts that for any *massive* object, no matter how small the mass, the answer is no. In fact, the daily activity of the particle accelerators around the world is a testimony to this impossibility. By providing more and more energy to an elementary particle, these machines can accelerate the particle to higher and higher speeds, just as the provision of fuel to the engine accelerates a car. The amount of energy delivered to particles has grown steadily several hundred thousand times over the last few decades. However, in no instance has a single particle been observed to have a speed greater than, or even equal to, the speed of light. In 2000, the Large Electron Positron Collider (LEP) at CERN was able to accelerate the electrons to a speed of $0.999999999997c$, corresponding to a γ of 409000.

"But light—which Einstein himself discovered to be composed of *particles* (photons)—does travel at the speed of light," you may remark, "How do you explain the speed of photons?" In the previous paragraph, I took care to use the word "massive." It turns out that photons are massless, and you can find a full account of this and other energy-related topics in Chapter 5.

Example 2.1.8. FASTER THAN LIGHT?

The astronomical object named 3C345 produces clouds that appear to be moving away from it faster than light. Astronomers have observed one particular cloud and have noted that over a period of approximately five years it moved 2 milliarcseconds (mas). An arcsecond is 1/3600 degree. So, a mas is 2.78×10^{-7} degree or about 4.85 nanoradians. Thus moving only twice this angle in five years does not seem very fast, except that 3C345 happens to be at an enormous distance from us, about 7.5 billion light years, or 7×10^{22} km. Therefore, noting that there are 3.16×10^7 seconds in a year, the transverse velocity of the cloud appears to be

$$v_{tr} = \frac{\text{distance} \times \text{angle}}{\text{time}} = \frac{(7 \times 10^{22}) \times (9.7 \times 10^{-9})}{5 \times 3.16 \times 10^7} = 4300000 \text{ km/s},$$

which is more than 14 times the speed of light! Does this cloud have a superluminal speed?

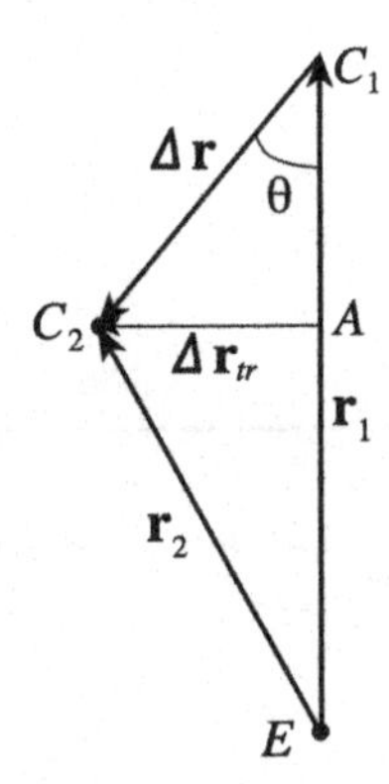

FIGURE 2.3 A distant object moving along $\Delta\mathbf{r}$ appears to have a transverse motion along $\Delta\mathbf{r}_{tr}$.

It turns out that if the cloud is moving *toward* Earth at a large enough speed, then it may *appear* that it has a superluminal transverse velocity. In Figure 2.3, an object moves from its location at C_1 at time t_1 to its later location C_2 at time t_2 along a path making an angle θ with $\mathbf{r}_1$ which connects the Earth E to C_1. The light rays from C_1 and C_2 reach the Earth at times $T_1 = t_1 + r_1/c$ and $T_2 = t_2 + r_2/c$, respectively. Therefore the apparent (transverse) speed of the object is

$$v_{tr} = \frac{|\Delta\mathbf{r}_{tr}|}{T_2 - T_1} = \frac{|\Delta\mathbf{r}|\sin\theta}{t_2 - t_1 + (r_2 - r_1)/c}.$$

Assuming that the object has a constant speed v between C_1 and C_2, it is clear that $|\Delta\mathbf{r}| = v(t_2 - t_1)$, and therefore,

$$v_{tr} = \frac{v\sin\theta}{1 + \dfrac{1}{c}\dfrac{r_2 - r_1}{t_2 - t_1}}. \tag{2.5}$$

With $\mathbf{r}_2 = \mathbf{r}_1 + \Delta\mathbf{r}$, you obtain

$$|\mathbf{r}_2|^2 \equiv r_2^2 = (\mathbf{r}_1 + \Delta\mathbf{r})\cdot(\mathbf{r}_1 + \Delta\mathbf{r}) = |\mathbf{r}_1|^2 + |\Delta\mathbf{r}|^2 + 2\mathbf{r}_1\cdot\Delta\mathbf{r}$$

or

$$r_2 = \sqrt{r_1^2 + |\Delta\mathbf{r}|^2 - 2r_1|\Delta\mathbf{r}|\cos\theta} = r_1\left(1 + \frac{|\Delta\mathbf{r}|^2}{r_1^2} - \frac{2|\Delta\mathbf{r}|\cos\theta}{r_1}\right)^{1/2}$$

$$\approx r_1\left(1 + \frac{|\Delta\mathbf{r}|^2}{2r_1^2} - \frac{|\Delta\mathbf{r}|\cos\theta}{r_1}\right) \approx r_1 - |\Delta\mathbf{r}|\cos\theta, \qquad r_1 >> |\Delta\mathbf{r}|,$$

and

$$\frac{r_2 - r_1}{t_2 - t_1} = -\frac{|\Delta\mathbf{r}|\cos\theta}{t_2 - t_1} = -v\cos\theta.$$

Inserting this in (2.5) you get

$$v_{tr} = \frac{v\sin\theta}{1 - \dfrac{v}{c}\cos\theta} = c\frac{\beta\sin\theta}{1 - \beta\cos\theta}. \tag{2.6}$$

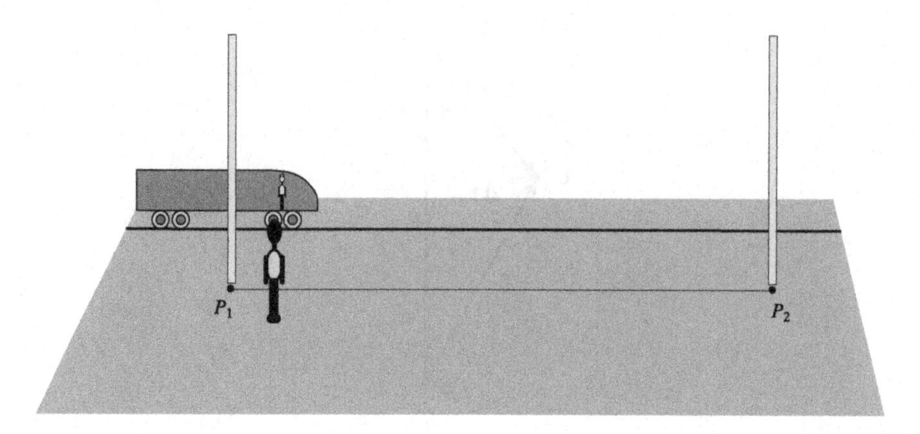

FIGURE 2.4 The distance between P_1 and P_2 is stationary relative to Sam, but in motion relative to Sonya.

If v is very close to c, and θ is small, v_{tr} can be many times the speed of light. For example, with $\theta = 5$ degrees and $\beta = 0.995$, $v_{tr} \approx 10c$! Problem 2.2 sheds more light on superluminal transverse speeds. ■

2.2 LENGTH CONTRACTION

Example 2.1.6 may seem to suggest a violation of the supremacy of the speed of light. It takes light 24.63 years to travel the Earth-planet distance, yet Enterprise covers it in 5 years. Is Enterprise going faster (almost five times faster) than light? Of course not! The key to this puzzle is in the notion of length contraction encountered before.

Deriving length contraction formula.

Consider two points P_1 and P_2 both at rest relative to Sam. These two points can be locations of two stars, locations of two cities, or merely the two ends of a meter stick. Sam measures the distance between the two points and calls it L_0. Sonya, moving relative to Sam with speed v (see Figure 2.4), sees L_0 moving along her direction of motion. Thus, the distance between P_1 and P_2 will appear smaller to Sonya. How much smaller? The time interval it takes Sonya to go from P_1 to P_2 is a proper time interval, because her clock is present at the two events "passing by P_1" and "passing by P_2." She thus concludes that the distance between P_1 and P_2 is $L = v\Delta\tau$. On the other hand, Sam measures Sonya's travel time from P_1 to P_2 to be Δt and concludes that $L_0 = v\Delta t$. Equation (2.4) now gives $L = v[\Delta t\sqrt{1-\beta^2}]$ or

$$L = (v\Delta t)\sqrt{1-\beta^2} = L_0\sqrt{1-\beta^2} = \frac{L_0}{\gamma}. \tag{2.7}$$

L_0 is the **rest length**, and L the **moving length**. Problem 2.3 shows another derivation of the length contraction formula (2.7).

"Explaining" why space travel is possible.

Example 2.2.1. Going back to Example 2.1.6, we can now understand why Enterprise can cover the Earth-planet distance in 5 years. On their way to their destination, the Enterprise passengers see this distance in motion, and they measure it to be

$$L = L_0\sqrt{1-\beta^2} = 24.63\sqrt{1-(0.98)^2} = 4.9 \text{ ly}.$$

Since they are moving with a speed of $0.98c$, their travel time is

$$\text{time} = \frac{\text{distance}}{\text{speed}} = \frac{4.9 \text{ ly}}{0.98c} = 5 \text{ years},$$

as given in the statement of Example 2.1.6.

On their way back, they measure the same distance as

$$L = L_0\sqrt{1-\beta^2} = 24.63\sqrt{1-(0.95)^2} = 7.69 \text{ ly},$$

and with a speed of $0.95c$, their travel time is

$$\text{time} = \frac{\text{distance}}{\text{speed}} = \frac{7.69 \text{ ly}}{0.95c} = 8.1 \text{ years},$$

which is what was found in Example 2.1.6 using the time dilation formula. ■

Example 2.2.2. Having learned about the length contraction, Sonya wants to try to make an ellipse out of a circle by moving it. First she compares lengths differing slightly from 1 meter to see how much a meter stick should shrink before she can actually notice the difference by merely eyeballing it. She decides that she can tell the difference between 100 cm and 95 cm by merely looking at them, i.e., if somebody showed her a meter stick (100 cm long) and a 95-cm stick, not next to the first one of course but separated by a large enough distance, she could tell that the first stick is longer.

Next, she tries to see how fast a circle with a diameter of 100 cm should move horizontally so she could see it as a vertical ellipse. Clearly, it has to move at such a speed that the horizontal diameter shrinks at least to 95 cm. She uses Equation (2.7) with $L = 95$ cm and $L_0 = 100$ cm: $95 = 100\sqrt{1-\beta^2}$, whose solution gives $\beta = 0.312$, or $v = 0.312c$, or $v = 9.4 \times 10^7$ m/s, or 209 million mph. Little wonder we do not see people and objects shrink as they move by us! ■

When I introduced length contraction in Section 1.3, I was careful—in fact I was forced—to measure the location of the two ends of the moving length *simultaneously according to the stationary observer*. A practical way of doing this was to have two fire crackers at the two ends of the length explode simultaneously and make marks *in their immediate vicinity* in the stationary reference frame. If the marks are not made immediately next to the fire crackers, the image will not be the true image of the length. The following example elucidates the point I'm trying to make.

Example 2.2.3. Let a rod $\overline{AB}$ of rest length $2L$ move with speed v along the x-axis of an observer, with A trailing B as shown in Figure 2.5. When the center C of the rod reaches the origin, a camera with its pinhole located on the z-axis a distance b from the origin takes a picture of the moving rod. What does the image of the rod look like?

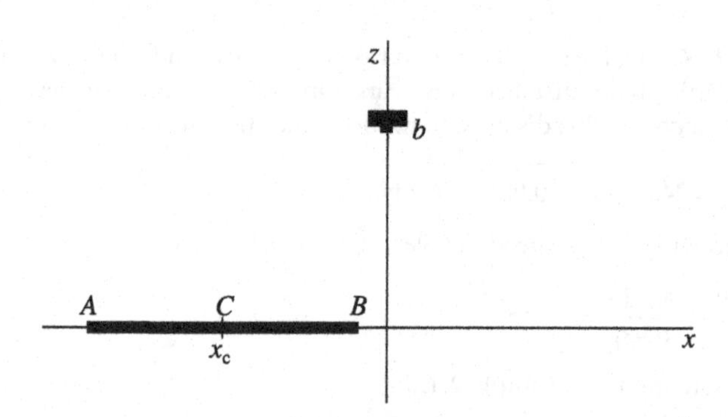

FIGURE 2.5 Location of the center of the rod when the ray from A is emitted. A little later, the center moves to x_c' and the light ray from B is emitted.

Let $t = 0$ be the time that the center reaches the origin *and* the time that all light rays forming the image reach the pinhole. For the light from A to reach the pinhole at $t = 0$, it must have been emitted earlier (at a negative time). Let x_c be the location of C when A emits its ray. When the center is at x_c, A is at $x_c - L/\gamma$ (remember length contraction). Therefore, its distance from the pinhole is $\sqrt{(x_c - L/\gamma)^2 + b^2}$, and it takes light $\sqrt{(x_c - L/\gamma)^2 + b^2}/c$ to reach the camera. This is also the time for the center to cover $|x_c|$. Therefore,

$$\frac{|x_c|}{v} = \frac{\sqrt{(x_c - L/\gamma)^2 + b^2}}{c} \quad \text{or} \quad |x_c| = \beta\sqrt{(x_c - L/\gamma)^2 + b^2}.$$

Squaring both sides gives

$$x_c^2 = \beta^2(x_c^2 + L^2/\gamma^2 - 2x_c L/\gamma + b^2) \quad \text{or} \quad \frac{x_c^2}{\gamma^2} + \frac{2\beta^2 L}{\gamma}x_c - \beta^2(b^2 + L^2/\gamma^2) = 0.$$

Solve this quadratic equation for x_c to obtain

$$x_c = -\beta^2\gamma L \pm \sqrt{\beta^4\gamma^2 L^2 + \beta^2\gamma^2(b^2 + L^2/\gamma^2)},$$

which you should be able to simplify to

$$x_c = -\beta^2\gamma L - \beta\gamma\sqrt{b^2 + L^2}, \tag{2.8}$$

where I chose the negative sign because $x_c < 0$ for all nonzero values of β. The coordinate of A can now be determined (you should supply all the missing steps):

$$x_A = x_c - L/\gamma = -\gamma\left(L + \beta\sqrt{b^2 + L^2}\right). \tag{2.9}$$

You can also find the time when the light from A was emitted. Denote this time by t_A, and note that

$$\begin{aligned} c^2 t_A^2 &= x_A^2 + b^2 = \gamma^2\left(L + \beta\sqrt{b^2 + L^2}\right)^2 + b^2 \\ &= \gamma^2\left(\beta L + \sqrt{b^2 + L^2}\right)^2. \end{aligned}$$

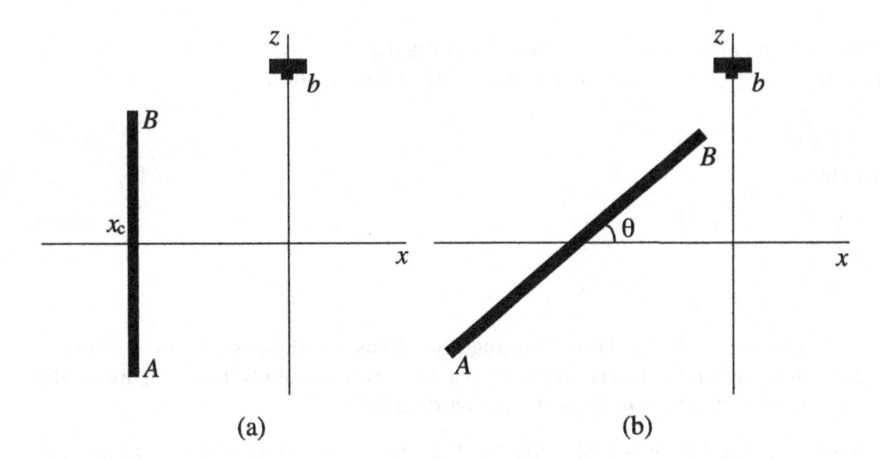

FIGURE 2.6 (a) The location of the rod when A emits its ray. (b) How the rod appears to the camera.

Therefore,

$$ct_A = -\gamma\left(\beta L + \sqrt{b^2 + L^2}\right). \tag{2.10}$$

Now let x_c' be the location of C when B emits its ray. You can calculate x_c' in exactly the same way I calculated x_c. I leave it for you to show that for the light ray from B to reach the pinhole at $t = 0$, we must have

$$x_B = x_c' + L/\gamma = \gamma\left(L - \beta\sqrt{b^2 + L^2}\right). \tag{2.11}$$

The photographic image is that of a rod that appears to have a length of

$$x_B - x_A = 2\gamma L,$$

which is *elongated* by the Lorentz factor!

Let me elaborate on this result. Suppose you have two identical cameras, one moving with the rod (in the rest frame of the rod), and one that I have been talking about so far. The image in the first camera is that of a rod of length $2L$. The image in the second camera—even though it is taking the picture of a rod that is *shorter* (its length being $2L/\gamma$)—is $2\gamma L$, which is *longer* than the image in the first camera by the Lorentz factor.

You can also find t_B from $c^2t_B^2 = x_B^2 + b^2$. You should be able to get

$$ct_B = \gamma\left(\beta L - \sqrt{b^2 + L^2}\right). \tag{2.12}$$

Clearly $t_B > t_A$, implying that A must emit its light earlier, as expected. ■

Example 2.2.4. What happens when the rod is parallel to the z-axis? Since the rod moves perpendicular to the direction of motion, its length does not change: It is $2L$. Figure 2.6(a) shows the rod at the moment that A sends its ray to the pinhole. For the ray to reach the pinhole at $t=0$, it must be emitted at $t_A = -\sqrt{(b+L)^2 + x_c^2}/c$, during which time the rod moves a distance of $|x_c|$. Solve the equation

$$|x_c| = \beta\sqrt{(b+L)^2 + x_c^2} \tag{2.13}$$

for x_c to get

$$x_A = x_c = -\gamma\beta(b+L). \tag{2.14}$$

Similarly,

$$x_B = x_c' = -\gamma\beta(b-L). \tag{2.15}$$

The z-coordinates of A and B do not change. So, A has coordinates $(x_A, 0, -L)$ when it sends its light ray, and B has coordinates $(x_B, 0, L)$ when it sends *its* ray. Figure 2.6(b) shows the rod as it *appears* to the camera.[4] Or does it?

To make sure that the rod actually appears as a *straight* line to the camera as shown in the figure, you have to show that all points between A and B lie on the straight line segment connecting the two end points. You should have no difficulty showing that if instead of the end points you choose any point along the rod with coordinates (x, z), you get

$$x = -\gamma\beta(b-z) \quad \text{or} \quad z = \frac{x+b}{\gamma\beta}. \tag{2.16}$$

This is the equation of a straight line with slope $\tan\theta = 1/(\gamma\beta)$, or $\cos\theta = \beta$. Furthermore, since $\cos\theta = (x_B - x_A)/\overline{AB}$, you get the interesting result:

$$\beta = \frac{-\gamma\beta(b-L) + \gamma\beta(b+L)}{\overline{AB}} \quad \text{or} \quad \overline{AB} = 2\gamma L.$$

The photographic image is that of a rod that is *stretched* by a factor of γ and *rotated* about the y-axis by an angle $\theta = \cos^{-1}\beta$! ■

You might have been surprised when I wanted to make sure that the rod was actually a *straight* line. "What else could it be?" you may ask. After all, motion should not bend a straight rod! However, while it is true that motion does not mechanically bend or deform a rod, the light rays that reach the pinhole may appear to have come from a curved rod! How is that possible?

Example 2.2.5. Reorient the rod so now it is parallel to the y-axis—which is perpendicular to the plane of Figure 2.6 and points into it. The light from an arbitrary point $(x, y, 0)$ of the rod reaches the pinhole located at $(0, 0, b)$ after traveling a distance of $\sqrt{x^2+y^2+b^2}$. This is also the time that the rod travels a distance of $|x|$. So, as before,

$$\frac{|x|}{v} = \frac{\sqrt{x^2+y^2+b^2}}{c} \quad \text{or} \quad |x| = \beta\sqrt{x^2+y^2+b^2}.$$

---

[4] The figure doesn't show the image itself, but what the camera is taking a picture of. The best way to understand this is to compare the picture with that of a stationary rod. If the rod does not move, the image is that of a line segment connecting $(0, 0, -L)$ and $(0, 0, L)$, which is just a single dot on the photographic plate as expected. When the rod moves, the picture will be that of a line segment connecting $(x_A, 0, -L)$ and $(x_B, 0, L)$, with x_A and x_B given by (2.14) and (2.15), respectively.

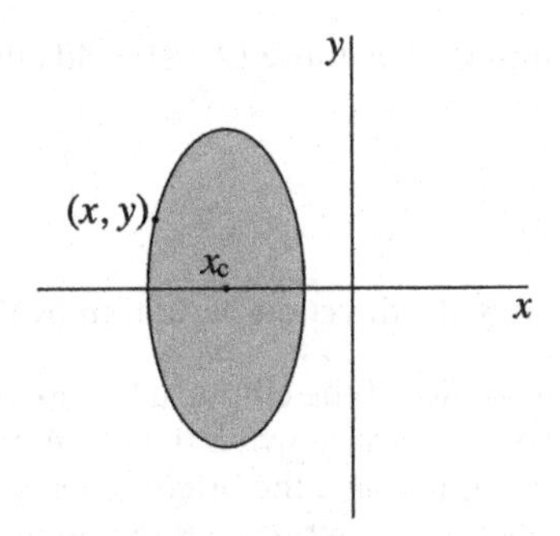

FIGURE 2.7 Location of the center of the circle when the ray from the point (x, y) is emitted. The z-axis is out of the page and the pinhole of the camera is on it at a distance b from the origin.

You can now easily show that

$$\frac{x^2}{\beta^2\gamma^2 b^2} - \frac{y^2}{b^2} = 1. \tag{2.17}$$

This is the equation of a hyperbola in the xy-plane! The branch $x = -\beta\gamma\sqrt{y^2+b^2}$ describes the shape of the rays coming to the camera from the rod. ■

Example 2.2.6. In Example 2.2.2, you saw the transformation of a circle into an ellipse whose axis along the direction of motion was shrunk by the Lorentz factor. In the meantime, you have seen that the *photograph* of a moving rod is not that of a shrunken rod, but that of an *elongated* one (see Example 2.2.3). So, it is instructive to see how a moving circle appears to a pinhole camera.

Let the circle of radius a be in the xy-plane, centered on the x-axis, and moving along that axis with speed v. As soon as the center of the circle reaches the origin, a pinhole camera located on the z-axis at a distance b from the origin, takes its picture. Figure 2.7 shows the location of the shrunken circle when the light ray from an arbitrary point (x, y) on it is emitted. This ray covers a distance of $\sqrt{x^2+y^2+b^2}$ at the same time that the center of the ellipse covers a distance of $|x_c|$. Therefore,

$$\frac{|x_c|}{v} = \frac{\sqrt{x^2+y^2+b^2}}{c} \quad \text{or} \quad x_c^2 = \beta^2\left(x^2+y^2+b^2\right).$$

Substitute for y^2 in this equation from the equation of the ellipse, which is

$$\frac{(x-x_c)^2}{(a/\gamma)^2} + \frac{y^2}{a^2} = 1, \tag{2.18}$$

and obtain

$$x_c^2 = \beta^2\left[x^2 + a^2 - \gamma^2(x-x_c)^2 + b^2\right]. \tag{2.19}$$

You should solve the quadratic equation in x_c that results from (2.19) and show that

$$x_c = \beta^2 x - \frac{\beta}{\gamma}\sqrt{a^2+b^2}. \tag{2.20}$$

Substituting this in the equation of the ellipse (2.18) yields the equation of what the camera sees:

$$\frac{\left(x+\gamma\beta\sqrt{a^2+b^2}\right)^2}{\gamma^2 a^2}+\frac{y^2}{a^2}=1, \tag{2.21}$$

which is an ellipse *elongated* along the direction of motion by the Lorentz factor.

It is interesting to note that the center of the ellipse at $(-\gamma\beta\sqrt{a^2+b^2}, 0)$ is farther and farther to the left as $v \to c$. This is because when the circle moves close to the speed of light, it will take less time for it to reach the origin. Therefore, for the rays to reach the pinhole at the same time that the center reaches the origin, they would have to be emitted earlier. ■

I have so far discussed how to find the location of the light-emitting points that form the image on the photographic plate. The *actual* image may not have any resemblance to the locus of these points,[5] because to obtain that image, you'll have to know how to map the three-dimensional points onto the two-dimensional plane of the photograph. You'll have to wait until Chapter 7 to see how to do that.

2.3 RELATIVISTICITY

All the preceding discussions dealt with speeds approaching light speed. How important are relativistic effects of time dilation and length contraction in daily motions? For $v << c$, the relativistic factor γ is very nearly 1. Therefore, a convenient measure of "relativisticity" is $\gamma - 1$, i.e., how much γ differs from its slow value. Approximate $(1+\epsilon)^x$ by $1+x\epsilon$, where both x and ϵ could be positive or negative and $|\epsilon| << 1$. Now note that

$\gamma - 1$ measures relativisticity.

$$\gamma - 1 = \frac{1}{\sqrt{1-\beta^2}} - 1 = (1-\beta^2)^{-1/2} - 1 \approx 1 + \tfrac{1}{2}\beta^2 - 1 = \tfrac{1}{2}(v/c)^2, \quad |\beta| << 1. \tag{2.22}$$

This result is useful when we investigate time dilation and length contraction of objects moving with speeds that are much smaller than the speed of light. For example the difference between L_0, the rest length, and L, the length in motion, to a very good approximation, is given by

$$L_0 - L = \gamma L - L = (\gamma - 1)L \approx \tfrac{1}{2}(v/c)^2 L. \tag{2.23}$$

This equation expresses the length shrinkage in terms of L. You can also write the same shrinkage in terms of L_0:

$$L_0 - L = L_0 - \frac{L_0}{\gamma} = L_0\left(1 - \sqrt{1-\beta^2}\right)$$

---

[5] As an example, you may be surprised to know that the image of the moving rod of Example 2.2.4 on the photographic plate is just a dot!

$$\approx L_0\left[1-\left(1-\tfrac{1}{2}\beta^2\right)\right]=\tfrac{1}{2}(v/c)^2L_0. \tag{2.24}$$

Therefore, it does not matter which length you use on the right-hand side because L_0 and L are almost equal for $v/c << 1$. Defining the fractional difference in length as

Defining the fractional difference.

$$\frac{\delta L}{L_0}=\frac{L_0-L}{L_0}\approx\frac{\delta L}{L},$$

you see that relativisticity is simply the fractional difference in length.

The same procedure can be used to find the difference between the proper and coordinate time intervals. Thus,

$$\Delta t-\Delta\tau=(\gamma-1)\Delta\tau\approx\tfrac{1}{2}(v/c)^2\Delta\tau\approx\tfrac{1}{2}(v/c)^2\Delta t. \tag{2.25}$$

The last approximation follows from the fact that $\Delta t\approx\Delta\tau$ when $v << c$. Defining the fractional difference in the time interval as in the case of length, you can write the last equation as

$$\frac{\delta(\Delta t)}{\Delta t}=\frac{\Delta t-\Delta\tau}{\Delta t}\approx\frac{\delta(\Delta t)}{\Delta\tau}=\tfrac{1}{2}(v/c)^2. \tag{2.26}$$

Example 2.3.1. Example 2.2.2 showed how hard (actually impossible) it was to shrink a meter stick by a mere 5 cm. Take the fastest and largest man-made object, a 100-meter-long satellite moving at about 8000 m/s. How much does it shrink as it passes by? Use Equation (2.24) to get

$$L_0-L\approx\frac{1}{2}\left(\frac{8000}{3\times10^8}\right)^2\times100=3.56\times10^{-8}\text{ m}=35.6\text{ nm},$$

which is the size of a (large) molecule! It is impossible to see the difference between 100 m and another length that differs from it by the size of a molecule.

What about time dilation? By how much does a clock slow down when placed on a fast vehicle and kept in motion for hours? A jet plane flies for 10 hours according to a clock on the ground. Assuming a speed of 270 m/s, Equation (2.25) gives

$$\Delta t-\Delta\tau\approx\frac{1}{2}\left(\frac{270}{3\times10^8}\right)^2\times(10\times3600)=1.46\times10^{-8}\text{ sec}=14.6\text{ ns}.$$

Although this is too small a time interval to measure by ordinary clocks, it is easily measurable by atomic clocks. In fact, a test very similar to this was done in 1975, and, to within experimental errors, the result agreed with the prediction of relativity. ■

Note 2.3.2. *A physical process for which $\gamma-1>>1$ is called* ***ultra-relativistic****; that for which $\gamma-1<<1$ is called* ***non-relativistic****.*

Definition of ultra- and non-relativistic processes.

While at ultra-relativistic speeds, strange phenomena, such as a daughter getting older than her father, are possible, even the fastest and the farthest round trip achievable today—an Apollo mission to the Moon, for example—yields only minute fractions of a second as a time dilation effect.

Example 2.3.3. The crew of Apollo 23 goes to the Moon with a speed of 15 km/s. It spends 10 hours exploring the Moon, and comes back with the same speed. The captain of the spaceship has just had a baby when he leaves on the mission. The whole trip takes 24 hours for the crew.

Time dilation for a Moon trip.

To feel the enormity of Apollo's speed, let's estimate its travel time from New York to Los Angeles (a distance of about 5000 km):

$$\text{time} = \frac{5000 \text{ km}}{15 \text{ km/s}} = 333.33 \text{ s} = 5.55 \text{ min},$$

an extraordinarily fast vehicle by any human standard!

How many hours does the captain spend on the way from Earth to the Moon? The round-trip travel time is $24 - 10 = 14$ hours. Since the speed is the same outbound and inbound, the time for each leg of the trip is half this amount, i.e., 7 hours or $7 \times 3600 = 25{,}200$ s.

How many more seconds does it take the baby for her father to reach the Moon? Here we want to find the difference between the proper and the coordinate time. Since the speed is small compared to light speed (the ratio v/c is 15/300,000 or 0.00005), we can use (2.25) with $\Delta\tau$ on the right-hand side, because we know $\Delta\tau$ (it is 7 hours or 25,200 s):

$$\Delta t - \Delta\tau \approx \tfrac{1}{2}(v/c)^2 \Delta\tau = \tfrac{1}{2}(0.00005)^2 \times 25200 = 0.0000315 \text{ s}.$$

So, it will take the captain 25,200 s to get to the Moon, during which time the baby has aged 25200.0000315 s.

How many more seconds has the baby aged than her father when Apollo 23 returns, assuming that the time difference develops only when the spaceship is in motion?[6] The aging difference is twice the difference for each leg of the trip, or $2 \times 0.0000315 = 0.000063$ s.

How far is the Moon from Earth according to the Earth observers? Since we know the speed and the time, we can find the distance by multiplying:

$$\text{distance} = \text{speed} \times \text{time} = 15 \times 25200.0000315 = 378000.00045 \text{ km}.$$

What is the Earth-Moon distance according to the crew? Use the same formula:

$$\text{distance} = \text{speed} \times \text{time} = 15 \times 25200 = 378000 \text{ km}.$$

The Earth-Moon distance has shrunk by only 0.00045 km or 45 cm (less than a foot and a half) for the crew! ■

Examples 2.3.1 and 2.3.3 show that even for humanly extraordinary speeds, the relativistic effects are immeasurably small. The dramatic relativistic effects of time dilation and length contraction demonstrated in Examples 2.1.6 and

[6] Strictly speaking, this is not true! While in ultra-relativistic cases we could ignore the motion of the planet, here the speed of the Moon is comparatively not as small as the speed of the planet relative to the spaceship. Therefore, the time interval spent on the Moon is also dilated, and should be taken into account. You should try incorporating this into the solution!

2.2.1 require enormous vehicular speeds, speeds unavailable to our species now or in the near future. Stated differently, ordinary phenomena, in which relevant speeds are much smaller than light speed, obey classical ideas. It does not mean that relativity gives wrong results when applied to ordinary phenomena, it just means that for ordinary phenomena the relativistic effects are so small as to be unobservable by crude measuring devices such as ordinary watches and meter sticks. In fact,

The Correspondence Principle

Note 2.3.4. *All the formulas and concepts of relativity reduce to the corresponding Newtonian formulas and concepts when all relevant speeds become much smaller than light speed.*

For example, if v/c is much smaller than 1, then $\Delta\tau \approx \Delta t$ and $L \approx L_0$, respectively, as expected in classical physics.

While it is fairly straightforward to show that *reducing* relativistic formulas leads to Newtonian formulas, *generalizing* the latter to relativistic formulas is nontrivial. For example, the innocent-looking generalization of the definition of acceleration as the rate of change of velocity, $a = dv/dt$, in a particular reference frame is wrong because its integration leads to $v = at$, which does not prevent v from surpassing the speed of light. You will see how acceleration is defined correctly in Section 6.8.

2.4 THE TWIN PARADOX

You may very well feel uneasy about time dilation, especially when the aging process comes into consideration. So it is worth examining this topic in some detail. Sam and Sonya are twins. As soon as they are born, Sam is put on the spaceship Marinarus heading to one of the planets of the star Epsilon Eridani about 11 light years away from Earth. The speed of Marinarus is $0.9c$. Both the ground control and the crew of the spaceship agree to send "happy birthday" signals to the twins on each anniversary of their birth.

On the first anniversary, the ground control sends the first "happy birthday" signal. They wait for a month, then 2 months, then the whole second year to receive Sonya's "happy birthday" message, all in vain. They send Sam his second happy birthday signal and wait another year. Still no signal! On the third anniversary, they repeat their annual message, and they urge the crew to respond. Still no response! Although it has given up hope by now, the ground control nevertheless sends a "Happy fourth, Sam!" knowing that the message is probably aimless. Marinarus is now considered "lost in space," but the communication channel is left open just in case. Exactly 131 days after the fourth message is sent the ground crew receives the signal saying "Happy first, Sonya!" Puzzled by this strange time warp, the ground control consults its theoretical physicist for an explanation. His crude response is as follows: There are two kinds of time delays involved. One is the relativistic time dilation which stretches the year in the spaceship to something more than a year. The second is the time delay due to the travel time of the radio signal from the spaceship to the Earth. His exact response is explained as follows.

Connecting traveling twin's birthday signals to ground twin's birthday.

The spaceship clock measures the proper time for the following two events: Lift-off, when Sam is zero year old, and departure of birthday signal, when Sam is one year old. Not aware of relativistic effects, the crew members assume that Sonya is also one; so, they send the message "Happy first, Sonya!" on Sam's first birthday. Denote the time passed according to the Earth due to the time dilation by Δt_1. Then

$$\Delta t_1 = \gamma \Delta \tau = \frac{\Delta \tau}{\sqrt{1-\beta^2}}. \tag{2.27}$$

Thus, Δt_1 is when—according to the Earth—the crew in Marinarus sends the birthday message. However, this message is sent at a very far distance. In fact, the spaceship is precisely at a distance of $v\Delta t_1$, when the signal is issued. This signal, being an EM wave, travels this distance with speed c. So, it takes the signal

$$\Delta t_2 = \frac{v\Delta t_1}{c} = \beta \Delta t_1 = \frac{\beta \Delta \tau}{\sqrt{1-\beta^2}}$$

to reach the Earth *after* it is issued. The Earth observers will, therefore, receive the signal after the time interval

$$\begin{aligned}\Delta t = \Delta t_1 + \Delta t_2 &= \frac{1+\beta}{\sqrt{1-\beta^2}}\Delta \tau \\ &= \frac{\sqrt{1+\beta}\sqrt{1+\beta}}{\sqrt{(1-\beta)(1+\beta)}}\Delta \tau = \sqrt{\frac{1+\beta}{1-\beta}}\,\Delta \tau.\end{aligned} \tag{2.28}$$

If you substitute $\beta = v/c = 0.9$ and $\Delta \tau = 1$ year, you get

$$\Delta t = \sqrt{\frac{1+0.9}{1-0.9}} \times 1 \text{ year} = 4.3589 \text{ years} = 4 \text{ years and } 131 \text{ days.}$$

How do we separate the delay due to the signal travel from the relativistic effect of time dilation? In other words, given Δt, the *total* elapsed time measured by the Earth clock, how can we calculate from it Δt_1 and Δt_2? This is a relevant question because in actuality, it is the total time that is measured by the Earth clocks, and we may be interested in the relativistic effect only. Furthermore, since *each reference frame has a unique time* (see Note 1.2.4), we can eliminate the delay time Δt_2 (once we have separated it from Δt) by relaying messages to outposts located at strategic distances from the main control room.

From Equations (2.27) and (2.28) and the fact that $\Delta t_2 = \beta \Delta t_1$, you can easily obtain

$$\Delta t_1 = \frac{\Delta t}{1+\beta}, \quad \Delta t_2 = \frac{\beta}{1+\beta}\Delta t.$$

For instance, the theoretical physicist, foreseeing the relativistic time dilation, can suggest outposts located approximately

$$c\Delta t_2 = \frac{0.9c}{1+0.9} 4.3589 \text{ years} \cdot c$$

or about 2.065 light years apart. These correspond to locations that Marinarus reaches just at the time that the crew sends their annual signals. In this case, the signal will be received immediately after it leaves the spaceship. The usefulness of these outposts comes from their ability to record the information immediately. Thus, when the crew of Marinarus sends the code "Happy first, Sonya!" the inhabitants of the outpost can immediately look at their clock and see that 2.29 years have passed (instead of one year), as expected from relativistic time dilation.

Is there something magical about the spaceship that keeps people young? Do we start staying young as soon as we jump on a spaceship headed into outer space? This seems to be contradictory to the first postulate of relativity which gives equal rights to all RFs. To understand the question, look at the same situation this time from the point of view of the spaceship RF. Marinarus's crew see Earth receding from them at the speed of $0.9c$ and after one year they send their signal and wait for the corresponding signal from the Earth. In fact, without going any further, we can switch Earth and the spaceship, and all their corresponding devices and people everywhere in the foregoing discussion of the trip, and come up with the spaceship version of the journey. Thus, for instance, the crew of Marinarus will receive the "Happy first, Sam!" 4 years and 131 days after departure. *In each RF, the moving twin ages less.*

How can this be? How can both twins see the other one grow less rapidly? What if we introduce a TV set that continually monitors the other twin? Then Sonya sees Sam as a 1 year old when she is more than 4 years old. On her 13th birthday she sees her brother as only 3. When she is married and has a child of her own at the age of 26, the monitor shows Sam as only 6. Finally, at the age of 91, in a retirement house where Sonya is spending the few remaining years of her life, she hardly recognizes her brother's face on the monitor appearing as a young 21-year-old man.

Such a state of affairs is hard to accept. What is worse, we have to accept the fact that Sam sees on *his* TV screen the youthfulness of his twin sister! Is there no way to determine once and for all who ages less? As long as the two are moving at constant speed relative to each other, no. Both RFs conclude exactly the same thing because time is a relative physical quantity. We are used to an absolute time because all the speeds we deal with are so small that we never experience the dilation of time. Thus, a watch in a car, a clock in a ship, and a timer in a plane all are treated the same as the watch on our wrist.

"Wait a minute," you may say. "There *is* a way of determining whether Sam stays younger or Sonya. Bring them face to face! Reverse the course of Marinarus on Sam's first birthday and bring it back to Earth. Then, when Sam is 2 years old, he will face his twin sister. Will she be younger, as the Marinarus crew may suspect, due to the motion of Earth clocks relative to them, or older as the ground control may calculate due to time dilation?" Stay tuned until Chapter 4 for the answer.

2.5 PROBLEMS

2.1. Take the most rigid rod you can find, and hit one end of it with a hammer. The rod *as a whole* starts to move because it is rigid.[7] Actually not! It takes time for the information that one end of the rod was hit to reach the other end, because of Note 2.1.7. Now go to the rest frame of the rod, which is now moving relative to the hammer. Assume that the hammer hits the rod in such a way as to cause (the front end of) it to stop. But the other end knows nothing about the hammer yet. So, it keeps moving! What does this say about the concept of "rigidity" in relativity?

2.2. In this problem you'll learn more about superluminal transverse speeds.

(a) Show that the angle that maximizes Equation (2.6) is given by $\cos\theta = \beta$.
(b) Substitute this in (2.6) to obtain $(v_{tr})_{\max} = c\beta\gamma$.
(c) Show that $(v_{tr})_{\max}$ is larger than c for any $\beta > 1/\sqrt{2}$.
(d) What speed makes $(v_{tr})_{\max}$ ten times faster than light? What is the angle corresponding to this speed?

2.3. Consider an MM clock moving horizontally with speed β relative to observer O. Denote its length in motion by L and at rest by L_0. Let Δt_1 be the time it takes light to go from the emitter to the mirror according to O. Let Δt_2 be the time it takes light to go from the mirror to the emitter according to O.

(a) Show that

$$c\Delta t_1 = L + \beta c\Delta t_1, \quad c\Delta t_2 = L - \beta c\Delta t_2.$$

(b) Show that a "tick" according to O is

$$\Delta t = \Delta t_1 + \Delta t_2 = \frac{2L/c}{1-\beta^2}.$$

(c) Now use the time dilation formula with $\Delta\tau = 2L_0/c$ to derive the length contraction formula.

---

[7] If you hit one end of a slinky, the other end does not move, at least not immediately.

2.4. The spaceship Enterprise goes to a planet in a star system far away with a speed of $0.9c$, spends 6 months on the planet, and comes back with a speed of $0.95c$. The entire trip takes 5 years for the crew.

(a) How far is the planet according to Earth observers?
(b) How long did it take the crew to get to the planet?
(c) How long did the entire trip take for the Earth observers?

2.5. A rocket ship leaves the Earth at a speed of $0.8c$. When a clock on the rocket says 1 hour has elapsed, the rocket ship sends a light signal back to Earth.

(a) According to Earth clocks, when was the signal sent?
(b) According to Earth clocks, how long after the rocket left did the signal arrive back on Earth?
(c) According to the rocket clock, how long after the rocket left did the signal arrive back on Earth?

2.6. A bicycle wheel of rest radius R is rotating in such a way that the rim has a linear speed of $0.866c$. What is the circumference of the rim? What is the length of the spokes in motion? But spokes are perpendicular to the direction of motion! Discuss whether in relativity anything can be considered "incompressible" or "rigid." See also Problem 2.1.

2.7. The spaceship Viking goes to a planet in a star system 30 light years away from Earth with a speed of $0.99c$, spends 1 year on the planet, and then returns home. The entire trip takes 10 years for the crew.

(a) How far is the planet according to the crew?
(b) How long does it take the crew to get to the planet?
(c) How long does it take the crew to return to Earth?
(d) What is the speed of the crew on return? Warning! The distance for the crew is *not* the same as the distance on their way to the planet.
(e) How far is the Earth from the planet according to the crew on their return?
(f) How long did the entire trip take for the Earth observers?

2.8. The spaceship Diracus goes to a planet in a star system with a speed whose Lorentz factor is γ, spends 1 year on the planet, and then returns home with a speed whose Lorentz factor is 4γ. The captain of the spaceship is 29 years old and has just had a newborn son. The entire trip takes 11 years for the crew. The odometer of the spaceship shows that the "milage" for the round trip is a quarter of the Earth-planet distance as measured by Earth observers.

(a) What is the outbound speed? The inbound speed?
(b) What is the Earth-planet distance according to the Earth observers?
(c) What is the Earth-planet distance according to the crew on their way to the planet? On their way back?
(d) How long does it take the crew to go to the planet? To return?

(e) Who is older, the son or the father when the ship lands on Earth? By how many years?

2.9. A rod of rest length L_0 moves with speed v along the positive x'-direction of observer O'. The rod makes an angle θ_0 with respect to the x-axis of its rest frame.

(a) Find the length of the rod as measured by O'.
(b) Find the angle θ the rod makes with the x'-axis as measured by O'.

2.10. A flasher produces a flash of light every second when at rest. It is moving away from you at $0.9c$.

(a) How frequently does it flash according to you?
(b) By how much does the distance between you and the flasher increase between consecutive flashes?
(c) How long after the emission of a given flash does it reach you?
(d) How often do you receive the flashes?

2.11. Charged pions are produced in many collisions in accelerators. They decay in their rest frame according to

$$N(t) = N_0 e^{-t/T},$$

where $T = 2.6 \times 10^{-8}$ s is their mean life. A burst of charged pions is produced at the target of an accelerator and it is observed that one percent of them decay after covering a distance of 1 m from the target. What is the Lorentz factor for pions and how fast are they moving?

2.12. Derive Equations (2.8), (2.9), and (2.10).

2.13. Derive Equations (2.11) and (2.12).

2.14. Derive Equations (2.14) and (2.15).

2.15. Derive Equation (2.16).

2.16. Find t_A and t_B, the times at which A and B emit their light rays when the rod is oriented along the z-axis as in Example 2.2.4.

2.17. Derive Equation (2.17).

2.18. In Example 2.2.5, find t_A and t_B, the times at which A and B emit their light rays when the rod is oriented along the y-axis.

2.19. Derive Equations (2.20) and (2.21).

CHAPTER 3

Lorentz Transformation

You have seen how time *intervals* and lengths warp as you go from one RF to another. Time intervals and lengths are time and space differences between *two* events. Is there a way of relating *individual* events as seen in two different RFs? To answer this question and find formulas that connect the time and position of an event in two different RFs, I'll first look at an analogous familiar case, from which you will get insight for the relativistic case.

3.1 "DERIVATION" OF EUCLIDEAN DISTANCE

Imagine creatures, *flatlanders*, who live in a three-dimensional world, but are aware of only two dimensions called *length* and *width*. Although they do not have a physiological experience with the third dimension, *height*, they have ways of measuring it with an *altimeter*. Carrying an altimeter, each flatlander can tell how "high" (s)he has "climbed" from some initial point.

Flatlanders know how to describe a point in terms of a coordinate system (they choose x and y for the two dimensions in which they live) drawn in their flatland and how to express the distance between them in terms of their coordinates: if points P_1 and P_2 have coordinates (x_1, y_1) and (x_2, y_2), respectively, then the distance between them is

$$d(P_1, P_2) = \sqrt{(x_2 - x_1)^2 + (y_2 - y_1)^2}. \tag{3.1}$$

Peculiar property of flatlanders' elevators.

The flatlanders' "elevators" have a strange but important property. Generally, an elevator in (invisible) vertical motion is seen to be moving horizontally as well. However, for some observers, a particular set of elevators, which they call *proper elevators*, do not move in the length and width directions, although the elevators are known to be moving because the riders inside can see the dials on their altimeter change. These same elevators move in the flatland of other observers. The first group of observers are said to belong to the same RF and the proper elevators designate that RF. Each RF has its own proper elevators, which are used to distinguish among different RFs. A quantity, which

Special Relativity. DOI:10.1016/B978-0-12-810411-8.00003-1

flatlanders call *epols*, differentiates one RF from another. One RF has an epols *relative* to another, which the flatlanders calculate as follows.

How flatlanders differentiate one RF from another.

Let Mas, a flatlander, ride on one of his proper elevators while Aynos, another flatlander, looks at the motion of this elevator in her RF. She keeps track of the position (x, y) and—using an altimeter—the height z of the elevator as it moves in her two-dimensional world. If the elevator is found at point P_1 with coordinates (x_1, y_1) and at height z_1, and a little later at point P_2 with coordinates (x_2, y_2) and at height z_2, then the epols of Mas relative to Aynos, which is a vector quantity, is defined as

$$\mathbf{e} \equiv \frac{(x_2 - x_1, y_2 - y_1)}{z_2 - z_1} \equiv \frac{(\Delta x, \Delta y)}{\Delta z} = \left(\frac{\Delta x}{\Delta z}, \frac{\Delta y}{\Delta z}\right). \tag{3.2}$$

If $\mathbf{e} = 0$, then $x_1 = x_2$ and $y_1 = y_2$. This means that Mas is not moving relative to Aynos, and their RFs coincide.

For simplicity, I'll assume that the relative epols between Mas and Aynos is constant. This means that the ratios of Δx and Δy to Δz do not depend on where Aynos started her observation. So, if the elevator moves from P_1' to P_2' and the difference between the coordinates and heights of these two points are $\Delta x'$, $\Delta y'$, and $\Delta z'$, then the epols $\mathbf{e} = (\Delta x', \Delta y')/\Delta z'$ is the same as before.

Since height seems to be a quantity that is present in all aspects of the flatlanders' physical universe and in a general way they seem to be able to "move" in it, they find it convenient to treat height as a "third dimension" and define *events* as points of this mysterious three-dimensional space. Thus, an event E is described by three numbers (x, y, z). Furthermore, they want to generalize their distance formula (3.1) to include this new dimension.

They start by ignoring the y-coordinate and concentrating on the new two-dimensional geometry of the xz-plane, because they know how to incorporate y in the distance formula. So, the big challenge for them is to find a generalized distance formula for two events having coordinates (x_1, z_1) and (x_2, z_2). The generalization can be dictated only by observation. As our experience with Euclidean geometry has shown, a cherished and "obvious" assumption such as Euclid's Fifth Postulate[1] can lead only to one geometry out of many possible geometries. So, the question of which geometry is the right geometry can be answered only by observation.

To make this observation, the flatlanders have Mas and Aynos perform a crucial experiment. Mas, standing at the origin ($X = 0$) of his coordinate system and riding his proper elevator, is to record the height when his elevator passes the events E_1 and E_2, while Aynos records both the heights and x-coordinates

[1] Recall that Euclid's Fifth Postulate states that from a point not located on a line, you can draw only one line parallel to the given line. In the nineteenth century, this postulate was changed in two different ways, giving birth to two new geometries: you can draw infinitely many lines, or none at all.

of the events. Let Z_1 and Z_2 be the heights as measured by Mas, and (x_1, z_1) and (x_2, z_2) be the coordinates measured by Aynos. The result of this crucial experiment can be summarized as

$$z_2 - z_1 = \frac{Z_2 - Z_1}{\sqrt{1+e^2}}, \tag{3.3}$$

where e is the epols of Mas's RF relative to Aynos's. Equation (3.3) is a very important relation, because as you'll see below, it alone determines the form that the generalized distance takes.

Flatlanders know that if (the generalization of the) distance is to have any meaning, they have to assign to it the following property:

Note 3.1.1. *If two events happen to have equal coordinates in all but one dimension, then the distance between them is the (absolute value of the) difference between the values of the unequal coordinates.*

Thus, for the two events E_1 and E_2 above, whose coordinates according to Mas are $(0, Z_1)$ and $(0, Z_2)$, the distance *must* be

$$d(E_1, E_2) = Z_2 - Z_1,$$

assuming that $Z_2 > Z_1$. Now they note that $e = (x_2 - x_1)/(z_2 - z_1)$.[2] Therefore, using (3.3), they obtain

Flatlanders discover a distance formula incorporating height.

$$\begin{aligned} d(E_1, E_2) &= Z_2 - Z_1 = (z_2 - z_1)\sqrt{1+e^2} \\ &= (z_2 - z_1)\sqrt{1 + [(x_2 - x_1)/(z_2 - z_1)]^2} \\ &= \sqrt{(x_2 - x_1)^2 + (z_2 - z_1)^2}. \end{aligned} \tag{3.4}$$

This is of course the familiar Euclidean distance, which we, the creatures familiar with the third dimension, have known for centuries. However, precisely because of this familiarity, we can appreciate the ingenuity of the flatlanders in arriving at the same result without having a physiological contact with the third dimension. You should keep in mind that the distance formula (3.4) depends crucially on the purely empirical result of Equation (3.3).

3.1.1 Coordinate Transformation

Equation (3.4), being the distance between two events, must not depend on the coordinates used to describe them. What kind of coordinate transformation leaves (3.4) unchanged? I pick two observers in Flatland, O and O'. O uses (x, z) for his coordinates and O' uses (x', z') for hers. I am looking for a relation between the primed and unprimed coordinates that respects (3.4). So,

[2] Remember that there is no y coordinate. So, e has only one component.

I start with

$$x' = a_0 + a_1 x + a_2 z, \quad z' = b_0 + b_1 x + b_2 z \tag{3.5}$$

where the a's and b's are unknown constants to be determined. I have to assume a linear relation (no x^2, z^3, or any other functions), because I don't want straight lines in O to change to curves other than straight lines in O' (see Problem 3.1).

Now I take any two events E_1 and E_2 with coordinates (x_1, z_1) and (x_2, z_2) in O and (x'_1, z'_1) and (x'_2, z'_2) in O' and write $\Delta x' \equiv x'_2 - x'_1$ and $\Delta z' \equiv z'_2 - z'_1$ in terms of $\Delta x \equiv x_2 - x_1$ and $\Delta z \equiv z_2 - z_1$:

$$\begin{aligned} \Delta x' &= a_1(\Delta x) + a_2(\Delta z) \\ \Delta z' &= b_1(\Delta x) + b_2(\Delta z). \end{aligned} \tag{3.6}$$

Next, I demand that the two distances calculated in O and O' be equal:

$$(\Delta x')^2 + (\Delta z')^2 = (\Delta x)^2 + (\Delta z)^2.$$

If I substitute (3.6) in the left-hand side (LHS) of this equation and rearrange the terms, I get

$$\text{LHS} = (a_1^2 + b_1^2)(\Delta x)^2 + (a_2^2 + b_2^2)(\Delta z)^2 + 2(a_1 a_2 + b_1 b_2)(\Delta x)(\Delta z).$$

If this is to be equal to the right-hand side (RHS) for *any* values of Δx and Δz (corresponding to *any* two points in the plane), then the coefficients must satisfy the following relations:

$$a_1^2 + b_1^2 = 1, \quad a_2^2 + b_2^2 = 1, \quad a_1 a_2 + b_1 b_2 = 0. \tag{3.7}$$

Since (3.6) holds for any Δx and Δz, I am free to use any convenient values to find the unknowns in (3.7). I take advantage of this freedom and let $\Delta x = 0$ in Equation (3.6) to obtain

$$e = \frac{\Delta x'}{\Delta z'} = \frac{a_2}{b_2} \quad \text{or} \quad a_2 = e b_2.$$

Now I substitute this in the second equation of (3.7) to get $b_2 = \pm 1/\sqrt{1+e^2}$. With b_2 so determined, I can find a_2: $a_2 = \pm e/\sqrt{1+e^2}$. Using the values of a_2 and b_2 (either plus or minus) in the last equation of (3.7), I get $b_1 = -e a_1$ (you should verify this).

Now I have to decide which sign to choose. If I choose the negative sign, then Equation (3.5) would yield $z' = -z$ for the special case in which $e = 0$ and $b_0 = 0$ (then $b_1 = -e a_1$ is also zero).[3] But this is nonsense, because for these parameters, the two coordinates should coincide! Therefore, for b_2 (as well as a_2), I have to choose the positive sign. Now I put $b_1 = -e a_1$ in the first equation of (3.7) and obtain $a_1 = \pm 1/\sqrt{1+e^2}$. Again, I have to choose the

[3] We know that it is possible for e to be zero. That b_0 can also be zero will become clear later.

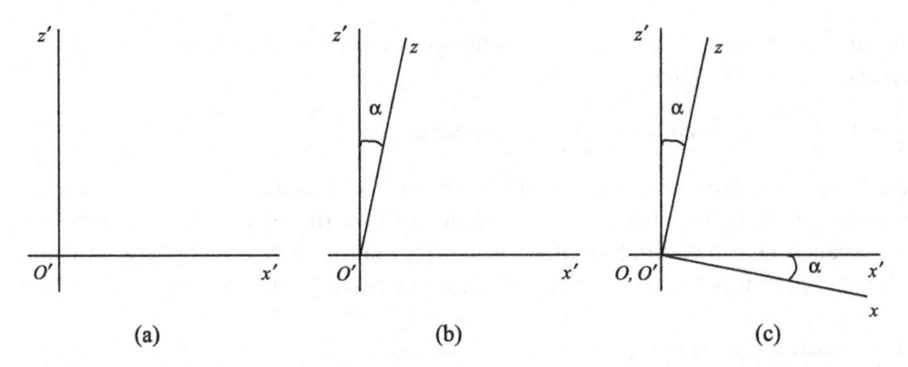

FIGURE 3.1 (a) The two perpendicular axes of O'. (b) The z-axis of O makes an angle α with the z'-axis of O'. (c) The axes x and z turn out to be perpendicular to each other.

plus sign, because otherwise Equation (3.5) would yield $x' = -x$ for $e = 0$ and $a_0 = 0$. Substituting these values in Equation (3.5) I get

$$\begin{aligned} x' &= a_0 + \frac{1}{\sqrt{1+e^2}}x + \frac{e}{\sqrt{1+e^2}}z \\ z' &= b_0 - \frac{e}{\sqrt{1+e^2}}x + \frac{1}{\sqrt{1+e^2}}z. \end{aligned} \tag{3.8}$$

These two equations give me the coordinates (x', z') of an "event" in O' if I know the coordinates (x, z) of that event in O, which has epols e relative to O'. In particular, if I substitute the coordinates $(x = 0, z = 0)$ of the "event" describing the origin of O in (3.8), I get (a_0, b_0) as the coordinates of the origin of O in O'. Therefore, a_0 and b_0 describe a "translation" of the origin of O relative to the origin of O'. I'll set them equal to zero (i.e., I assume that the two origins coincide) and write

$$\begin{aligned} x' &= \frac{1}{\sqrt{1+e^2}}x + \frac{e}{\sqrt{1+e^2}}z \\ z' &= -\frac{e}{\sqrt{1+e^2}}x + \frac{1}{\sqrt{1+e^2}}z. \end{aligned} \tag{3.9}$$

What is the geometric meaning of Equation (3.9)? Aynos, observer O', draws her coordinates perpendicular to each other as shown in Figure 3.1(a). How does the z-axis of Mas, observer O, look in the $x'z'$-plane? The z-axis is a line; so it must have an equation in terms of x' and z'. In fact, since z is the set of all points whose x-values are zero, setting $x = 0$ in Equation (3.9), gives the parametric equation of the z-axis with z as the parameter. Eliminate z in the two equations to get $z' = x'/e$, which is a line passing through the origin O' and having $1/e$ as its slope.[4] Designate the angle between z and z' by α and

[4] So, flatlanders' "epols" is just the inverse of our "slope!"

recall that the slope of a line is the tangent of the angle between the line and the *horizontal* axis. Then

$$\frac{1}{e} = \tan\left(\frac{\pi}{2} - \alpha\right) = \cot\alpha = \frac{1}{\tan\alpha} \Rightarrow e = \tan\alpha$$

[see Figure 3.1(b)]. You can similarly obtain the equation of the x-axis by setting $z = 0$. This yields $z' = -ex'$, showing that the slope of the x-axis is $-e = \tan(-\alpha)$. Now conclude that the x-axis makes an angle of $-\alpha$ with the x'-axis, i.e., that *the x-axis is perpendicular to the z-axis* [Figure 3.1(c)], and

Orthogonal transformation leaves the distance (3.4) invariant.

$$\begin{aligned} x' &= \frac{1}{\sqrt{1+e^2}}x + \frac{e}{\sqrt{1+e^2}}z = x\cos\alpha + z\sin\alpha \\ z' &= -\frac{e}{\sqrt{1+e^2}}x + \frac{1}{\sqrt{1+e^2}}z = -x\sin\alpha + z\cos\alpha. \end{aligned} \tag{3.10}$$

These equations describe a rigid rotation of the coordinate system, the **orthogonal transformation** in two dimensions.

Figure 3.1 also sheds some light on the mysterious properties of "height" and the strange behavior of elevators: Mas's proper elevator moves along the z-axis, and thus has no projection on the x-axis. Therefore, it does not move in the O reference frame. However, as the figure shows, the same elevator moves in Aynos's x' direction.

3.2 SPACETIME DISTANCE

We are three-dimensional creatures living in a four-dimensional *spacetime* and having no physiological contact with the fourth dimension. However, just like the flatlanders, we have instruments, our clocks, which can measure the invisible fourth dimension. We can tell if something occurred before or after something else, but cannot picture these the same way we do two buildings that are 500 meters apart.

Just like the flatlanders, we want to find a formula for the distance between two events (points in the spacetime continuum having four coordinates) and the transformation that leaves that distance unchanged. And just like flatlanders, we know that the formula can be dictated only by experimental observation. Furthermore, as in the flatlanders' case, we confine ourselves to two dimensions, one of which is time. So, what is that crucial experiment that gives us the invariant distance formula for (the two-dimensional) spacetime? It consists of comparing the rate of the ticking of two special identical clocks in relative motion. We have already performed this experiment with the MM clocks and have obtained the equivalent of Equation (3.3). It is given in Equation (2.4), which I rewrite here:

$$\Delta t = \frac{\Delta\tau}{\sqrt{1 - v^2/c^2}}. \tag{3.11}$$

To be more transparent, I let O move in the positive x'-direction of O'. Two events E_1 and E_2 occur in such a way that O measures the proper time interval $\Delta\tau \equiv \tau_2 - \tau_1$ between them. Since proper time is measured by a clock that is present at both events, the coordinates of E_1 and E_2 according to O are $(0, c\tau_1)$ and $(0, c\tau_2)$. The same events have coordinates (x_1, ct_1) and (x_2, ct_2) according to O'. Now I use a notation that is very common in relativity, namely $s \equiv c\tau$, and write Equation (3.11) as

$$ct_2 - ct_1 = \frac{s_2 - s_1}{\sqrt{1 - v^2/c^2}}. \tag{3.12}$$

This is the crucial relation that will dictate the formula for the distance in spacetime. Its similarity to Equation (3.3) is striking. Retracing exactly the same steps following Equation (3.3), with ct replacing z, and invoking Note 3.1.1, I can identify $s_2 - s_1$ as the distance $d(E_1, E_2)$. Then, with $v = (x_2 - x_1)/(t_2 - t_1)$, Equation (3.12) gives me

$$ct_2 - ct_1 = \frac{d(E_1, E_2)}{\sqrt{1 - (x_2 - x_1)^2/(ct_2 - ct_1)^2}},$$

or

$$d(E_1, E_2) = \sqrt{c^2(t_2 - t_1)^2 - (x_2 - x_1)^2}.$$

Because of the negative sign under the radical and the possibility of getting an imaginary number, it is more common to square both sides and write

We discover spacetime distance!

$$d^2(E_1, E_2) = c^2(t_2 - t_1)^2 - (x_2 - x_1)^2. \tag{3.13}$$

This is the invariant **spacetime distance**, which is sometimes called the **Minkowskian distance** and is applicable to the two-dimensional spacetime.

3.3 LORENTZ TRANSFORMATION

Now that you have the formula for the distance, you can find the transformation that leaves this distance invariant. As in Section 3.1, start with the linear relations

$$x' = a_0 + a_1 x + a_2 t, \quad t' = b_0 + b_1 x + b_2 t. \tag{3.14}$$

Take any two events E_1 and E_2, where E_1 has coordinates (x_1, t_1) in O and (x'_1, t'_1) in O' and E_2 has coordinates (x_2, t_2) in O and (x'_2, t'_2) in O'. Substitute the coordinates of E_1 and E_2 in (3.14), subtract the resulting equations, and use the obvious notations $\Delta x = x_2 - x_1$, etc., to obtain

$$\begin{aligned}\Delta x' &= a_1(\Delta x) + a_2(\Delta t)\\ \Delta t' &= b_1(\Delta x) + b_2(\Delta t).\end{aligned} \tag{3.15}$$

Now demand that the distance defined in (3.13) be the same for O and O', i.e., that

$$(c\Delta t')^2 - (\Delta x')^2 = (c\Delta t)^2 - (\Delta x)^2. \tag{3.16}$$

Next substitute for $\Delta x'$ and $\Delta t'$ from (3.15) in terms of Δx and Δt on the LHS of (3.16) and rearrange to get

$$\text{LHS} = (c^2b_1^2 - a_1^2)(\Delta x)^2 + (c^2b_2^2 - a_2^2)(\Delta t)^2 + 2(c^2b_1b_2 - a_1a_2)(\Delta x)(\Delta t).$$

If the two sides of Equation (3.16) are to be equal for *any* values of Δx and Δt (corresponding to *any* two events in the spacetime plane), then the following relations must hold:

$$c^2b_1^2 - a_1^2 = -1, \quad c^2b_2^2 - a_2^2 = c^2, \quad c^2b_1b_2 - a_1a_2 = 0. \tag{3.17}$$

When $\Delta x = 0$, i.e., when E_1 and E_2 occur at the same point on the x-axis, you must get $\Delta x'/\Delta t' = v$, the speed of that point relative to O'.[5] Setting $\Delta x = 0$ in Equation (3.15) gives $v = \Delta x'/\Delta t' = a_2/b_2$ or $a_2 = b_2 v$. Furthermore, since E_1 and E_2 occur at the same point, Δt is the proper time. It follows from Equation (3.11) and the second equation in (3.15) that $b_2 = 1/\sqrt{1-(v/c)^2}$. With a_2 and b_2 so determined, the last equation in (3.17) yields $b_1 = a_1 v/c^2$. Substitute this in the first equation of (3.17) to get $a_1 = \pm 1/\sqrt{1-(v/c)^2}$. For similar reasons as in the flatlanders' case, only the positive sign is acceptable.

Using the symbols β and γ introduced earlier, rewrite Equation (3.14) as

$$x' = a_0 + \gamma x + \gamma\beta ct, \quad t' = b_0 + \gamma(\beta/c)x + \gamma t, \tag{3.18}$$

or

$$\begin{aligned} x' &= x'_0 + \gamma(x + \beta ct) \\ ct' &= ct'_0 + \gamma(\beta x + ct), \end{aligned} \tag{3.19}$$

where the time coordinate has been multiplied by c to give it the same dimension as the space coordinate. In (3.19) $x'_0 \equiv a_0$ and $ct'_0 \equiv cb_0$ are the coordinates of the origin of O relative to O'. It is common to set $x'_0 = 0 = ct'_0$, corresponding to the convention that the two origins coincide, and to which convention I adhere for almost all applications in the book. Then we obtain the celebrated **Lorentz transformation**:

Lorentz transformation.

$$\begin{aligned} x' &= \gamma(x + \beta ct) \\ ct' &= \gamma(\beta x + ct). \end{aligned} \tag{3.20}$$

If you solve these two equations for x and t, you get

$$\begin{aligned} x &= \gamma(x' - \beta ct') \\ ct &= \gamma(-\beta x' + ct'). \end{aligned} \tag{3.21}$$

Inverse Lorentz transformation.

This is called the Lorentz transformation **inverse** to Equation (3.20). You should have guessed (3.21) because the only difference between O and O' is that O moves in the *positive* x'-direction of O', while O' moves in the *negative* x-direction of O.

[5] For example, E_1 could be $(0, 0)$, corresponding to O being at the common origin of the two observers, and E_2 could be $(0, t_2)$, corresponding to when O measures his time to be t_2.

Note 3.3.1. *Let O and O' be two observers in motion relative to one another. Let E be an event with coordinates (x, ct) in O and (x', ct') in O'. Assume that the origin of time for both O and O' is when the origins of their space axes coincide. Then the coordinates of E are related via the* **Lorentz transformation (LT)**

$$x' = \gamma(x + \beta ct)$$
$$ct' = \gamma(\beta x + ct), \tag{3.22}$$

where β, the relative fractional speed of the two observers, is positive (negative) if O moves along the positive (negative) x'-axis of O'.

Equation (3.22) is sometimes written in matrix form:

$$\begin{pmatrix} x' \\ ct' \end{pmatrix} = \begin{pmatrix} \gamma & \gamma\beta \\ \gamma\beta & \gamma \end{pmatrix} \begin{pmatrix} x \\ ct \end{pmatrix} \tag{3.23}$$

What is the geometric meaning of Equation (3.22)? Let O' draw her coordinates perpendicular to each other as shown in Figure 3.2(a). How does the ct-axis of O look in the $x'ct'$-plane? Since ct is the set of all points whose x are zero, setting $x = 0$ in Equation (3.22) gives the equation of the ct-axis: $ct' = x'/\beta$, which is a line passing through the origin O' and having a slope $1/\beta$. Designate the angle between ct and ct' by α and note that

$$\frac{1}{\beta} = \tan\left(\frac{\pi}{2} - \alpha\right) = \cot\alpha = \frac{1}{\tan\alpha} \quad \Rightarrow \quad \beta = \tan\alpha. \tag{3.24}$$

If $\beta > 0$, then O is moving in the positive x'-direction and the ct-axis should be to the right of the ct'-axis. This is consistent with the slope of ct being positive: $\tan\left(\frac{\pi}{2} - \alpha\right) > 0$. Therefore, positive angle α corresponds to a clockwise rotation of the ct-axis away from the ct'-axis. Similarly, negative angle corresponds to a counterclockwise rotation away from the ct'-axis, i.e., having a slope $\tan\left(\frac{\pi}{2} + |\alpha|\right) < 0$ [see Figure 3.2(b) in which I assume that β is positive].

You can similarly obtain the equation of the x-axis by setting $t = 0$. This yields $ct' = \beta x'$, showing that the slope of the x-axis is $\beta = \tan\alpha$. The conclusion is that the x-axis makes an angle of α with the x'-axis and therefore an angle of $\pi/2 - 2\alpha$ with the ct-axis, i.e., that the *x-axis is not perpendicular to the ct-axis* [Figure 3.2(c)]. For $\beta > 0$, the x-axis is above the x'-axis and the ct-axis is to the right of the ct'-axis. Therefore, the angle between the unprimed axes is acute. On the other hand, for $\beta < 0$, the x-axis is below the x'-axis and the ct-axis is to the left of the ct'-axis. Therefore, the angle between the unprimed axes is obtuse.

If the x'-axis is perpendicular to the ct'-axis, then the x-axis is not perpendicular to the ct-axis.

Most often we are interested in space and time *intervals* between two events. If events E_1 and E_2 have respective coordinates (x_1, ct_1) and (x_2, ct_2) relative to O and (x_1', ct_1') and (x_2', ct_2') relative to O', then even for the more general case

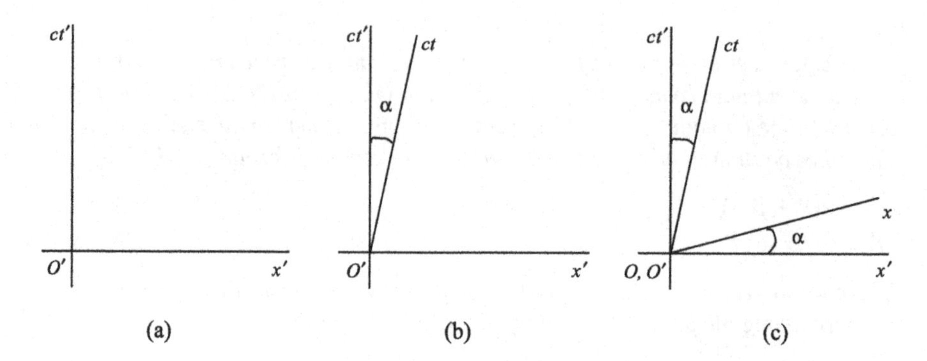

FIGURE 3.2 (a) The two perpendicular axes of O'. (b) The ct-axis of O makes an angle α with the ct'-axis of O'. Here $\beta > 0$. If β were negative, then the ct-axis would be to the left of the ct'-axis. (c) The axes x and ct are *not* perpendicular to each other. The angle between them is acute if $\beta > 0$ and obtuse if $\beta < 0$.

of (3.19), you get

$$\begin{aligned} \Delta x' &= \gamma(\Delta x + \beta c \Delta t) \\ c\Delta t' &= \gamma(\beta \Delta x + c\Delta t), \end{aligned} \tag{3.25}$$

where Δ denotes the difference between the coordinates of the two events. The order of subtraction is irrelevant as long as the *same* order is applied to all Δ's. For example, if $\Delta x = x_2 - x_1$, then all the six Δ's must correspond to coordinates of E_1 being subtracted from those of E_2.

Lorentz transformations are sometimes used to simplify the analysis of physical processes. Because of the motion of an object, it may be hard to answer questions regarding processes in which the object may be involved. For example, when various parts of an object emit light rays that arrive at a certain point at a given time, it may be easier to go to the rest frame of the object, identify the events corresponding to the emission and arrival of light signals, and then Lorentz transform those events to the original RF.

A corollary of this procedure, which also makes use of Note 1.2.1, is the following:

> Note 3.3.2. *If a physical process is impossible in one reference frame, then it is impossible in **all** reference frames.*

3.4 EXAMPLES OF THE USE OF LT

When y and z coordinates are involved, you should keep in mind that they don't change as you go from one RF to another, because they are perpendicular to the direction of motion—which is always assumed to be x—and perpendicular lengths are not contracted.

Example 3.4.1. SIMULTANEITY
Lorentz transformation quantifies all the qualitative examples I discussed in the previous chapters. In this example I calculate the time difference between the explosion of the two firecrackers on the train as measured by Sam when Sonya sees them simultaneously (see Figure 1.4 of Chapter 1). I let Sam be O' and Sonya O, and assume that right is positive. Thus, Sonya is moving in the positive direction, implying a positive β. Suppose E_1 is the explosion of A and E_2 the explosion of B. Let the origin of Sonya's coordinate system be where she is located, i.e., the middle of the train, whose length is L. Also let the time of explosion be when Sonya's clock starts ticking, i.e., at time zero. Under these assumptions, E_1 and E_2 have respective coordinates $(-L/2, 0)$ and $(+L/2, 0)$ according to Sonya.

Finding time interval for one observer when simultaneous for another.

If you take Δx to be $x_2 - x_1$ then $\Delta x = L$ and $\Delta t = t_2 - t_1 = 0$. The second equation of (3.25) yields

$$c\Delta t' = ct'_2 - ct'_1 = \gamma\beta\Delta x = \gamma\beta L.$$

For example, if the train moves at half the speed of light and the car is 30 m long, then $\beta = 0.5$, $\gamma = 1/\sqrt{1-(0.5)^2} = 1.155$, and

$$c\Delta t' = 1.155 \times 0.5 \times 30 = 17.32 \text{ m}$$

or

$$\Delta t' = t'_2 - t'_1 = \frac{17.32}{3 \times 10^8} = 5.8 \times 10^{-8} \text{ s} = 58 \text{ ns},$$

i.e., $t'_2 = t'_1 + 58$ ns. This shows that t'_2, the time of the occurrence of B according to Sam, is larger than t'_1, the time of the occurrence of A. Thus, Sam sees A 58 ns before B, as I explained qualitatively in Chapter 1. ■

Example 3.4.2. TIME DILATION AND LENGTH CONTRACTION
You can use LT to derive both time dilation and length contraction formulas. If $\Delta x = 0$, then Δt is the proper time between the two events. The second equation of (3.25) gives

Time dilation and length contraction from Lorentz transformation.

$$c\Delta t' = \gamma(0 + c\Delta t) \quad \text{or} \quad \Delta t' = \gamma\Delta t = \frac{\Delta t}{\sqrt{1-\beta^2}},$$

which is the relation between proper and coordinate time as given in Equation (2.4).

For length contraction, first you have to recall from the beginning of Section 1.3 that to measure the length of a moving object, you have to spot the two ends of the object *at the same time*, i.e., you can call $\Delta x'$ "the length of the object" only if $\Delta t' = 0$. This happens only if $\beta\Delta x + c\Delta t = 0$ or $c\Delta t = -\beta\Delta x$. If you substitute this in the first equation of (3.25), you get

$$\Delta x' = \gamma(\Delta x - \beta^2\Delta x) = \frac{1}{\sqrt{1-\beta^2}}(1-\beta^2)\Delta x = \sqrt{1-\beta^2}\,\Delta x,$$

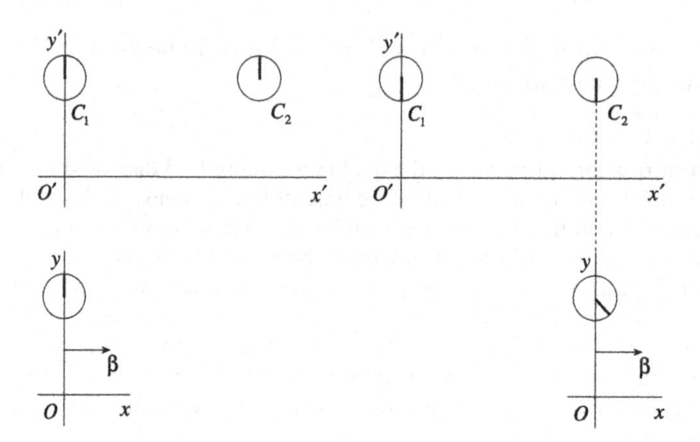

FIGURE 3.3 Comparison of two synchronized clocks in Sam's RF with the clock in Sonya's RF. When Sonya's clock reaches C_2, a camera in each RF takes a picture of the two clocks. This picture should look the same whether Sam looks at it or Sonya.

which is the length contraction formula (2.7). You can obtain the same formula more easily if you start with the LT having the O' coordinates on the right-hand side, i.e., the inverse LT. ■

Lorentz transformation relates coordinates of *events* as seen by two observers. Events are points in a spacetime continuum. We know how to represent an event when we explicitly have a ct-axis. But there is another way of representing an event.

Note 3.4.3. *The presence of a clock at a point makes an event E with spacetime coordinates (x, ct), where x is the spacial coordinate of the clock and t is the reading of the clock.*

Example 3.4.4. COMPARISON OF CLOCKS

Two clocks C_1 and C_2 are separated by a distance L and synchronized in Sam's rest frame O'. Sonya is moving with speed β from left to right as shown in Figure 3.3. Take the origin of time for both observers to be the event at which Sonya's clock passes C_1.

Q: When Sonya's clock reaches C_2, how do the readings of the two clocks compare?

A: Sam sees Sonya move from C_1 to C_2 with speed β, so he concludes that it took Sonya $\Delta t = L/(c\beta)$ to cover that distance. Therefore, C_2 must read $L/(c\beta)$. On the other hand, he knows that Sonya's clock slows down by a factor γ, so he reads Sonya's clock and sees that it shows $L/(c\beta\gamma)$.

Q: How do you derive the same result using Lorentz transformation?

A: Let E_1 and E_2 be the events at which Sonya's clock coincides with C_1 and C_2, respectively. The coordinates of E_1 are $(0, 0)$ in both RFs. The coordinates of E_2 in Sam's

RF are $(L, L/\beta)$. Using the *inverse* LT, you get

$$x_2 = \gamma(L - \beta L/\beta) = 0$$
$$ct_2 = \gamma(-\beta L + L/\beta) = \frac{\gamma}{\beta}(1-\beta^2)L = \frac{L}{\gamma\beta},$$

which agrees with the previous derivation.

Now consider the sequence of events from the point of view of Sonya. She sees the two clocks, but *they are not synchronized* according to her! Problem 3.11 shows the details of how the final result is that C_2 reads $L/(c\beta)$ while Sonya's clock reads $L/(c\beta\gamma)$, as before. ■

A peculiar paradox!

I started the quantitative discussion of relativity with time dilation, in which a Michelson-Morley clock was held *vertically* and moved horizontally with high speed (see Section 2.1). Then I used the Pythagorean theorem to obtain the formula (2.4) connecting proper time and the coordinate time and clearly indicating that the moving clock *slows down*. In the meantime, we have discovered the phenomenon of length contraction. This creates a dilemma.

Suppose Sonya and Sam are on a spaceship moving at 86.6% light speed, and each carries a Michelson-Morley clock. Sonya holds her clock vertically while Sam keeps his horizontal. As they travel from Earth to a distant planet, they both age the proper time interval of $\Delta t = 5$ years. You know how to calculate Earth time interval $\Delta t'$ from Sonya's clock: Equation (2.4) gives how much the Earth inhabitants age:

$$\Delta t' = \frac{\Delta t}{\sqrt{1-(v/c)^2}} = \frac{5}{\sqrt{1-(0.866)^2}} = 10 \text{ years}.$$

Sam's clock, on the other hand, moving along its length, is shrunk for the Earth people, and thus must *tick faster* than Sonya's. Are the Earth people, therefore, to conclude that the proper time is more than 5 years? Which clock is right? Have the Earth people aged 10 years or something else? The following example resolves this strange double timing.

Example 3.4.5. MM CLOCKS PERPENDICULAR TO EACH OTHER

To begin with, assume a length of 3 m for the MM clock and note that each tick of Sonya's clock is 20 ns according to Sonya, and

$$\Delta t' = \frac{20}{\sqrt{1-(0.866)^2}} = 40 \text{ ns}$$

MM clock orientation and its time measurement.

according to Earth. Next, place Sam's clock so that the emitter is at the origin, and the mirror is +3 m away from the origin. Let E_1 be the emission of light and E_2 its reflection. Sam measures Δx_{21} to be 3 m and Δt_{21} to be the time it takes light to go from one end of the clock to the other, i.e., $3/(3 \times 10^8)$ or 10^{-8} s.[6] It follows that $c\Delta t_{21}$ is also 3 m. What is $c\Delta t'_{21}$, the time interval between the same two events according

[6] I am employing the convention that $\Delta x_{21} = x_2 - x_1$.

to Earth? With $\beta = 0.866$, you get $\gamma = 1/\sqrt{1-0.866^2} = 2$, and the second equation in (3.25) yields

$$c\Delta t'_{21} = \gamma(\beta\Delta x_{21} + c\Delta t_{21}) = 2(0.866 \times 3 + 3) = 11.196 \text{ m},$$

giving $\Delta t'_{21} = 11.196/(3 \times 10^8) = 3.732 \times 10^{-8}$ s, or 37.32 ns. So, the Earth people measure the trip of the light signal from the bottom (or left end) of Sam's clock to its top (or right end) to be only slightly less than an entire tick of Sonya's clock. Is the second half of the light trip as long as the first? That would be disastrous, because it would imply that Sam's clock is almost twice *slower* than Sonya's, and the Earth people must age almost 20 years, rather than 10 years, as Sonya's clock suggests! Before jumping to any conclusion, let's calculate the second half of the light signal's trip.

Denote the event of the arrival of the signal back to the bottom of the clock as E_3. Then using the same equation as above (with due consideration to signs), you get

$$c\Delta t'_{32} = \gamma(\beta\Delta x_{32} + c\Delta t_{32}) = 2[0.866 \times (-3) + 3] = 0.804 \text{ m}$$

yielding $\Delta t'_{32} = 0.804/(3 \times 10^8) = 0.268 \times 10^{-8}$ s, or 2.68 ns. Adding the two flight times, you get $37.32 + 2.68 = 40$ ns, exactly the same as the Earth measurement of the tick of Sonya's clock! ■

In Examples 2.2.3, 2.2.4, and 2.2.5, I examined how the image of a straight rod would look in a pinhole camera for various orientations of the rod. The discussion was slightly complicated because of the motion of the rod. The analysis of the shape of the photograph of a stationary rod is much simpler, of course. But a stationary rod and a moving one are connected via Lorentz transformation. Can we derive everything in the rest frame of the rod and Lorentz transform the result to the frame in which it is moving? The following examples give an affirmative answer to the question.

Example 3.4.6. ROD ALONG THE x-AXIS
The rod is as in Example 2.2.3. Let O be the rest frame of the rod, whose center is at the origin of O as shown in Figure 3.4(a). When the two origins coincide at $t = t' = 0$, the pinhole camera located at P_0 on the z'-axis in O' takes a picture, i.e., the rays from the two ends of the rod reach the pinhole of the camera on the z'-axis at $t = t' = 0$. Since the z'-axis is perpendicular to the direction of motion, P_0 has the same coordinates in O as in O'. So, at time $t = 0$, observer O sees the rays from the two ends reach P_0 on the z-axis. Let E_A and E_B be, respectively, the events at which the rays from A and B leave toward the pinhole.

Write the spacetime coordinates of these events in O: The coordinates of E_A are $(-L, ct_A)$, and of E_B are (L, ct_B), where $|ct_A|$ and $|ct_B|$ are just the length of the line segments $\overline{AP_0}$ and $\overline{BP_0}$, respectively. Therefore,

$$ct_A = ct_B = -\sqrt{L^2 + b^2}.$$

With time and x-coordinates provided, you can now find the coordinates in O':

$$x'_A = \gamma\left(-L - \beta\sqrt{L^2 + b^2}\right)$$
$$x'_B = \gamma\left(L - \beta\sqrt{L^2 + b^2}\right).$$

These are identical to Equations (2.9) and (2.11).

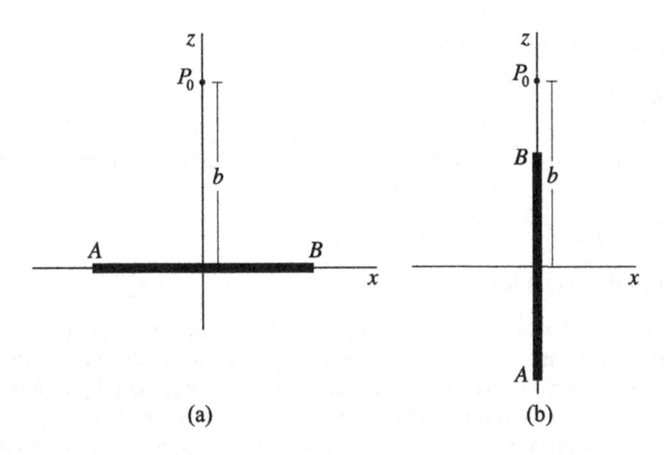

FIGURE 3.4 The rod is stationary at the origin of O. The point P_0 is where the pinhole of the camera will be instantaneously located as it passes by. The light rays from A and B reach P_0 at time $t = 0$.

Note that E_A and E_B occur simultaneously according to O. According to O',

$$ct'_A = \gamma\left(-\beta L - \sqrt{L^2 + b^2}\right)$$
$$ct'_B = \gamma\left(\beta L - \sqrt{L^2 + b^2}\right),$$

which agree with Equations (2.10) and (2.12). Since A is farther away from the pinhole than B, E_A must occur earlier than E_B so its light can reach the pinhole at the same time as that of B. ■

Example 3.4.7. ROD ALONG THE z-AXIS

The rod is as in Example 2.2.4. In the reference frame O it is as shown in Figure 3.4(b). The two-dimensional spacetime coordinates of E_A are $(0, ct_A)$, and of E_B are $(0, ct_B)$, where

$$ct_A = -(b + L), \quad ct_B = -(b - L).$$

With time and x-coordinates provided, you can now find the coordinates in O':

$$x'_A = \gamma\left[0 - \beta(b + L)\right] = -\gamma\beta(b + L)$$
$$x'_B = \gamma\left[0 - \beta(b - L)\right] = -\gamma\beta(b - L).$$

These are identical to Equations (2.14) and (2.15). Instead of its two ends, you can use an arbitrary point of the rod and obtain (2.16). I leave this as an exercise for you (see Problem 3.13). ■

Example 3.4.8. PHOTOGRAPH OF A CIRCLE

To illustrate the power of LTs, I reexamine Example 2.2.6, in which a moving circle (in the xy-plane of its rest frame O and centered at the origin) is photographed by a pinhole camera located on the z'-axis of O' a distance b from the origin. In the rest frame of the circle, all rays from it are emitted at the same time: $ct = -\sqrt{a^2 + b^2}$. Therefore, the x-coordinates are transformed to

$$x' = \gamma\left(x - \beta\sqrt{a^2+b^2}\right) \quad \text{or} \quad x = \frac{x'}{\gamma} + \beta\sqrt{a^2+b^2}.$$

Substitute this in the equation of the circle $x^2 + y^2 = a^2$, and note that $y = y'$ to get

$$\left(\frac{x'}{\gamma} + \beta\sqrt{a^2+b^2}\right)^2 + y'^2 = a^2,$$

which is identical to Equation (2.21). ■

Example 3.4.9. Observer O is moving in the positive x'-direction of observer O' with speed β. At the moment that the two origins coincide, observer O' sends a light signal (event E_1) to an object located on the x'-axis at a distance L. The light gets reflected from the object (event E_2) and the reflection is received by O (event E_3) and O' (event E_4). O reaches the object at event E_5. Assume that both observers are located at their respective origins. The aim of the exercise is to find the coordinates of all events in both RFs.

Q: What are the coordinates of E_1, E_2, and E_4 in O'?

A: This should be a no-brainer: $(0,0)$, (L,L), and $(0,2L)$, respectively.

Q: What are the coordinates of E_5 in O'?

A: The x' coordinate is L and the ct' coordinate is the time it takes O to reach the object. Therefore, the coordinates are $(L, L/\beta)$.

Q: What are the coordinates of E_1, E_2, E_4, and E_5 in O?

A: From inverse LT (or noting that O' is moving in the negative x-direction), you can find the following coordinates:

$$
\begin{aligned}
\text{for } E_1: &\quad (0,0);\\
\text{for } E_2: &\quad x_2 = \gamma(L - \beta L) = \sqrt{\frac{1-\beta}{1+\beta}}\,L,\\
&\quad ct_2 = \gamma(-\beta L + L) = \sqrt{\frac{1-\beta}{1+\beta}}\,L;\\
\text{for } E_4: &\quad x_4 = \gamma(0 - 2\beta L) = -2\gamma\beta L,\\
&\quad ct_4 = \gamma(0 + 2L) = 2\gamma L;\\
\text{for } E_5: &\quad x_5 = \gamma(L - \beta L/\beta) = 0,\\
&\quad ct_5 = \gamma(-\beta L + L/\beta) = L/\gamma.
\end{aligned}
$$

Note that $x_2 = ct_2$ just as $x'_2 = ct'_2$ because light must travel from E_1 to E_2 with the same speed for both observers. Note also that $x_5 = 0$ because the object and O will have the same space coordinates when O reaches the object.

The coordinates of E_3 are given neither in O nor in O'. However, you know that in O, the event is the reception of the light signal coming from E_2, which is at a distance of x_2 from O. Since O is at the origin, the reception occurs at $x_3 = 0$.

Q: What is the time coordinate of E_3 in O?

A: It takes light x_2/c to reach O *after* it is reflected. Therefore,

$$t_3 = x_2/c + t_2 = 2\sqrt{\frac{1-\beta}{1+\beta}}\frac{L}{c} \quad \text{or} \quad ct_3 = 2\sqrt{\frac{1-\beta}{1+\beta}}L$$

Q: What are the coordinates of E_3 in O'?

A: Use LT:

$$x'_3 = \gamma(0 + \beta ct_3) = 2\gamma\beta\sqrt{\frac{1-\beta}{1+\beta}}L = \frac{2\beta}{1+\beta}L$$

$$ct'_3 = \gamma(0 + ct_3) = 2\gamma\sqrt{\frac{1-\beta}{1+\beta}}L = \frac{2}{1+\beta}L.$$

Note that $x'_3/t'_3 = c\beta$ as expected, because, according to O', O goes from the origin to x'_3 in time t'_3. ■

Example 3.4.10. Two clocks C_1 and C_2 separated by a fixed distance L' are placed on the x'-axis of a reference frame O' with C_1 at the origin. The two clocks are synchronized in O'. Reference frame O is moving relative to O' in the positive direction of x' with speed β. Observer O_1 is at the origin of O and observer O_2 is placed strategically on the x-axis in such a way that O_1 and O_2 can read, respectively, C_1 and C_2 *at the same time.* O_1 records the reading as 12:00 (the zero time).

Q: What time does O_2 record and where is O_2 located in O?

A: Note that O_2 records *the time in* O' of a clock with x' coordinate L'.[7] Hence, the event of "recording," which in O occurs at $t = 0$, in O' has the coordinates (L', ct'). Using the inverse LT, you can obtain

$$x = \gamma(L' - \beta ct')$$

$$0 = \gamma(-\beta L' + ct') \quad \text{or} \quad t' = \frac{\beta L'}{c}.$$

Substituting t' in the first equation, you get $x = L'/\gamma$, which is the length contraction formula, as expected.

This rather confusing example has a very simple graphical representation in terms of spacetime geometry discussed in Chapter 4 (see Problem 4.15). ■

3.5 RELATIVISTIC LAW OF ADDITION OF VELOCITIES

Relativity prohibits any object to go faster than light speed. What happens, then, when a passenger on a super train fires a super gun, shooting bullets at almost light speed while the train is also moving at almost light speed? Would a platform observer see the bullet moving faster than light? Your intuition may lead you to believe that if the train moves at $0.9c$ and the bullet is fired in

---

[7] O_2 can take a picture of the face of C_2 as soon as it reaches O_2 and read the picture at his/her convenience.

the forward direction with a speed of $0.95c$ with respect to the train, then the platform observer should see the bullet move at $1.85c$. Your intuition is wrong, of course! Lorentz transformation is right and will give you the right answer.

Relativistic law of addition of velocities.

Suppose that the speed of an object (a bullet) relative to observer O is v_b. How does she measure this speed? She locates the bullet at two different times, measures the distance between those locations, and divides by the time interval. In other words, she picks two events $E_1 = (x_1, ct_1)$ and $E_2 = (x_2, ct_2)$ and takes the ratio of $\Delta x = x_2 - x_1$ to $\Delta t = t_2 - t_1$.

Observer O' looks at O as she makes her speed measurement. He sees the two events as (x'_1, ct'_1) and (x'_2, ct'_2) and concludes that the speed of the bullet v'_b is the ratio of $\Delta x'$ to $\Delta t'$, where $\Delta x'$ and $\Delta t'$ are related to Δx and Δt via the LT. Thus, by Equation (3.25) you can write

$$\begin{aligned}\frac{v'_b}{c} &= \frac{\Delta x'}{c\Delta t'} = \frac{\gamma(\Delta x + \beta c\Delta t)}{\gamma(\beta\Delta x + c\Delta t)} = \frac{\Delta x + \beta c\Delta t}{\beta\Delta x + c\Delta t} \\ &= \frac{\Delta x/(c\Delta t) + \beta c\Delta t/(c\Delta t)}{\beta\Delta x/(c\Delta t) + c\Delta t/(c\Delta t)} = \frac{v_b/c + \beta}{\beta(v_b/c) + 1}.\end{aligned}$$

Introduce two obvious notations: $\beta'_b \equiv v'_b/c$ and $\beta_b \equiv v_b/c$ and rewrite the relation above as

$$\beta'_b = \frac{\beta_b + \beta}{1 + \beta\beta_b}, \quad \text{or} \quad v'_b = \frac{v_b + v}{1 + v v_b/c^2}. \tag{3.26}$$

The law of addition of velocities (3.26) never violates relativity theory as you may verify by inserting some large values (but smaller than c, of course) for v and v_b and noting that v'_b comes out smaller than c as well. However, you can show quite generally that it is impossible to surpass the speed of light c by adding two speeds that are less than c, regardless of how close they are to c (see Problem 3.32). Thus, although the bullet is moving close to the speed of light in the train and the train is also moving close to light speed, the observer on the platform will measure a combined speed for the bullet that is smaller than light speed.

Equation (3.26) also agrees with the second postulate of relativity. If instead of a bullet, O sends a light beam so that $\beta_b = 1$, (3.26) gives

$$\beta'_b = \frac{1 + \beta}{1 + \beta} = 1.$$

Hence, O' also measures the speed of light to be c.

Example 3.5.1. LAV FROM COMPOSITE LORENTZ TRANSFORMATIONS

The relativistic law of addition of velocities can also be obtained from composite Lorentz transformation. Observer O moves relative to observer O' with speed β in the positive x'-direction. Observer O' moves relative to observer O'' with speed β' in the positive x''-direction. Therefore, O moves relative to O'' with some speed β'' in the positive x''-direction. What is β'' in terms of β and β'?

I'll use the matrix form of the Lorentz transformation (3.23). I can write

$$\begin{pmatrix} x' \\ ct' \end{pmatrix} = \begin{pmatrix} \gamma & \gamma\beta \\ \gamma\beta & \gamma \end{pmatrix} \begin{pmatrix} x \\ ct \end{pmatrix} \quad \text{and} \quad \begin{pmatrix} x'' \\ ct'' \end{pmatrix} = \begin{pmatrix} \gamma' & \gamma'\beta' \\ \gamma'\beta' & \gamma' \end{pmatrix} \begin{pmatrix} x' \\ ct' \end{pmatrix},$$

or

$$\begin{pmatrix} x'' \\ ct'' \end{pmatrix} = \begin{pmatrix} \gamma' & \gamma'\beta' \\ \gamma'\beta' & \gamma' \end{pmatrix} \begin{pmatrix} \gamma & \gamma\beta \\ \gamma\beta & \gamma \end{pmatrix} \begin{pmatrix} x \\ ct \end{pmatrix}.$$

Multiplying the matrices, I get

$$\begin{pmatrix} x'' \\ ct'' \end{pmatrix} = \begin{pmatrix} \gamma\gamma'(1+\beta\beta') & \gamma\gamma'(\beta+\beta') \\ \gamma\gamma'(\beta+\beta') & \gamma\gamma'(1+\beta\beta') \end{pmatrix} \begin{pmatrix} x \\ ct \end{pmatrix}.$$

If O moves relative to O'' with some speed β'', then I also have

$$\begin{pmatrix} x'' \\ ct'' \end{pmatrix} = \begin{pmatrix} \gamma'' & \gamma''\beta'' \\ \gamma''\beta'' & \gamma'' \end{pmatrix} \begin{pmatrix} x \\ ct \end{pmatrix}.$$

Therefore,

$$\begin{pmatrix} \gamma'' & \gamma''\beta'' \\ \gamma''\beta'' & \gamma'' \end{pmatrix} = \begin{pmatrix} \gamma\gamma'(1+\beta\beta') & \gamma\gamma'(\beta+\beta') \\ \gamma\gamma'(\beta+\beta') & \gamma\gamma'(1+\beta\beta') \end{pmatrix}.$$

Dividing $\gamma''\beta'' = \gamma\gamma'(\beta+\beta')$ by $\gamma'' = \gamma\gamma'(1+\beta\beta')$, I readily obtain (3.26). It is interesting that this procedure gives the combined γ factor directly in terms of the individual ones, whose derivation requires some algebra (see Problem 3.37). ■

Example 3.5.2. FIZEAU'S EXPERIMENT

The universal speed of light c assumes the vacuum as its medium of propagation. When light moves in a transparent medium, it slows down by a factor $n > 1$ called the *index of refraction* of the medium. Thus, in a *stationary* transparent medium of index of refraction n, light travels with the speed of c/n. What is the speed v' of light in that medium when the medium moves with speed v? Equation (3.26) holds the answer to this question. With $v_b = c/n$, it gives

$$v' = \frac{c/n + v}{1 + v/nc} = \left(\frac{c}{n} + v\right)\left(1 + \frac{v}{nc}\right)^{-1}.$$

For small v, you can expand the second parentheses using binomial series and keep terms only up to first power in v/c. Then you obtain

$$v' \approx \left(\frac{c}{n} + v\right)\left(1 - \frac{v}{nc}\right) \approx \frac{c}{n} + v\left(1 - \frac{1}{n^2}\right). \tag{3.27}$$

A transparent medium "drags" light with it.

This shows that the medium "drags" light with it. The non-relativistic law of addition of velocities predicts a complete drag, namely $v' = c/n + v$. But, when Fizeau performed an experiment in 1851 to measure the drag, he found that it disagreed with the classical prediction. His experiment showed that the drag was not 100%, but was lower by $1/n^2$, drag in agreement with (3.27). Although physicists found a way of explaining the discrepancy using electrodynamic arguments, it wasn't until after the discovery of the special theory of relativity that a simple satisfactory explanation was offered. ■

Example 3.5.3. The speed limit in Metropolis is $0.8c$. A driver is moving at $0.9c$ (with respect to the ground) as she passes a policeman without noticing him. The policeman, accelerating to $0.99c$ (also with respect to the ground) *instantaneously*, catches up with the speeder one hour later according to the speeder's clock.

Q: According to the speeder's clock, when did the policeman start chasing her and how far away from her was the policeman?

A: Assume that the origin of the speeder's coordinate system is when and where she passes the policeman. Let $(x_{\text{chase}}, ct_{\text{chase}})$ be the coordinates in the speeder's RF of the event at which the policeman starts his motion. According to the speeder, the policeman is moving in the negative x-direction at the rate of $0.9c$ before he starts chasing her. Therefore, $x_{\text{chase}} = -0.9ct_{\text{chase}}$. Then at t_{chase}, the policeman starts moving in the positive x-direction with a *relative* speed β, which *is not* $0.99c - 0.9c$. In fact, from the relativistic law of addition of velocities,

$$\beta = \frac{0.99 - 0.9}{1 - 0.99 \times 0.9} = 0.826.$$

Therefore, the speeder concludes that $x_{\text{chase}} = -0.826c(T - t_{\text{chase}})$, where T is one hour. Thus

$$0.9ct_{\text{chase}} = 0.826c(T - t_{\text{chase}}) \quad \text{or} \quad t_{\text{chase}} = \frac{0.826}{1.726}T$$

$$\text{or} \quad t_{\text{chase}} = 0.479 \text{ hour} = 1722.8 \text{ s},$$

and

$$x_{\text{chase}} = -0.9ct_{\text{chase}} = -0.9 \times 3 \times 10^8 \times 1722.8 = -4.65 \times 10^{11} \text{ m}.$$

This, by the way, is more than three times Earth-Sun distance!

Q: When did the policeman start chasing the speeder according to the policeman's clock?

A: Note that t_{chase} is the time *interval* between two events: speeder passes the policeman and policeman starts chasing her. The policeman measures the proper time between these two events. Thus,

$$\Delta\tau_1 = 0.479\sqrt{1 - 0.9^2} = 0.21 \text{ hour}.$$

Q: How long did it take the policeman to catch up with the speeder after she passed him?

A: The time between the start of the policeman's motion and the time that he catches up with the speeder is also proper time for the policeman, but the speed is now $0.826c$. For the speeder this time is $1 - 0.479 = 0.521$ hour. Therefore,

$$\Delta\tau_2 = 0.521\sqrt{1 - 0.826^2} = 0.29 \text{ hour},$$

and the total time for the policeman is $0.21 + 0.29 = 0.5$ hour.

Q: What are the coordinates (in the ground RF) of the event at which the policeman catches up with the speeder?

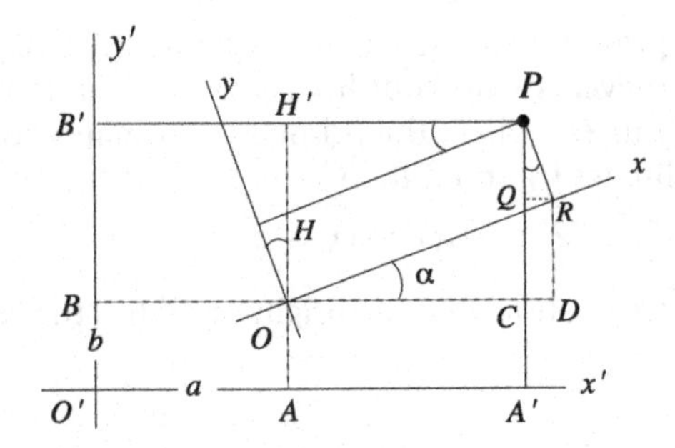

FIGURE 3.5 The same point P has different pairs of coordinates in different coordinate systems. Note that $a = \overline{O'A}$ and $b = \overline{O'B}$.

A: The coordinates in the speeder's RF are $x = 0$ and $t = T = 1$ hour. The speeder is moving in the positive x'-direction of the ground with speed $\beta = 0.9$. Therefore, $\gamma = 2.294$ and LT gives

$$x' = 2.294(0 + 0.9cT) = 2.065cT = 2.065 \text{ light hour} = 2.23 \times 10^{12} \text{ m}$$
$$ct' = 2.294(cT + 0) = 2.294cT = 2.294 \text{ light hour} = 2.478 \times 10^{12} \text{ m.}$$

A *light hour* is the distance light travels in one hour. ■

3.6 PROBLEMS

3.1. A line in the xz-plane has slope m and intercept b.

(a) Show that Equation (3.5) maps this line onto a line in the $x'z'$-plane.

(b) What are the slope and the intercept of the line in the $x'z'$-plane?

(c) Show that the transformation

$$x' = a_0 + a_1 x + a_2 z, \quad z' = b_0 + b_1 x^2 + b_2 z$$

transforms a straight line in the xz-plane into a parabola in the $x'z'$-plane.

3.2. Start with Equation (3.6) and provide all the missing steps that lead to Equation (3.8).

3.3. Derive Equation (3.10).

3.4. Use Figure 3.5 to prove the coordinate transformation

$$x' = a + x\cos\alpha - y\sin\alpha$$
$$y' = b + x\sin\alpha + y\cos\alpha.$$

3.5. In Euclidean space, the locus of points equidistant from the origin of a plane is a circle. What is the locus of points equidistant (in the spacetime distance sense) from the origin of a spacetime plane?

3.6. Verify that LT preserves the spacetime distance (3.13). In other words, if E_1 and E_2 are two events with coordinates (x_1, ct_1) and (x_2, ct_2) in O and (x'_1, ct'_1) and (x'_2, ct'_2) in O', where the primed coordinates are obtained from the unprimed coordinates by an LT, then

$$c^2(t'_2 - t'_1)^2 - (x'_2 - x'_1)^2 = c^2(t_2 - t_1)^2 - (x_2 - x_1)^2.$$

3.7. Starting with (3.19), show that the origin of O' has the coordinates

$$x_0 = \gamma(-x'_0 + \beta ct'_0)$$
$$ct_0 = \gamma(\beta x'_0 - ct'_0)$$

relative to O. Show also that the inverse of Equation (3.19) can be written as

$$x = x_0 + \gamma(x' - \beta ct')$$
$$ct = ct_0 + \gamma(-\beta x' + ct').$$

3.8. A ruler of length L is at rest in O with its left end at the origin. O moves from left to right with speed β relative to O' along the length of the ruler. The two origins coincide at time zero for both, at which time a photon is emitted toward the other end of the ruler. What are the coordinates in O' of the event at which the photon reaches the other end?

3.9. Start with the inverse LT and derive the time dilation (corresponding to $\Delta x = 0$) and length contraction (corresponding to $\Delta t' = 0$) formulas.

3.10. Two rockets A and B of rest length L_0 travel toward each other along their x-axes with relative speed β.[8] According to A, when the tail of B passes the nose of A, a missile is fired from the tail of A toward B. It will clearly miss due to length contraction of B as seen by A. But B will see the length of A contracted. So, when the nose of A coincides with the tail of B, the tail of A is in the middle of B's rocket and it will hit B. Who is right? Consider five events: nose of A passes nose of B (call this the origin of spacetime for both), tail of B passes nose of A, missile is fired, nose of B passes tail of A, and tail of B passes tail of A. Find the coordinates of these five events in both frames to see who is right.

3.11. This problem is the continuation of Example 3.4.4. It is important for you to remember that LT applies only to *events*. The coincidence of Sonya's clock with C_1 is an event. But what is the event corresponding to the reading of C_2? Note 3.4.3 gives the answer: There is an event E_0 at the location of C_2 with coordinates $(L, 0)$ in Sam's RF.

(a) What are the coordinates of E_0 in Sonya's RF? Note that although C_2 is at the 12 o'clock position for Sam, it is not for Sonya!

(b) How long does it take her to reach C_2? Warning: The length Sonya has to travel to reach C_2 *is not* L/γ, because when Sonya is at C_1, C_2 is not at $x = L/\gamma$ for her!

---

[8] Assume that their path is slightly shifted in the y-direction so they do not collide.

(c) What time does Sonya's clock show when she reaches C_2? This should be $L/(c\beta\gamma)$ as in Example 3.4.4.
(d) What are the coordinates (in her RF) of the event at which Sonya reaches C_2?
(e) Using LT, show that the time coordinate of the same event in Sam's RF is $L/(c\beta)$.

3.12. To gain more insight into the clock comparisons of Example 3.4.4 and Problem 3.11, consider the synchronization process of the two clocks C_1 and C_2 in Sam's RF as seen by Sonya. To synchronize the clocks, let two light signals be sent simultaneously to the two clocks from the midpoint between them. Call this event E.
(a) Show that E has coordinates $(L/2, -L/2)$ in Sam's RF.
(b) Show that E has coordinates

$$x = \sqrt{\frac{1+\beta}{1-\beta}}\frac{L}{2}, \quad ct = -\sqrt{\frac{1+\beta}{1-\beta}}\frac{L}{2}$$

in Sonya's RF.
(c) How long does it take this signal to arrive at the (common) origin? What is the time at which it arrives at the origin? Is that what you expect?
(d) What is the location of C_2 relative to Sonya? What is the distance between E and C_2 as measured by Sonya? Show that it takes $\gamma(1-\beta)L/2c$ for the signal to cover this distance.
(e) From (b) and (d), show that the time of arrival of the signal at C_2 is $-\gamma\beta L/c$ according to Sonya.
(f) From (d), show that it takes Sonya $\gamma L/(c\beta)$ to reach C_2. Therefore, her time of arrival at C_2 is $L/(c\gamma\beta)$ *after* the arrival of the light signal at C_2, as in Example 3.4.4.
(g) Using (f) and LT show that the time shown on C_2 is $L/(c\beta)$.

3.13. Obtain Equation (2.16) using Lorentz transformation.

3.14. Generalize the discussion of Example 3.4.5 to any clock length L and any speed β by showing that

$$c\Delta t'_{21} = \gamma L(1+\beta), \quad \text{and} \quad c\Delta t'_{32} = \gamma L(1-\beta).$$

3.15. Reorient the rod of Example 3.4.6 so that now it lies along the y-axis. Let t be the event at which the light from an arbitrary point of the rod is emitted in such a way that it reaches P_0 of Figure 3.4 at $t = 0$.
(a) Find ct in terms of y and b.
(b) Find the x'-coordinate of the event according to O', with respect to which the rod is moving.
(c) Show that the locus of source points whose light rays are collected simultaneously at the pinhole is a hyperbola in the $x'y'$-plane as in Example 2.2.5.

(d) Do you expect the rays from the two ends of the rod to have been emitted simultaneously according to O? According to O'?
(e) Find t_A, t_B, t'_A, and t'_B, the times of occurrence of the emission of the light rays from the two ends of the rod according to the two observers.

3.16. The Earth and Alpha Centauri are 4.3 light years apart. Ignore their relative motion. Events A and B occur at $t = 0$ on Earth and at 1 year on Alpha Centauri, respectively.
(a) What is the time difference between the events according to an observer moving at $\beta = 0.98$ from Earth to Alpha Centauri?
(b) What is the time difference between the events according to an observer moving at $\beta = 0.98$ from Alpha Centauri to Earth?
(c) What is the speed of a spacecraft that makes the trip from Alpha Centauri to Earth in 2.5 years according to the spacecraft clocks?
(d) What is the trip time in the Earth RF?

3.17. All muons in a group move toward Earth with the same speed. After covering a distance of 2911 m, half of them survive. Recall that for any decay process, $N(t) = N_0 e^{-t/\tau}$, where τ is the mean life of the decaying particles as measured in their rest frame, and for muons $\tau = 2.2\ \mu s$.
(a) What is the speed of the muons?
(b) What is the mean life of the muons in the Earth frame?
(c) According to the muons, how far did they travel?
(d) A spaceship is launched from Earth with a speed of $0.95c$. What is the mean life of the muons in the frame of the spaceship?

3.18. A ruler moves past another ruler of rest length L and parallel to it with speed β to the right. Two clocks are attached to the ends of the moving ruler and synchronized in the RF of that ruler. When the left end of the moving ruler reaches the left end of the stationary ruler, the left clock reads zero. When the right end of the moving ruler reaches the right end of the stationary ruler, the right clock reads Δt.
(a) Find the rest length of the moving ruler in terms of β, L, and Δt.
(b) Relative to the stationary ruler, how long after the left coincidence does the right coincidence occur?

The pole-and-barn paradox.

3.19. Sonya (observer O) is holding a long pole of length L in the middle as she runs with speed β toward a barn. Sam (observer O'), standing in the barn, sees that the pole fits exactly between the entrance and exit doors of the barn. On the other hand, Sonya sees her pole—which is at rest relative to her—longer than the barn and concludes that there is no way that the pole can fit in the barn. Who is right? To find out, let E_1 be the event when the front end of the pole exits the barn and E_2 the event when the rear end of the pole enters the barn. Write the general Lorentz transformation for these events, and answer the following questions:
(a) What are x_1 and x_2?

(b) How does t'_1 compare to t'_2?
(c) How does t_1 compare to t_2? Conclude that Sam and Sonya are both right!
(d) What is $x'_1 - x'_2$? Is that what you expect?

3.20. Sam sees a firecracker explode at $x = 1.5$ km (event E_1), then after 5 μs, sees another firecracker explode at $x = 100$ m (event E_2).
(a) What is the velocity relative to Sam of Sonya for whom the events occur at the same place?
(b) Which event occurs first according to Sonya and what is the time interval between the explosions for her?

3.21. Sonya is in the middle of a train car of length L moving relative to Sam with speed β along Sam's positive x'-direction. Firecrackers A in the back and B in the front of the car explode simultaneously according to Sam at the same time that Sonya passes him. Let $t = t' = 0$ be the time that Sam and Sonya pass each other, and assume that both observers are at the origins of their coordinate systems. Without using length contraction or time dilation formulas, find the coordinates of all the following events for both observers: explosion of A, explosion of B, reception of light from A by Sam, reception of light from B by Sam, reception of light from A by Sonya, reception of light from B by Sonya.

3.22. Two spaceships of rest length L_0 are approaching the Earth from opposite directions at velocities $\pm 0.8c$. How long does one of them appear to the other?

3.23. Spaceship A is twice as long as spaceship B when they are at rest. Spaceship B is moving at three quarters the speed of light relative to Earth. As A overtakes B, an Earth observer notices that they both have the same length.
(a) How fast is A moving?
(b) What is the length of A relative to B?
(c) What is the length of B relative to A?

3.24. Sam and Pat move in the same direction at $0.8c$ and $0.6c$ relative to Earth, respectively.
(a) How fast should Sonya move relative to Earth so that she observes Sam and Pat approaching her at the same speed?
(b) What is the speed of Sam (or Pat) relative to Sonya?

3.25. Sonya (observer O) is approaching Sam (observer O') with speed β from the negative values of his x'-axis. Sonya has a source of light of wavelength λ sending signals to Sam. Consider two events: E_1, the source sends a wave crest, and E_2, the source sends the next wave crest. So, $t_2 - t_1$ is the period of the wave according to Sonya.

Doppler effect.

(a) Write the Lorentz transformation for these two events assuming that Sonya (at her origin) is holding the source of light.
(b) What is the period according to Sam? Hint: It is *not* $t'_2 - t'_1$!

(c) Show that λ', the wavelength as measured by Sam, is related to λ via the following formula:

$$\lambda' = \sqrt{\frac{1-\beta}{1+\beta}}\,\lambda.$$

3.26. Sonya (observer O) is moving away from Sam (observer O') with speed β along his positive x'-axis. At time t', Sam sends a light signal to Sonya (event E), which gets reflected from a reflector at Sonya's origin (event E_{ref}) and received by a detector at Sam's origin (event E_{det}).

(a) Find (x, t), the coordinates of E according to Sonya in terms of t'.
(b) Find t_{ref}, the time of E_{ref} according to Sonya in terms of t'.
(c) Find $(x'_{\text{ref}}, t'_{\text{ref}})$, the coordinates of E_{ref} according to Sam in terms of t'.
(d) Show that

$$t'_{\text{det}} = \frac{1+\beta}{1-\beta}\,t'.$$

3.27. Speeder O is in a spacecraft moving away from a policeman (observer O') with speed β. The policeman sends an EM signal of wavelength λ to O and receives the reflected wave of wavelength λ_{ref}.

(a) Use the result of Problem 3.26 for the emission of two successive wave crests to show that

$$\lambda_{\text{ref}} = \frac{1+\beta}{1-\beta}\,\lambda.$$

(b) The speed limit in Metropolis is half the speed of light. Will the speeder get a ticket if the policeman sends a violet signal of 400 nm and receives an infrared signal of 1.3 μm?

3.28. Sam is on a train of rest length L_0 moving at speed β relative to Sonya standing on the ground. Consider the time interval between the front of the train coinciding with Sonya and the back of the train coinciding with her.

(a) Find this time interval in Sonya's frame by calculating in her frame.
(b) Find this time interval in Sonya's frame by calculating in Sam's frame.
(c) Find the time interval in Sam's frame by calculating in Sonya's frame.
(d) Find the time interval in Sam's frame by calculating in his frame.

3.29. Sam and Sonya sit in two identical rockets of length L. Sam approaches Sonya from behind with relative speed β along the length of their rockets. Assume that at time zero for both, the nose of Sam's rocket coincides with the tail of Sonya's. Now consider three events occurring in succession: Event A is when the tail of Sam's rocket coincides with the tail of Sonya's; B is when the nose of Sam's rocket coincides with the nose of Sonya's; and C is when the tail of Sam's rocket coincides with the nose of Sonya's. Using only Lorentz transformation and inverse Lorentz transformation (no length contraction or time dilation) find the coordinates of these three events in Sam's and Sonya's RFs paying attention to signs.

3.30. Sam is on a train of rest length L_0 moving at speed β relative to Sonya standing on the ground. He fires a gun from the back of the train to the front. The speed of the bullet relative to the train is β_b.

(a) How much time does the bullet spend in the air before hitting the front of the train according to Sonya?
(b) What is the distance that the bullet travels according to Sonya?

3.31. Sam and Sonya are newborn twins. Sam is placed on a rocket that moves at speed $0.99c$ to a star system 25 light years away. At the moment of his departure a light signal is sent to the star system and gets reflected.

(a) How old is Sam when he receives the reflected signal? How old is Sonya?
(b) How old is Sonya when *she* receives the reflected signal? How old is Sam?
(c) How old is Sam when he reaches the star system?
(d) How old is Sonya when Sam reaches the star system?

Hint: See Example 3.4.9.

3.32. Multiply both sides of the inequality $\beta_b < 1$ by the positive quantity $1 - \beta$ and show that $\beta + \beta_b < 1 + \beta\beta_b$. From this conclude that the relativistic law of addition of velocities never yields a speed larger than the speed of light.

3.33. A super-ball has the property that when it hits a wall with a given speed *relative to the wall*, it bounces back with the same speed in the opposite direction. What do you measure the speed of the bounced ball to be if you throw it at speed β_s toward a wall that is moving toward you at speed β_w?

3.34. In an intergalactic race, team A is moving at speed $0.8c$ relative to the finish line. They notice that a faster team B passes them at $0.9c$. Team B observes another team C to pass them at $0.95c$. What are the speeds of teams B and C relative to the finish line?

3.35. Sam sees Sonya and Pat flying in opposite directions with a speed $\beta = 0.995$. What is the speed of Pat relative to Sonya?

3.36. Two spaceships approach each other with relative speed $0.9c$. What are the velocities of the spaceships relative to Earth assuming that they move with the same speed relative to Earth?

3.37. Observer O moves in the positive x'-direction of observer O' with speed β. Observer O' moves in the positive x''-direction of observer O'' with speed β'. Use the relativistic law of addition of velocities (3.26) to obtain γ'' in terms of β, β', γ, and γ'.

3.38. All motions are in the positive x-direction. Observer O_1 moves relative to observer O with speed β_1. Observer O_2 moves relative to O_1 with speed β_2. Let $P_2^+ \equiv (1 + \beta_1)(1 + \beta_2)$ and $P_2^- \equiv (1 - \beta_1)(1 - \beta_2)$.

(a) Show that O_2 moves relative to O with speed β_2' given by

$$\beta_2' = \frac{P_2^+ - P_2^-}{P_2^+ + P_2^-}.$$

(b) Now introduce another observer O_3 moving relative to O_2 with speed β_3, and let

$$P_3^+ = (1+\beta_1)(1+\beta_2)(1+\beta_3)$$
$$P_3^- = (1-\beta_1)(1-\beta_2)(1-\beta_3).$$

Use the relativistic law of addition of velocities for β_2' and β_3 in the form given in (a) to prove that O_3 moves relative to O with speed β_3' given by

$$\beta_3' = \frac{P_3^+ - P_3^-}{P_3^+ + P_3^-}.$$

(c) Can you predict what happens if you introduce a fourth observer?

(d) If you are familiar with mathematical induction, prove the formula for N observers.

3.39. Recall that **hyperbolic** functions are defined by

Hyperbolic functions.

$$\cosh\phi = \frac{e^\phi + e^{-\phi}}{2}, \quad \sinh\phi = \frac{e^\phi - e^{-\phi}}{2}.$$

(a) Using these definitions, show the following properties:

$$\cosh^2\phi - \sinh^2\phi = 1$$
$$\sinh(\phi_1+\phi_2) = \sinh\phi_1\cosh\phi_2 + \sinh\phi_2\cosh\phi_1$$
$$\cosh(\phi_1+\phi_2) = \cosh\phi_1\cosh\phi_2 + \sinh\phi_1\sinh\phi_2$$
$$\tanh(\phi_1+\phi_2) = \frac{\tanh\phi_1 + \tanh\phi_2}{1+\tanh\phi_1\tanh\phi_2}$$

Rapidity.

(b) Now define the **rapidity** ϕ by $\tanh\phi = \beta$ and show that

$$\cosh\phi = \gamma, \quad \sinh\phi = \beta\gamma, \quad e^\phi = \left(\frac{1+\beta}{1-\beta}\right)^{1/2}.$$

3.40. Show that in terms of rapidity of Problem 3.39, the LT can be written as

$$x' = x\cosh\phi + ct\sinh\phi$$
$$ct' = x\sinh\phi + ct\cosh\phi.$$

3.41. Redo Problem 3.37 using LTs in terms of rapidities. How is the rapidity ϕ'' of O'' relative to O related to the rapidity ϕ' of O'' relative to O' and the rapidity ϕ of O' relative to O? The identities in Problem 3.39 may be useful.

3.42. Derive the following relations among the coordinates of the RFs O and O':

$$x' + ct' = e^\phi(x+ct), \quad x' - ct' = e^{-\phi}(x-ct),$$

where ϕ is the rapidity as defined in Problem 3.39. Now define two new coordinates (ξ, η) as

$$\xi = x + ct, \quad \eta = x - ct,$$

and show that they are perpendicular to each other (in the Euclidean sense). Show that under a LT, ξ and η do not change direction, but their calibration changes. By how much?

3.43. The speed limit in Metropolis is $0.8c$. A driver is moving at $0.92c$ (with respect to the ground) as he passes a policeman without noticing him. The policeman, accelerating to $0.996c$ (also with respect to the ground) instantaneously, catches up with the driver 10 minutes later according to his own clock. The origins of the RFs of both observers are when and where they pass each other.

(a) What are the coordinates (in the policeman's RF) of the event at which the policeman starts chasing the driver?
(b) What are the coordinates of that event in the driver's RF?
(c) How long after the driver passes the policeman does the policeman catch up with him according to the driver's clock?

3.44. The acceleration of a "bullet" in O' is $a'_b = dv'_b/dt'$, and in O, moving relative to O' with constant speed β in the positive direction of the x'-axis, is $a_b = dv_b/dt$. Use (infinitesimal) LT and Equation (3.26) to show that

$$a'_b = \frac{a_b}{\gamma^3(1+\beta\beta_b)^3}.$$

3.45. In Problem 3.44, let $\beta'_b = \beta$, that is let O be moving with the bullet. Furthermore, assume that the acceleration a_b is constant. Call it a.

(a) Show that (ignoring the prime on t)

$$a dt = \frac{c d\beta}{(1-\beta^2)^{3/2}}.$$

(b) Integrate the equation above and show that if $\beta = 0$ at $t = 0$, then

$$\beta = \frac{at}{\sqrt{c^2 + a^2t^2}}.$$

How long do you have to wait for the bullet to reach a speed of $0.99c$ if the acceleration is 5 m/s^2? If you wait long enough, can you accelerate the bullet to a speed larger than c?

(c) Ignoring the prime on x as well and noting that $c\beta = v = dx/dt$, find x as a function of time assuming that $x = 0$ at $t = 0$. What kind of a curve do you get on the xt-plane?
(d) How long does it take the "bullet" to reach Alpha Centauri 4 light years away if the acceleration of the bullet is 5 m/s^2? What is its speed when it reaches there?

CHAPTER 4

Spacetime Geometry

Relativity theory abolishes the notion of absolute time and space. An event, specified by a location in space and an instant in time, is described differently by different RFs. This is the essence of Lorentz transformation derived in the previous chapter. The most elegant way of relating an event's space and time properties assigned by two observers is to use geometry.

You studied *solid* geometry of lines, surfaces, and volumes by assigning three numbers to each point of those figures. If a certain collection of points happen to lie on a geometric figure, then those points are related via a mathematical formula. For instance, a surface is described by an equation giving one coordinate of a general point of the surface in terms of the other two. All points in space whose coordinates satisfy the equation lie on that surface. All curves and surfaces are determined by equations. That is how solid geometry turns into coordinate (or *analytic*) geometry.

Many of the features of solid geometry can be derived by studying the easier **plane geometry**, as is done in a typical elementary geometry course. As shown in Figure 4.1(a), a Euclidean point is specified by only two numbers (x, y). Lines and curves and other geometric figures are objects for which x and y are related via equations. For instance, any point whose coordinates satisfy the equation $x^2 + y^2 = 1$ lies on a circle of unit radius centered at the origin.

Solid geometry is hard; simplify to *plane geometry.*

4.1 PLANE SPACETIME GEOMETRY

Coordinate geometry came after coordinate-free geometry. Euclid and other Greek mathematicians did not use coordinates. The entire Euclidean geometry was based on certain postulates and properties of points and straight lines. Coordinates came much later, at the end of the 16th and the beginning of the 17th centuries. On the other hand, relativity *starts* with coordinates by defining an event—a point in spacetime—by a collection of four numbers (x, y, z, t). Therefore, the geometry of spacetime is inherently an analytic geometry.

If it is only hard to picture three-dimensional geometric figures, it is impossible to picture four-dimensional objects. In fact, the only way to work with

Special Relativity. DOI:10.1016/B978-0-12-810411-8.00004-3

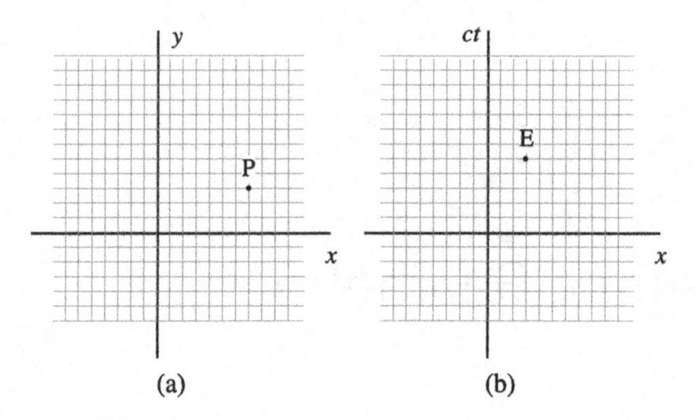

FIGURE 4.1 (a) A point in the plane is specified by a pair of numbers or coordinates (x, y); for instance P has coordinates $(7, 3)$. (b) An event in the plane spacetime is specified by a pair of numbers or coordinates (x, ct); for instance E has coordinates $(3, 5)$.

spacetime objects is through coordinates and mathematical relations between them. Fortunately, just as plane geometry captures many of the important features of solid geometry, **plane spacetime geometry**, shown in Figure 4.1(b), captures the subtle features of the full four-dimensional spacetime geometry. Now, in the case of Euclidean geometry, it does not matter which two coordinates we choose to represent the plane. We usually choose (x, y); but any other pair would be just as good. For spacetime, we *must* include time as one of the coordinates,[1] because time is not equivalent to other (space) coordinates. As the second coordinate, we take x, although y and z are just as good. Thus, our plane spacetime geometry consists of a set of events described by pairs like (x, ct), where the factor of c is introduced as in the last chapter.

4.1.1 Events and Worldlines

The building blocks of spacetime geometry are *events* just as the building blocks of Euclidean geometry are *points*. Events of spacetime geometry and points of Euclidean geometry exist independently of the observers plotting them on their coordinatized sheets of paper. In Euclidean geometry one picks a point of the plane, calls it the origin, erects his axes from there, and measures the coordinates of all points relative to this coordinate system. In plane spacetime geometry one picks an event, calls it the origin, erects her axes from it, and measures the coordinates of all events relative to that coordinate system. These origins and axis orientations are quite arbitrary and are chosen to make calculations as simple as possible.

Worldlines are curves in spacetime.

A collection of points arranged and connected in a certain succession constitutes a curve in Euclidean plane. Similarly, a collection of events arranged and

[1] This is the reason that Δt was called the *coordinate time* in Chapter 2.

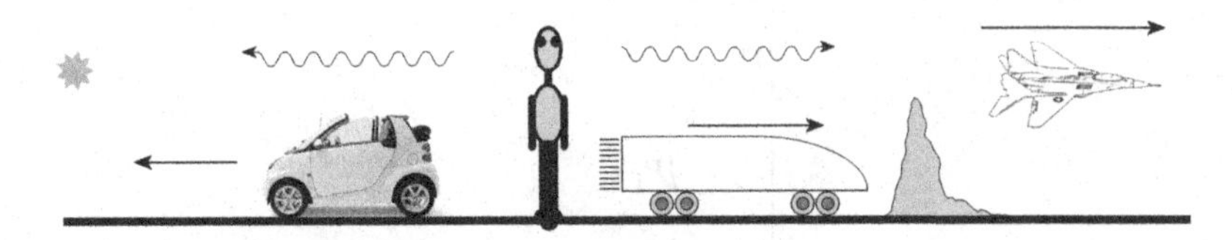

FIGURE 4.2 Some moving objects in Sonya's RF.

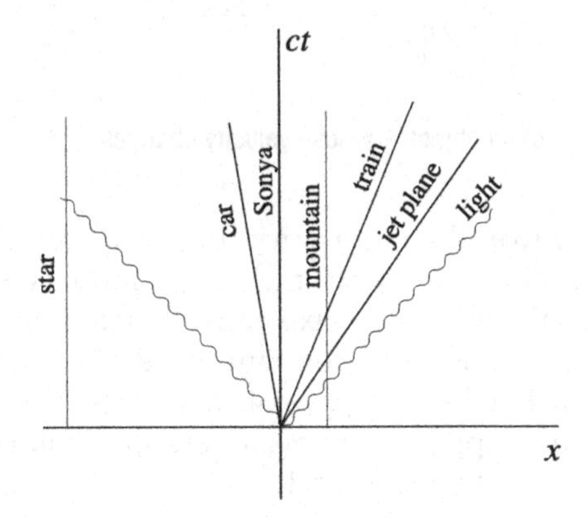

FIGURE 4.3 The worldline of the objects moving in Sonya's RF. The positions of the mountain and star are not to scale.

connected in succession makes up a spacetime curve. This curve, which can be thought of as the location of some moving object at different times, is called the **worldline** of that object.

Figure 4.2 shows Sonya, standing at the origin of her coordinate system, as she watches some objects, each assumed to be moving with a constant velocity that is comparable to light speed. In Figure 4.3 are drawn the worldlines of these objects. All the worldlines are straight lines, because all speeds are assumed constant. The slope[2] of an object moving with speed v_b relative to Sonya[3] is

$$\frac{c\Delta t}{\Delta x} = \frac{1}{\Delta x/c\Delta t} = \frac{1}{v_b/c} = \frac{1}{\beta_b},$$

where $\beta_b = v_b/c$ is the fractional speed of the object. For light, this is equal to 1; therefore, on a spacetime coordinate system whose axes are perpendicular to each other, light has a worldline that makes a 45° angle with the positive x-axis.

---

[2] As usual, slope is measured relative to the horizontal axis. However, later we will find it convenient to use slopes relative to the vertical axis as well.

[3] The subscript "*b*" reflects my choice of a *b*ullet or a *b*all as a typical moving object. You will see this subscript later as well. I add the subscript so you don't confuse the speed of an *object* with the relative speed of two RFs.

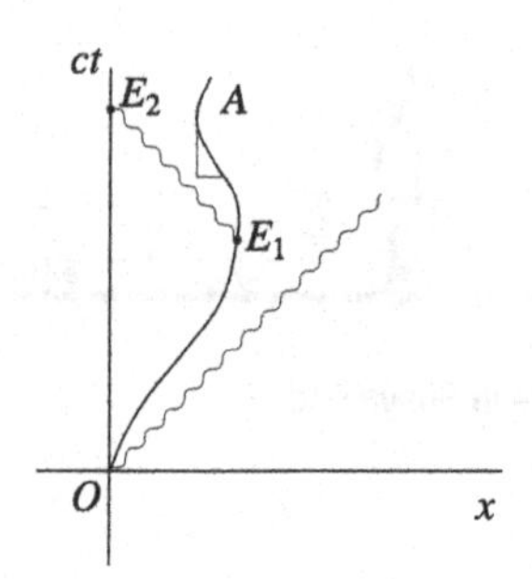

FIGURE 4.4 The worldline of an object A whose velocity changes.

All objects have a worldline whose slope is larger than one.

I often draw the light worldline as a wavy line, or occasionally as a dashed line. Since $\beta_b < 1$ for all other objects, their slope must be *larger* than 1, i.e., the angle that they make with the x-axis must be larger than 45°. This goes not only for objects that move with constant speed as in Figures 4.2 and 4.3, but also for objects such as A in Figure 4.4, whose speed is changing. In the latter case, the restriction applies to *instantaneous* speed: at every event of the worldline the slope must be larger than 1.

The worldline of any observer is his/her time axis.

An object that is stationary relative to Sonya has a vertical worldline, because its distance (x-coordinate) does not change. The mountain to her left and the star to her right are such objects (their positions in Figure 4.3 are not to scale). Sonya herself has a worldline that is vertical (her time axis), because she is not moving (and her x-coordinate is always zero). This is true for all observers: *The worldline of any observer is his/her time axis.* In fact, you can think of the worldline of any observer as the curve that connects the ticks of the clock carried by the observer, each tick being one event.

When an observer O' is present at two events, his worldline, which may not be a straight line in the coordinate system of another (inertial) observer O, passes through (or connects) those two events. The clock of O', therefore, measures the proper time between those events (see Note 2.1.2). If the events are very close together on the worldline of O', the arc of this worldline connecting those events is almost a straight line and can be identified as the instantaneous *time axis* of O'. If O' is not moving with constant speed relative to O, this time axis keeps changing, but if O' has constant speed relative to O, the worldline of O' is his (permanent) time axis.

Inertial observer.

Many observers can be present at any given two events, but only one of them, the **inertial observer**, has constant speed: it is the observer whose worldline is a *straight line* passing through the two events. This straight line must have a slope larger than 1 in the (perpendicular) spacetime coordinate system of any other observer. When an observer can be present at two events, we say that those events are **causally connected**. In fact, causal connection is generalized to also include the case when the worldline of a *light signal* connects the two

Causally and luminally connected events.

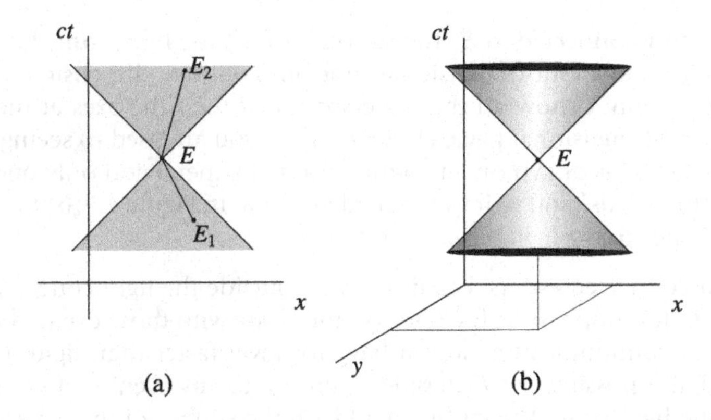

FIGURE 4.5 (a) Events E_1 in the past and E_2 in the future are causally connected to E. (b) One more axis of the space is added to show why the shaded region is called the light "cone."

events. When a light signal is present at the two events, we say that they are **luminally connected**. Let me summarize what we have learned so far:

> **Note 4.1.1.** *Relative to an observer O, every object moves on a worldline whose* instantaneous *slope at every event of the worldline is larger than* 1, *except for light, whose slope is* 1. *The instantaneous time axis of an observer O' is the tangent to his/her worldline. If O' moves with constant speed relative to O, the worldline of O' is his/her time axis. Two events that can be connected by a straight worldline whose slope is larger than (or equal to)* 1 *are called* **causally connected,** *otherwise they are* **causally disconnected.**

The intersection of the worldlines of two objects is an *event* (just as the intersection of two lines is a *point* in Euclidean geometry) at which both objects are present. If these objects happen to be observers, then the *observers* are present at that event. For example, observers O and A in Figure 4.4 are both present at the event with coordinates $(0, 0)$ in the RF of O. For the same reason, both A and a light signal are present at E_1, with the light signal initiating on the worldline of A and moving away from it. We call E_1 the *emission* of a light signal by observer A. Similarly, E_2 is the *reception* of the same light signal by O.

4.1.2 Light Cone

Given any event E, how do you identify the collection of all events that are causally connected to E? By definition, this collection consists of all the events that can be connected to E by a straight line whose slope is larger than or equal to 1. Some of these, such as E_1 in Figure 4.5(a), occurred in the past of E; some, such as E_2, will occur in the future. The shaded region is the set of all events

Light cone as the collection of all events causally connected to a given event.

that are causally connected to E. The region is called the **light cone**, because it is the projection of a four-dimensional cone onto our two-dimensional spacetime plane. I cannot show all the three space *and* the time axes at the same time on a two-dimensional page. However, since you are used to seeing three-dimensional objects drawn on a two-dimensional paper, I add only one more space axis (the y-axis) and redraw the shaded region in Figure 4.5(b) to explain the word "cone" in the *light cone*.

Causally disconnected events, i.e., those lying outside the light cone, are inaccessible to E. It is impossible for E to communicate with those events, because the signal of communication would have to travel faster than light. On the other hand, it *is* possible for E to send a "probe" to any event that lies within (and on) the light cone. The events that lie on the surface of the cone require a light probe (signal); those inside require a probe traveling slower than light.

4.2 RULES OF SPACETIME GEOMETRY

Geometry is the most elegant and powerful way of studying relativity. As I have restricted this geometry to two dimensions, I'll be working in a plane. Every spacetime plane has an origin, which is both a space origin and a *time origin*. You are familiar with the space origin from your ordinary geometry course. It is the point at which positive and negative distances are separated from each other. But what exactly is the origin of time? It is really analogous to the origin of space (coordinates). When we draw our x- and y-axes in a plane, we are doing it quite arbitrarily, and some random point in the plane of the sheet of paper—the point at which the two axes cross—receives the "honor" of being the *origin*. The next time we draw the axes, another random point becomes the origin. Similarly, the origin of time is an arbitrary instant—usually called "now"—at which future and past events are separated.[4]

Coordinate axes need not be horizontal and vertical.

Not only is the origin of a coordinate system (CS) arbitrary, but so is the *orientation* of its axes. In the case of plane geometry, this causes the *same* point to have *different* pairs of coordinates in different CSs. As you saw in Section 3.1.1, the coordinates of a point P in O were (x, z), which were different from its coordinates (x', z') in O'. These coordinates were tied together via Equation (3.10), where α was the angle between the axes x and x' or z and z' (see also Problem 3.4 and Figure 3.5).

Single points and their transformations are important in studying coordinated geometry. What is equally important is the properties of objects consisting of a collection of points; objects such as lines and curves. Since any curve (or

[4] As I am writing these lines, I am 2014 years, 24 days, 14 hours, and 27 minutes away from the *arbitrary* origin of time chosen by a certain population of the Earth. This same instant (specified to within a minute) of time is described by another population—for example, the Moslems or the Chinese—differently.

even surface) can be built up from very small (infinitesimal) line segments, the nature of such a line segment determines the shape of lines and surfaces and, therefore, the character of the geometry. For plane Euclidean geometry, the length of the line segment is given by (3.4), and as you saw, the *invariance* of this formula for different observers determines the transformation rule (3.10).

Since Euclidean geometry was discovered before the notion of coordinates was introduced in mathematics, the idea of the distance being independent of the coordinate system used is deeply ingrained in our mind. We never question the invariance of the Euclidean distance, and when we transform the coordinates of points from a system O to another system O', we expect that the distance between any two points remains the same.

Our knowledge of spacetime is restricted to coordinates; we don't have a visual image of the spacetime geometry as we do of the Euclidean geometry. This does not mean that we cannot draw pictures in spacetime; it just means that the pictures we draw have to be interpreted differently than the familiar Euclidean pictures.

4.2.1 Invariant Distance

At the heart of the spacetime geometry (like any other geometry) is the *invariant* distance between two infinitesimally close events, which you already encountered in Section 3.2. Denoting this distance by s, I rewrite the result here:

Spacetime distance and its invariance.

Note 4.2.1. *The* ***spacetime distance (line segment)*** Δs *between two close events* E_1 *and* E_2*, with respective coordinates* (x_1, ct_1) *and* (x_2, ct_2) *in some reference frame* O*, is related to proper time by* $\Delta s = c\Delta\tau$ *and is given by*

$$(\Delta s)^2 = (c\Delta\tau)^2 = (c\Delta t)^2 - (\Delta x)^2,$$

where $\Delta t = t_2 - t_1$ *and* $\Delta x = x_2 - x_1$*. This spacetime distance is an* ***invariant quantity****, meaning that it is independent of the observer.*

It is important that you understand the nature of Δs.[5] If observer O' assigns coordinates (x'_1, ct'_1) and (x'_2, ct'_2) to E_1 and E_2, respectively, and $\Delta t' = t'_2 - t'_1$ and $\Delta x' = x'_2 - x'_1$, then

$$(\Delta s)^2 = (c\Delta t')^2 - (\Delta x')^2.$$

Both observers calculate exactly the same number for $(\Delta s)^2$, even though they assign different sets of coordinates to the two events E_1 and E_2.

The negative sign in the expression for $(\Delta s)^2$ has a dramatic physical consequence: It allows for the spacetime distance between two *distinct* events to be

---

[5] Because of the minus sign in the definition of $(\Delta s)^2$ and the possibility of getting an imaginary number for its square root, it is common to use $(\Delta s)^2$ instead of Δs.

Spacetime distance of zero.

zero! Observer O spots a light beam at x_1 at time t_1 (event E_1). A little later he finds the beam at x_2 at time t_2 (event E_2). Since light travels from x_1 to x_2 with speed c,

$$x_2 - x_1 = c(t_2 - t_1) \quad \text{or} \quad \Delta x = c\Delta t.$$

Therefore,

$$(\Delta s)^2 = (c\Delta t)^2 - (\Delta x)^2 = (c\Delta t)^2 - (c\Delta t)^2 = 0,$$

which holds for any light signal, as the two events above are quite general.

Another observer O' sees these two events as (x_1', ct_1') and (x_2', ct_2'). The invariance of the spacetime distance implies that for her, $(\Delta s)^2$ is also zero. Therefore,

$$0 = (\Delta s)^2 = (c\Delta t')^2 - (\Delta x')^2 \Rightarrow c\Delta t' = \Delta x' \quad \text{or} \quad \frac{\Delta x'}{\Delta t'} = c,$$

i.e., she measures the speed of the light signal to be c, consistent with the second postulate of relativity.

Spacetime distance for light.

Note 4.2.2. *For any light signal $(\Delta s)^2 = 0$. In other words, if two events are connected by a light signal (i.e., if a light signal is present at the two events), then the spacetime distance between those two events is zero.*

The notion of causal connection can be stated succinctly in terms of $(\Delta s)^2$. For the speed of the observer present at two events to be less that the speed of light, $c\Delta t$ must be larger that Δx. Therefore, two events are causally connected if $(\Delta s)^2 > 0$. Such two events are also called **timelike separated**. Similar arguments apply to two causally disconnected events.

Note 4.2.3. *Let two events E_1 and E_2 be separated by Δx and Δt with respect to an observer O so that $(\Delta s)^2 = (c\Delta t)^2 - (\Delta x)^2$. Then these two events are*

causally connected or **timelike** if $(\Delta s)^2 > 0$,
luminally connected or **lightlike** if $(\Delta s)^2 = 0$,
causally disconnected or **spacelike** if $(\Delta s)^2 < 0$.

Invariance of $(\Delta s)^2$ implies that this classification is universal.

When two events are causally connected (luminal connection included), the notion of "later" or "earlier" is also universal. You can see this by noting that in such a case, $|c\Delta t| \geq |\Delta x|$ for the observer O. For any other observer O', the LT yields

Invariance of "later" and "earlier."

$$c\Delta t' = \gamma(c\Delta t + \beta\Delta x)$$

and since $\beta < 1$, the sign of $\Delta t'$ is the same as that of Δt. However, when E_1 and E_2 are causally *disconnected*, one observer may see E_1 earlier than E_2 and another observer later (see Example 4.2.4). This is another reason that causally disconnected events are not physical: no observer can see a child born before his/her parents are born!

Example 4.2.4. In the reference frame O, let events E_1 and E_2 have coordinates (x_1, ct_1) and (x_2, ct_2), respectively. Assume that E_1 and E_2 are causally disconnected, i.e., that $|\Delta x_{12}| > |c\Delta t_{12}|$. Introduce the symbol ϵ_t by writing $\Delta t_{12} = \epsilon_t |\Delta t_{12}|$, so that $\epsilon_t = \pm 1$: it is $+1$ if $\Delta t_{12} > 0$ and -1 if $\Delta t_{12} < 0$. Use similar symbols for other quantities that could be positive or negative. Write the Δ form of the second equation in LT as

$$c\epsilon_t' |\Delta t_{12}'| = \gamma \left(c\epsilon_t |\Delta t_{12}| + \epsilon_\beta \epsilon_x |\beta| |\Delta x_{12}|\right).$$

Multiply both sides by ϵ_t and note that $\epsilon_t^2 = 1$. Then

$$c\epsilon_t \epsilon_t' |\Delta t_{12}'| = \gamma \left(c|\Delta t_{12}| + \epsilon_t \epsilon_\beta \epsilon_x |\beta| |\Delta x_{12}|\right).$$

The notion of "earlier" or "later" is switched in O' if $\Delta t_{12}'$ and Δt_{12} have different signs, i.e., if $\epsilon_t \epsilon_t' = -1$. This can happen only if the right-hand side of the last equation is negative, that is, if

$$\epsilon_t \epsilon_\beta \epsilon_x = -1 \quad \text{and} \quad |\beta| > \frac{c|\Delta t_{12}|}{|\Delta x_{12}|},$$

and this is always possible because $c|\Delta t_{12}| < |\Delta x_{12}|$ and the first equation can be satisfied by choosing an appropriate direction for β. Thus for any two causally disconnected events, there are infinitely many reference frames in which the order of occurrence is switched. ■

An observer who is present at both E_1 and E_2 measures the proper time interval $\Delta\tau$ between those events. For this observer, $x_1 = x_2$ and $(\Delta s)^2 = (c\Delta t)^2$. Therefore,

Note 4.2.5. *A clock that is present at two events directly measures $\Delta\tau$ and thus the spacetime distance $\Delta s = c\Delta\tau$ between those events.*

4.2.2 The Rules

Because of the difference between the formula for the distance in the Euclidean and spacetime geometries, distortions take place, not unlike distortions resulting from mapping the spherical geometry of the globe onto a flat piece of paper. One main distortion is that the time and space axes need not be perpendicular, as you saw in Section 3.2 and shown in Figure 3.2. Despite these distortions, you can understand and manipulate them by resorting to a set of rules derived from the invariant distance formula of Note 4.2.1. Based on these rules, you can draw geometric figures and extract algebraic relations governing events and worldlines.

The first two rules have to do with the orientation of axes. Equation (3.24) and comments after it imply that the angle α between the axes of two observers

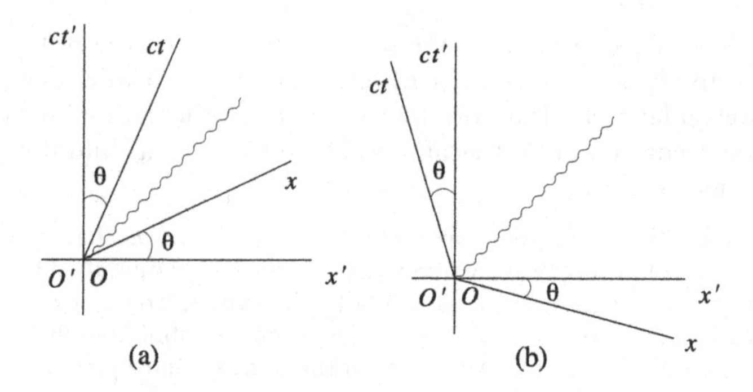

FIGURE 4.6 (a) O is moving in the positive direction of O'. (b) O is moving in the negative direction of O'. The relative speed is larger in (a) than in (b). Note how the light worldline (represented by the wavy line) makes equal angles with both axes of O and O'.

O and O' moving with fractional speed β relative to one another is given by $\beta = \tan\alpha$. The third rule tells us how to determine the coordinates of a point in a non-perpendicular system of axes. Here are the first three rules of spacetime geometry:

Rules 1, 2, and 3 of spacetime geometry.

Note 4.2.6. (Rules of spacetime geometry) *Suppose that O moves relative to O' with fractional speed β, and the axes of O' are drawn perpendicular.*

1. *The axes of O form an acute angle if O moves in the positive direction, and an obtuse angle if O moves in the negative direction (Figure 4.6).*

2. *The angle between the axes of O and the corresponding axes of O' is given by* $\tan\alpha = \beta$.

3. *Suppose that the lines drawn parallel to the axes from an event E intersect the time axis at T and the space axis at X; then $\overline{ET}$ is the space coordinate and $\overline{EX}$ is the time coordinate of E [Figure 4.7 (a)].*

Consider events E_1 and E_2 in Figure 4.7(b), which occur at the same point *relative to* observer O, i.e., for which her clock (or any other clock in *her* RF) is present at both events. It follows that $t_2 - t_1$ is the *proper time,* and must be related to $t_2' - t_1'$—the time interval between the same two events according to observer O'—via (2.4):

$$t_2' - t_1' = \gamma(t_2 - t_1).$$

Lengths are not what they appear to be!

Here you encounter another strange phenomenon of spacetime geometry: Although $\overline{E_1E_2}$ [or, equivalently, $c(t_2 - t_1)$] *appears* longer to O' than its projection, the projection is actually longer! In general, you can apply the rules of ordinary geometry only to lengths *measured by a single observer.* Lengths

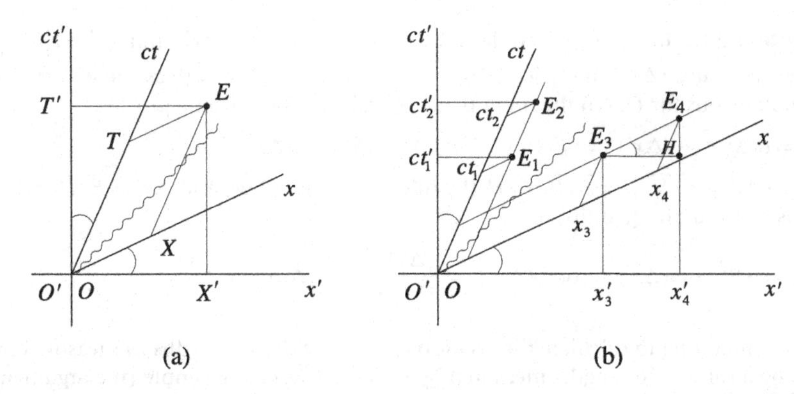

FIGURE 4.7 (a) Coordinates (X, cT) and (X', cT') of E in O and O'. (b) Events E_1 and E_2 occur at the same location relative to O, which is moving in the positive direction of O'. Events E_3 and E_4 occur simultaneously relative to O.

measured by two *different* observers *are not* related by the rules of ordinary geometry.

What about lengths along (or parallel to) the space axes? The events E_3 and E_4 in Figure 4.7(b) occur simultaneously according to O, i.e., $\overline{E_3E_4} = x_4 - x_3$ is parallel to O's x-axis, and as such represents a length (say of her spaceship if E_3 and E_4 are explosions of two firecrackers at the ends of the spaceship). The same two events are separated by $x'_4 - x'_3$ according to O', and you might think that, since moving lengths shrink, $x'_4 - x'_3 = (x_4 - x_3)/\gamma$. But as Example 4.2.8 below shows, $x'_4 - x'_3 = \gamma(x_4 - x_3)$. The reason for the discrepancy is that $x'_4 - x'_3$ *does not represent the length* of the spaceship for O'. For two explosions to represent the length of the spaceship as measured by O', they must occur simultaneously *for* O'. E_3 and E_4 are not simultaneous for O'. Example 4.2.8 also derives a rule that connects the length of the line segment $\overline{E_1E_2}$ (or $\overline{E_3E_4}$) as measured by O' (using a ruler, for example) and that measured by O. These two results complete the rules of spacetime geometry.

Rules 4 and 5 of spacetime geometry.

Note 4.2.7. (Rules of spacetime geometry) *Suppose that O moves relative to O' with fractional speed β and the axes of O' are drawn perpendicular.*

4. If E_1 and E_2 occur on a worldline parallel to an axis of one observer—for whom $\overline{E_1E_2}$ is the interval on that axis—and A_1 and A_2 are their (parallel) projections onto the **corresponding** *axis of the other observer, then $\overline{A_1A_2} = \gamma\overline{E_1E_2}$.*

5. The **Euclidean** *length $\overline{E_1E_2}$, as measured by O', is longer than the interval itself, as measured by O, by a factor of $\gamma\sqrt{1+\beta^2}$. This factor is called the* **stretch factor***.*

Example 4.2.8. In Figure 4.7(b), let $\Delta x = x_4 - x_3$, $\Delta x' = x'_4 - x'_3$, and $\Delta t' = t'_4 - t'_3$. Then O calculates $(\Delta s)^2$ and gets $(\Delta s)^2 = -(\Delta x)^2$, because $\Delta t = 0$, as the two events are simultaneous for O. On the other hand, O' calculates $(\Delta s)^2$ and gets

$$(\Delta s)^2 = (c\Delta t')^2 - (\Delta x')^2 = (\beta \Delta x')^2 - (\Delta x')^2 = (\beta^2 - 1)(\Delta x')^2$$

because $c\Delta t' = \overline{E_4 H}$, $\beta = \overline{E_4 H}/\overline{E_3 H}$ (by rule 2 of Note 4.2.6), and $\Delta x' = \overline{E_3 H}$. Since $(\Delta s)^2$ is an invariant quantity,

$$-(\Delta x)^2 = (\beta^2 - 1)(\Delta x')^2 \quad \text{or} \quad (\Delta x')^2 = \frac{(\Delta x)^2}{1 - \beta^2} = \gamma^2 (\Delta x)^2.$$

It is also interesting to calculate the relation between *Euclidean* lengths as measured by O' (using a ruler) and lengths measured by O. Take $\overline{E_3 E_4}$ for example (the argument for $\overline{E_1 E_2}$ is identical). O' sees this length as the hypotenuse of a right triangle and uses the Pythagorean theorem to find its length:

$$(\overline{E_3 E_4})^2_{O'} = \overline{E_3 H}^2 + \overline{H E_4}^2 = (\gamma \overline{E_3 E_4})^2 + [\beta(\gamma \overline{E_3 E_4})]^2 = (1 + \beta^2)\gamma^2 \overline{E_3 E_4}^2.$$

Therefore,

$$(\overline{E_3 E_4})_{O'} = \gamma \sqrt{1 + \beta^2}\, \overline{E_3 E_4}.$$

Keep in mind that $(\overline{E_3 E_4})_{O'}$ is the *Euclidean* length measured by O'—by placing a ruler on the two *points* E_3 and E_4. On the other hand, $\overline{E_3 E_4}$ is the difference between the x-coordinates of E_3 and E_4 as measured by O. ■

4.3 EXAMPLES OF SPACETIME DIAGRAMS

The best way to understand the implications of Notes 4.2.6 and 4.2.7 is to look at some examples. First let me take a closer look at the last item in the second Note. Sonya (reference frame O) is moving at $0.866c$ in the positive direction of Sam (reference frame O'). At the very moment that Sonya passes Sam, they both start their clocks; thus the origins of the two spacetime coordinate systems coincide (see Figure 4.8). Event E_1 occurs two years later according to Sonya. So on her time axis, E_1 is two years away from the origin. This time interval is proper. The interval that Sam measures is given by

$$\Delta t' = \frac{\Delta t}{\sqrt{1 - \beta^2}} = \frac{2}{\sqrt{1 - (0.866)^2}} = 4 \text{ years}.$$

This verifies the first part of rule 4 (which here translates to $\overline{O'T} = \gamma \overline{O E_1}$), because $\gamma = 1/\sqrt{1 - (0.866)^2} = 2$ and $\overline{O E_1} = 2$ years.

From $\beta = \tan\theta = \overline{T E_1}/\overline{O'T}$, you get

$\overline{T E_1} = \beta \overline{O'T} = 8.66 \times 4 = 3.464$ light years.

Now you can calculate the *Euclidean* length of $\overline{O' E_1}$ by the Pythagorean theorem:

$$\overline{O' E_1} = \sqrt{(\overline{T E_1})^2 + (\overline{O'T})^2} = \sqrt{(3.464)^2 + (4)^2} = 5.29.$$

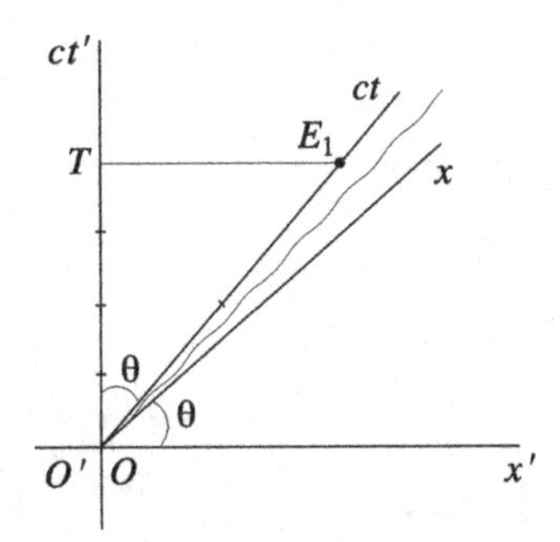

FIGURE 4.8 The event E_1 occurs 2 years after departure according to O, but 4 years according to O'. The *Euclidean* length of the line segment $\overline{O'E_1}$ is 5.29 light years.

So the actual length of the line segment $\overline{O'E_1}$ that Sam measures (when he places a ruler on the page of the book) is 5.29 light years. The real length (as measured by Sonya) is 2 light years, of course. Therefore, the stretch factor is 5.29/2=2.645, which is identical to what Note 4.2.7 says it should be: $\gamma\sqrt{1+\beta^2} = 2\sqrt{1+0.866^2} = 2.645$.

4.3.1 Simultaneity Revisited

The diagrammatic approach to relativity can elucidate some of the notions discussed earlier. Take the relativity of simultaneity, which was one of the first topics I introduced. Chapter 1 showed a picture identical to Figure 4.9(a), in which Sam (observer O') detects a simultaneous explosion of two firecrackers A and B. Sonya, on the other hand, describes the situation as B happening before A. Let's see if we can further unravel the succession of these events.

For simplicity, I'll assume that the explosions occur—according to Sam—at exactly the same time that Sonya passes him [corresponding to the picture at the top of Figure 4.9(a)],[6] and all these happen at Sam's time zero. This zero of time is also Sonya's zero of time, because I'm assuming that at the moment that they pass each other, Sonya and Sam start their clocks. How do we describe these events in Sam's and Sonya's spacetime coordinate systems (CS)?

I start with Sam and draw his axes perpendicular. Since A, B, and the passage of Sonya all occur at time $t' = 0$, they must all lie on the x'-axis. Sonya's worldline is the ct-axis, which crosses the x'-axis at the origin. The light signals (the wavy lines) from A and B reach Sam at a later time; this time is shown as a solid triangle on the ct'-axis. Sam, being in the middle of the two events, receives the two signals at the same time, as he should.

[6] Don't confuse the *reception* of the light signals at the bottom picture with the *occurrence* of the explosions.

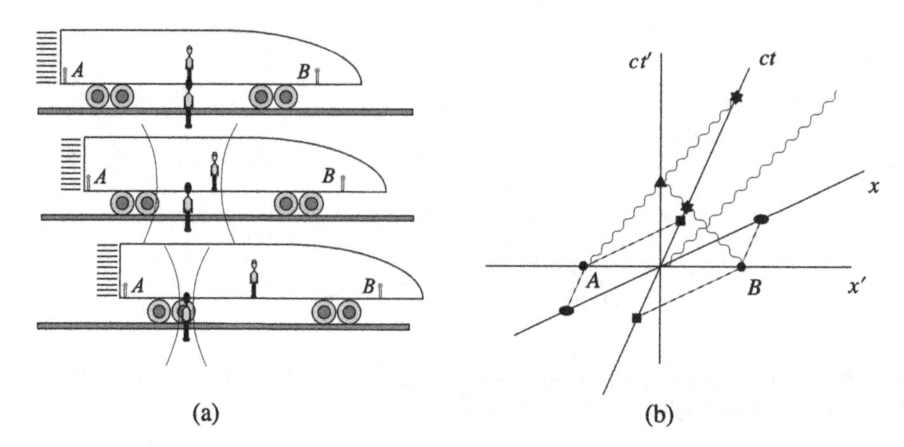

FIGURE 4.9 (a) To Sam, the explosion of the two firecrackers occurs at the same time that Sonya passes him by. (b) The spacetime geometry of the events as seen by Sam (O') and Sonya (O).

How does Sonya perceive the occurrence of the events and the reception of their signals? Draw parallel lines to Sonya's axes from A and B to find the coordinates of the two events. The locations on her x-axis are designated as ovals, and she is in the middle of them as is also clear in Figure 4.9(a). The times are represented by squares on the ct-axis. Note that B occurs first, as also indicated in the middle picture of Figure 4.9(a). But Figure 4.9(b) tells us more: that B occurred *before* Sonya reached Sam. The middle picture of Figure 4.9(a) shows the *reception* of B's light signal, not its explosion. Sonya's reception of the light signals is denoted by stars on her ct-axis. You see that although B explodes *before* Sonya reaches Sam, its signal reaches Sonya *after* she passes Sam, consistent with the middle picture of Figure 4.9(a). Finally, the signal from A reaches Sonya after the reception of the signal from B.

The discussion above, although qualitative, sheds some light on the notion of simultaneity as perceived by two different observers. If you know Sonya's speed relative to Sam and Sam's measurement of the length $\overline{AB}$, you can apply the rules of spacetime geometry to the figure to calculate the time difference between the explosions, the separation between the two firecrackers, and the time of the reception of the two signals all according to Sonya (see Problem 4.19).

4.3.2 The Train and the Tunnel

In the early days of relativity, there appeared to be a "paradox" having to do with the contraction of length; it was called the *pole-and-the-barn paradox* (see Problem 3.19). I'll consider a more modern version of it and call it the *train-and-the-tunnel paradox*. In Figure 4.10, Sonya's 756-meter-long train moves at $0.75c$ as it approaches a tunnel. Sam measures the contracted length of the train to be 500 m, and concludes that it should snuggly fit the 500-meter tun-

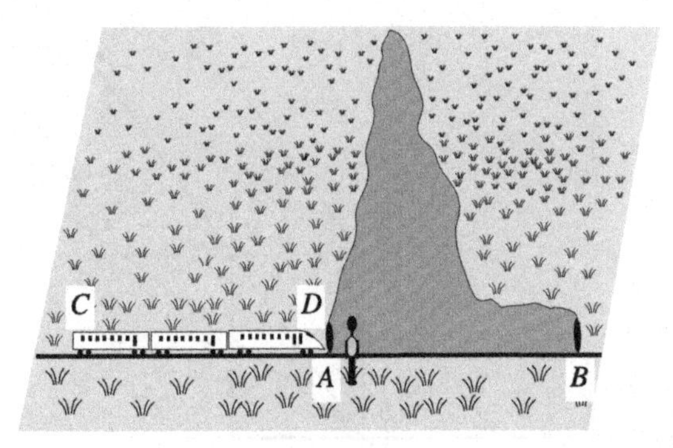

FIGURE 4.10 Sonya's train and the tunnel have the same length according to Sam.

nel he is standing by. Sonya, on the other hand, sees a contracted *tunnel*[7] of only 331 m and concludes that there is no way she can fit a 756-meter long train in that tunnel. What is going on?

With $\beta = 0.75$, I get $\gamma = 1.51$. To analyze the relative motion of the two RFs on a spacetime diagram, I'll label the ends of the tunnel and the train as A, B, C, and D, as shown in Figure 4.10. There are three conspicuous events in the diagram: the coincidence of A and D corresponding to the front of the train entering the tunnel (call it E_1, assumed to be the origin of the two RFs), the coincidence of D and B corresponding to the front of the train exiting the tunnel (call it E_2), and the coincidence of A and C corresponding to the end of the train entering the tunnel (call it E_3). To construct these events in a spacetime diagram, proceed as follows.

Draw Sam's axes as two perpendicular lines on a sheet of paper. From Sam's origin draw a line making a slope of 0.75 with Sam's x'-axis. This is Sonya's x-axis. Sonya's ct-axis is the line passing through the origin and making the same slope with Sam's ct'-axis (see Figure 4.11). Now draw the worldlines of A and B as two vertical lines, with A's worldline coinciding with the ct'-axis (assuming that Sam is standing at A). These worldlines are separated by 500 m, the length of the tunnel *according to Sam*. The worldlines of C and D are two parallel lines (because, being the two ends of the train, they move with the same velocity, i.e., same slope), with D's worldline coinciding with the ct-axis (assuming that Sonya is sitting in the front of the train at D).

How do I draw C's worldline? From the intersection of the ct-axis and B's worldline (event E_2) draw a line parallel to the x'-axis. This line meets A's worldline (the ct'-axis) at E_3 (this is because Sam sees E_2 and E_3 as simultaneous). I know that C's worldline must pass through E_3. I also know that

[7] Because the tunnel is moving relative to Sonya. So, its length should shrink for her.

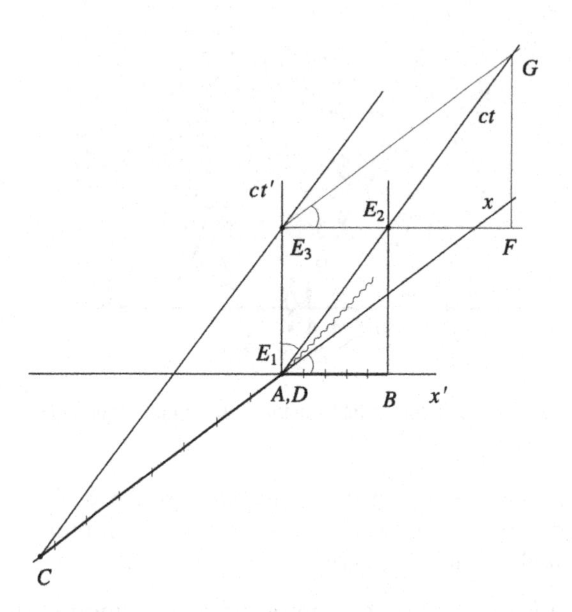

FIGURE 4.11 The spacetime diagram of the paradox of the train and the tunnel.

C's worldline must be parallel to D's (the ct-axis). So, draw a line through E_3 parallel to the ct-axis. This is C's worldline.

C's worldline meets the x-axis at a point, which you can naturally call C. The points C and D on the x-axis are separated by 756 m *as measured by Sonya*. The reason that the line segment $\overline{CD}$ appears much longer than 756 m (remember that $\overline{AB}$ is 500 m) is due to the stretch factor mentioned in rule 5 of Note 4.2.7.

It is clear from Figure 4.11 that E_3 has a time coordinate in Sonya's RF equal to $\overline{DG}$, which is obviously larger than $\overline{DE_2}$, the time of the coincidence of B and D according to Sonya. Therefore, the coincidence of C and A occurs *after* the coincidence of D and B (event E_2). This means that, although Sam sees the train completely in the tunnel, with the end points of the two coinciding at the same time, Sonya notices that the back of the train is outside the tunnel when the front of the train has reached the end of the tunnel. This is a diagrammatic representation of the relativity of simultaneity. The quantitative analysis of this discussion can be found in the following example.

Quantitative analysis of tunnel and train.

Example 4.3.1. The time of the occurrence of event E_2 (or E_3) according to Sam is the segment $\overline{E_1E_3}$ (see Figure 4.11). Furthermore, rule 2 of Note 4.2.6 gives $\overline{E_2E_3}/\overline{E_1E_3} = \beta$. But $\overline{E_2E_3}$ is just $\overline{AB} = 500$ m. Thus, $\overline{E_1E_3} = 500/0.75 = 667$ m. This is the projection of $\overline{E_1E_2}$—a segment on Sonya's time axis—onto Sam's time axis; so by rule 4 of Note 4.2.7,

$$\overline{E_1E_2} = \overline{E_1E_3}/\gamma = 667/1.51 = 441 \text{ m}.$$

This is the time of occurrence of the coincidence of B and D (times the speed of light) according to Sonya.

What is the time of occurrence of the coincidence of A and C, i.e., the time coordinate of E_3 (times the speed of light) according to Sonya? It is the line segment $\overline{DG}$, which is the projection of $\overline{E_1E_3}$ onto the ct-axis. Again, by rule 4 of Note 4.2.7,

$$\overline{DG} = \gamma \overline{E_1E_3} = 1.51 \times 667 = 1007 \text{ m}.$$

Therefore, E_3 occurs $1007 - 441 = 566$ m or $\frac{566}{3\times 10^8} = 1.9 \times 10^{-6}$ s later than E_2 according to Sonya.

The stretch factor is $1.51\sqrt{1+0.75^2} = 1.89$. This means that the ticks used on Sonya's axes should be 1.89 times farther apart than those used for Sam. The stretched units (still of 100 m each!) used in Figure 4.11 to measure $\overline{AC}$, the length of the train, take this fact into account. ■

4.3.3 The Doppler Effect

Sonya is moving with speed v away from Sam along his positive direction. She sends a light signal of wavelength λ to him, which is Doppler shifted to λ'. What is the relation between λ, λ', and v?

Figure 4.12(a) shows Sam's (the primed axes) and Sonya's (the unprimed axes) RFs, whose origins are assumed to have coincided at $t = 0 = t'$. Event E_1 is the emission of a crest of the light wave by Sonya's light source. After T seconds—where T is the period of the light wave—the next crest is emitted at event E_2. These two crests travel along the two wavy worldlines (making a 45° angle with the axes as they should) and are received by Sam at C and D. Therefore, $\overline{CD} = cT'$, where T' is the period of the light signal as measured by Sam. Since $\lambda = cT$ and $\lambda' = cT'$, what is left to do to obtain the Doppler formula is to find $\overline{CD}$ in terms of v and T.

Refer to Figure 4.12(b) and note that

$$\overline{CD} = \overline{BD} + \overline{CB} = \overline{BD} + \overline{AB} - \overline{AC}.$$

But, because light travels on a worldline that makes an angle of 45° with the axes, $\overline{BD} = \overline{BE_2}$ and $\overline{AC} = \overline{AE_1}$. So

$$\overline{CD} = \overline{BE_2} + \overline{AB} - \overline{AE_1} = (\overline{BE_2} - \overline{AE_1}) + \overline{AB},$$

or

$$\overline{CD} = (\overline{BE_2} - \overline{BF}) + \overline{AB} = \overline{FE_2} + \overline{AB}.$$

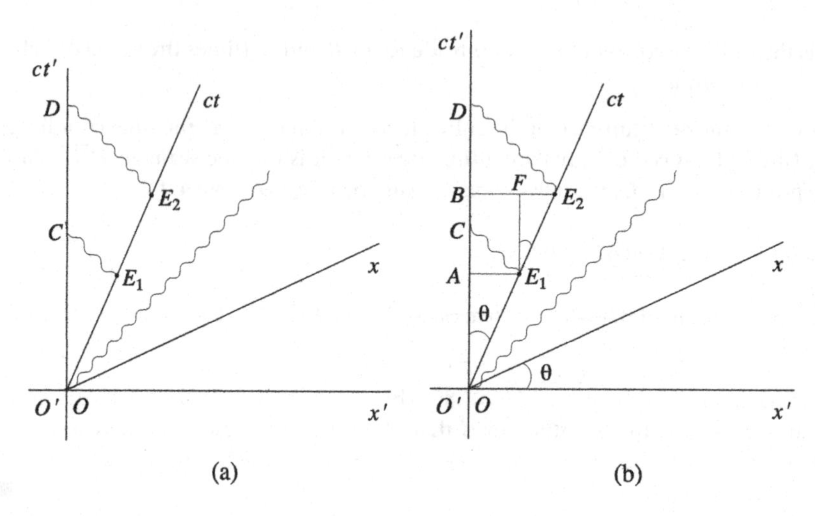

FIGURE 4.12 (a) The spacetime diagram of the Doppler effect. (b) Diagram used in the derivation of the Doppler effect.

Finding $\overline{CD}$ of Figure 4.13(a) in terms of v and T.

Furthermore, $\overline{FE_2}/\overline{FE_1} = \beta$ by rule 2 of Note 4.2.6, and $\overline{FE_1} = \overline{AB}$, yielding $\overline{FE_2} = \beta\,\overline{AB}$. Therefore,

$$\overline{CD} = \beta\,\overline{AB} + \overline{AB} = (1+\beta)\overline{AB} = (1+\beta)\gamma\,\overline{E_1E_2},$$

by rule 4 of Note 4.2.7. But $\overline{CD} = cT' = \lambda'$ and $\overline{E_1E_2} = cT = \lambda$; therefore

$$\lambda' = (1+\beta)\gamma\lambda = \frac{1+\beta}{\sqrt{1-\beta^2}}\lambda = \frac{\sqrt{1+\beta}\sqrt{1+\beta}}{\sqrt{(1-\beta)(1+\beta)}}\lambda,$$

or

$$\lambda' = \sqrt{\frac{1+\beta}{1-\beta}}\,\lambda. \tag{4.1}$$

This equation shows that if β is positive (Sonya is receding from Sam), the light is red-shifted ($\lambda' > \lambda$), and if she is approaching Sam (β is negative), the light is blue-shifted ($\lambda' < \lambda$). Problem 3.25 directly derives the equation for the case of approach.

If $\beta << 1$, then you can approximate (4.1) as follows:

$$\lambda' = (1+\beta)^{1/2}(1-\beta)^{-1/2}\lambda \approx (1+\tfrac{1}{2}\beta)[1-\tfrac{1}{2}(-\beta)]\lambda$$
$$\approx (1+\beta)\lambda = \left(1+\frac{v}{c}\right)\lambda, \quad \text{or} \quad \frac{\Delta\lambda}{\lambda} \equiv \frac{\lambda'-\lambda}{\lambda} = \frac{v}{c}, \tag{4.2}$$

which is the familiar non-relativistic Doppler formula.

Example 4.3.2. In the observation of the spectral lines of a certain element coming from a galaxy, it is seen that the green light, whose wavelength is 0.55 μm, has shifted

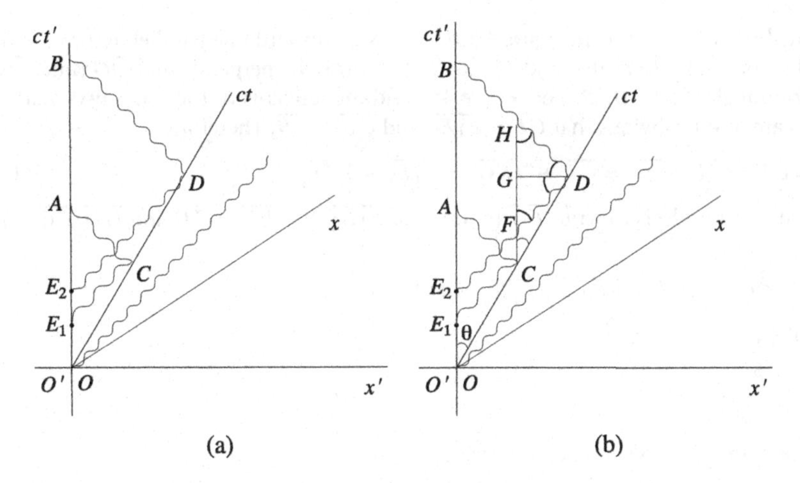

FIGURE 4.13 (a) The spacetime diagram of the Doppler effect including reflection from a moving object. (b) Diagram used in the derivation of the Doppler effect with reflection.

to 1.455 μm. What is the speed with which this galaxy is moving away from us? If you were to use the classical result of Equation (4.2), you would get

$$\lambda' = \lambda\left(1 + \frac{v}{c}\right), \quad \text{or} \quad 1.455 = 0.55\left(1 + \frac{v}{c}\right) \Rightarrow 1 + \frac{v}{c} = \frac{1.455}{0.55} = 2.645$$

or $v/c = 1.645$, a nonsensical result! Relativistic Doppler formula, on the other hand, gives

$$1.455 = \sqrt{\frac{1+\beta}{1-\beta}}\, 0.55 \Rightarrow \sqrt{\frac{1+\beta}{1-\beta}} = \frac{1.455}{0.55} = 2.645.$$

Squaring both sides gives

$$\frac{1+\beta}{1-\beta} = 6.996 \quad \text{or} \quad \beta = \frac{6.996 - 1}{6.996 + 1} \approx 0.75,$$

a reasonable result. ■

Spacetime diagrams can also be used to derive the Doppler shift of an electromagnetic wave reflected from a moving object. Figure 4.13(a) shows Sonya moving relative to Sam. Sam sends a radar signal to Sonya and measures the wavelength of its reflection, thereby determining Sonya's speed. How does he do it? Let E_1 be the sending of a wave crest, and E_2 that of the next crest, so that $\overline{E_1E_2}$ is cT, where T is the period of the radar. These waves intersect Sonya's worldline (are received by her) at C and D, whereupon they get reflected and are received by Sam at A and B. Note that the radar's worldlines $\overline{E_1C}$ and $\overline{E_2D}$ and their reflections $\overline{CA}$ and $\overline{DB}$ make 45° angles with Sam's axes. Our task is to find $\overline{AB}$ in terms of $\overline{E_1E_2}$. The following example derives the result.

Finding $\overline{AB}$ in terms of $\overline{E_1E_2}$ in Figure 4.13(b).

Example 4.3.3. From C in Figure 4.13(b) draw an upward line parallel to the ct'-axis. This line intersects $\overline{E_2D}$ at F and $\overline{DB}$ at H. Also drop the perpendicular $\overline{DG}$ onto $\overline{CH}$. The four angles at F, D, G, and H are 45° and the tangents of the other two marked angles are β. It is obvious that $\overline{CF} = \overline{E_1E_2}$ and $\overline{CH} = \overline{AB}$. Therefore,

$$\overline{AB} = \overline{CH} = \overline{CF} + \overline{FH} = \overline{CF} + 2(\overline{FG}) = \overline{E_1E_2} + 2(\overline{FG}). \tag{4.3}$$

All that is left to do is to write $\overline{FG}$ in terms of $\overline{E_1E_2}$. But $\overline{FG} = \overline{GD}$ and $\overline{GD}/\overline{CG} = \beta$. Thus,

$$\overline{FG} = \beta\overline{CG} = \beta(\overline{CF} + \overline{FG}) = \beta\overline{E_1E_2} + \beta\overline{FG}.$$

This gives

$$\overline{FG} = \frac{\beta}{1-\beta}\overline{E_1E_2}.$$

Substitute this in (4.3) and get the desired result:

$$\overline{AB} = \overline{E_1E_2} + \frac{2\beta}{1-\beta}\overline{E_1E_2} = \frac{1+\beta}{1-\beta}\overline{E_1E_2}.$$

Noting that $\overline{AB} = cT_{\text{ref}} = \lambda_{\text{ref}}$, with λ_{ref} the wavelength of the reflected light, and $\overline{E_1E_2} = cT = \lambda$, you obtain

$$\lambda_{\text{ref}} = \frac{1+\beta}{1-\beta}\lambda. \tag{4.4}$$

You can derive this formula also by noting that λ_{ref} is the Doppler-shifted wavelength λ' of (4.1). Thus,

$$\lambda_{\text{ref}} = \sqrt{\frac{1+\beta}{1-\beta}}\lambda' = \sqrt{\frac{1+\beta}{1-\beta}}\sqrt{\frac{1+\beta}{1-\beta}}\lambda = \frac{1+\beta}{1-\beta}\lambda.$$

When Sonya's speed becomes much smaller than light speed, β is very small and you can make the following approximations:

$$\frac{1}{1-\beta} \approx 1+\beta \quad \text{and} \quad (1+\beta)^2 \approx 1+2\beta.$$

Then Equation (4.4) reduces to $\lambda_{\text{ref}} = (1+2\beta)\lambda$. This is the equation used in determining the speed of a speeding car. ■

4.3.4 Time Travel?

Humankind have been traveling in three dimensions for millions of years. The two-dimensional aspect of this travel became frequent when man invented the wheel, horse and buggy, and particularly the automobile. Travel in the third dimension intensified with the invention of the airplane. Now you have learned that relativity makes time a "fourth dimension." So, you may ask: Can we travel in time?

A typical situation is the occurrence of an event E in the past of an observer O' as shown in Figure 4.14. Since O' has no "time machine," he exploits the

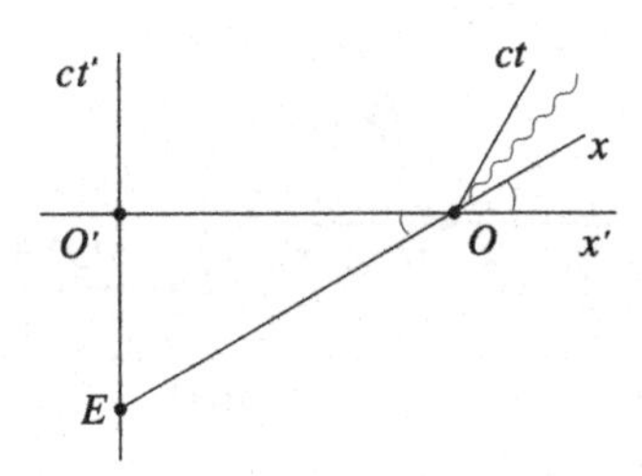

FIGURE 4.14 Spacetime diagram for rescuing Bruno by finding an observer whose present time is Bruno's execution.

variation of time as measured by others, and finds an observer O for whom E occurs now. O must lie on the x'-axis, because O' wants to do the "time traveling" now, i.e., at $t' = 0$. E must lie on the x-axis, because E must be taking place now for O. This determines the x-axis as the line EO. By drawing a light worldline at O, which makes a 45° angle with the O' axes, and reading the angle it makes with the x-axis, we can draw the ct-axis as well.

Can we stop the execution of Giordano Bruno?

Giordano Bruno was a fiercely independent Renaissance thinker whose ideas of Copernican astronomy and infinite worlds with infinite (perhaps similar) histories were blasphemy in the eyes of the sixteenth-century church. After a long period of imprisonment and defiance, the church decided to silence Bruno for good. On February 17, 1600 Giordano was burned at the stake. Now that we are familiar with the geometry of spacetime, let's see if we could have stopped the execution of this free thinker on the fifth centennial of his death, February 17, 2100.

We need to find an observer for whom "now" (17 February 2100) and Bruno's execution are simultaneous, i.e., they both lie on the x-axis of the observer. We look among our Galactic Explorers and find observer O 600 light years away, far enough that with the proper speed will have an x-axis that passes through the event E, Bruno's death [see Figure 4.14].

Q: What speed should O have? In which direction?

A: Figure 4.14 shows that O should be moving *away from* us, because the ct-axis makes an acute angle with the x'-axis. The same figure shows that $\beta = \overline{EO'}/\overline{O'O}$ or

$\beta = c \times 500 \text{ years}/(600 \text{ ly}) = 0.833.$

Q: How far away is Bruno's execution taking place from O?

A: By rule 4 of Note 4.2.7, $\overline{O'O} = \gamma \overline{EO}$. Therefore,

$$\overline{EO} = \sqrt{1-\beta^2}\,\overline{O'O} = 0.553 \times 600 \approx 332 \text{ light years},$$

too far away to prevent the execution.

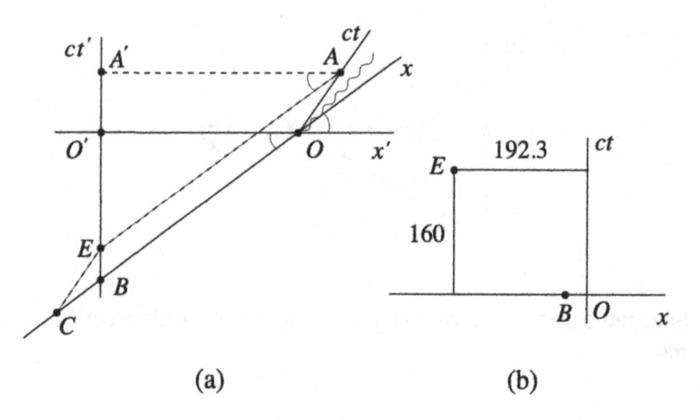

FIGURE 4.15 (a) Finding the coordinates of E in the non-perpendicular coordinate system O. (b) The events B and E as seen by O.

It appears that β determines how close we can get to the execution; the larger the β, the closer we might get. So we call on another observer who is 502 light years away. The required β—that which ensures that the x-axis of the observer passes through E—is $\beta = 500/502 = 0.996$, giving $\gamma = 11.21$. In this case, $\overline{EO} = \overline{O'O}/\gamma = 502/11.21 = 44.8$ light years, still too far away to prevent the execution.

Q: Is it even possible to get to E?

A: To answer this question, write $\overline{EO}$ in terms of $\overline{EO'}$. You can do this by simply noting that $\overline{EO} = \overline{O'O}/\gamma$ and $\overline{O'O} = \overline{EO'}/\beta$. Then,

$$\overline{EO} = \frac{\overline{O'O}}{\gamma} = \frac{\overline{EO'}/\beta}{\gamma} = \frac{\overline{EO'}}{\beta\gamma} = \frac{\sqrt{1-\beta^2}}{\beta}\overline{EO'} = \sqrt{\frac{1}{\beta^2} - 1}\,\overline{EO'}$$

and thus, for $\overline{EO}$ to be very small, β has to be very close to 1; and unless O moves *at the speed of light*, you can never shrink $\overline{EO}$ to zero! Although you *can* find observers for whom Bruno's death occurs *NOW*, you can never find an observer who is present *at the location* of the execution.

Since we can "go back" in time, why not go further back, to a time *before* the event of interest happened and "wait" for the event? Let's go back 10 years prior to Bruno's execution, to event B of Figure 4.15(a). In this case, the x-axis is the line OB, with the corresponding ct-axis drawn at equal angle from the light worldline, as done before. Assume that O is 511 light years away. Then rule 2 of Note 4.2.6 gives

$$\beta = \frac{\overline{BO'}}{\overline{O'O}} = \frac{510}{511} = 0.998 \quad \text{and} \quad \gamma = \frac{1}{\sqrt{1-0.998^2}} = 16.$$

To see if E can be prevented, we need to know where the location of E is in the spacetime plane of O.

Q: When is E happening according to O?

A: Draw a line from E parallel to the x-axis to cut the ct-axis at A. Then $\overline{OA}$ is the time coordinate of E in O. By rule 4 of Note 4.2.7,

$\overline{OA} = \gamma \overline{BE} = 16 \times 10 = 160$ ly.

So the execution is happening 160 years from now according to O. That is a long time into the future and you might think that the mission can be accomplished by the future generation of free thinkers. But wait!

Q: What is the x-coordinate of E according to O?

A: Draw a line from E parallel to the ct-axis to cut the x-axis at C. Then $-\overline{OC}$ is the x-coordinate of E in O. But $\overline{OC} = \overline{EA}$. To find $\overline{EA}$, draw a line from A parallel to the x'-axis to cut the ct'-axis at A'. Then the length of $\overline{EA}$ *as measured by an ordinary ruler* by O' is given by the Pythagorean theorem:

$$\overline{EA} = \sqrt{(\overline{EA'})^2 + (\overline{A'A})^2} = \sqrt{(\overline{EA'})^2 + (\overline{EA'}/\beta)^2} = \frac{\overline{EA'}}{\beta}\sqrt{1+\beta^2},$$

where the second equality follows from rule 2 of Note 4.2.6. By rule 5 of Note 4.2.7, the *real length* of $\overline{EA}$ as measured by O is obtained by dividing $\overline{EA}$ by the stretch factor $\gamma\sqrt{1+\beta^2}$. Therefore, denoting the x-coordinate of E by x_E, we get

$$x_E = -\frac{\overline{EA}}{\gamma\sqrt{1+\beta^2}} = -\frac{\overline{EA'}}{\gamma\beta}. \tag{4.5}$$

The only thing that is left now is to find $\overline{EA'}$, which is the sum of $\overline{EO'}$ and $\overline{O'A'}$. The first one is given, and we can get the second one from $\overline{OA}$ by another application of rule 4 of Note 4.2.7:

$\overline{O'A'} = \gamma \overline{OA} = 16 \times 160 = 2560$ ly.

Equation (4.5) now yields

$$x_E = -\frac{510 + 2560}{(16)(0.998)} = -192.3 \text{ ly}.$$

Thus, according to O, the execution will happen 160 years from now at a distance of 192.3 ly (in the negative x-direction). Any probe sent from O must have a fractional speed of 192.3/160; i.e., it must travel faster than light! Even a laser pulse sent with the purpose of stunning the executioner will not be able to make it in time!

For the sake of completeness, let's calculate the coordinates of B relative to O. Clearly, B has zero time coordinate. Its x-coordinate x_B is obtained by rule 4

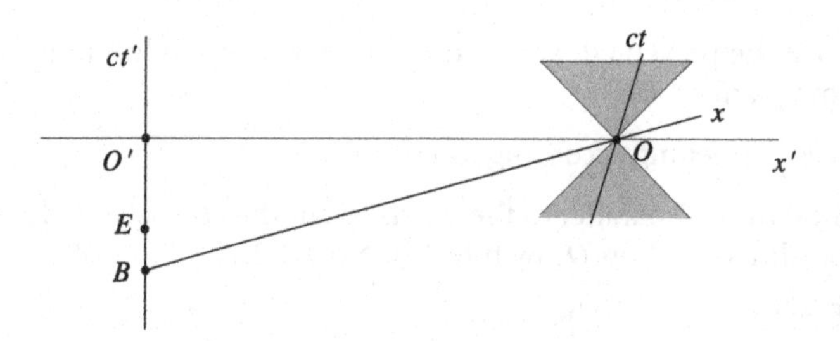

FIGURE 4.16 We want to get to E, but we go to B to "wait."

of Note 4.2.7: $\overline{OO'}$ is 511 ly and is γ times $\overline{OB}$. Therefore,

$$x_B = -\overline{OB} = -\frac{\overline{OO'}}{\gamma} = -\frac{511}{16} = -31.94 \text{ ly}.$$

Figure 4.15(b) shows the two events B and E in the coordinate system O.

Bruno's execution took place in the past. What about time traveling in the opposite direction? Can we stop *future* mishaps from happening? Problem 4.18 has yet another disappointing answer!

All these examples have shown that *using the numbers given in those examples,* it is impossible to travel back and forth in time. Can this be generalized?

Note 4.3.4. *It is impossible to reach events either in the past or in the future using the laws of special relativity.*

The following general argument shows that the past is indeed out of our reach. A similar argument shows that the future is also inaccessible.

Proof of impossibility of going back in time.

The reference frame O that we are seeking lies on the Earth's x'-axis (see Figure 4.16). It could be any point of this axis with one condition: the resulting coordinate system for O must have its *time* axis in the light cone at O, because the time axis of a RF is its worldline and no worldline can be outside the light cone. It follows that the x-axis must be to the right of the cone. But the x-axis is the line OB, because B is happening NOW (at the same time as the origin of time) for O. So we must choose O on the horizontal line in such a way that the line BO is outside the light cone. This makes B causally disconnected from O. The line $O'O$ is also outside the line cone. Since the line EO lies between $O'O$ and BO, it too must lie outside the light cone. Therefore, E is causally disconnected from O, and no probe (even light) can reach E from our RF. This conclusion is independent of the location of B and E in the coordinate system. So, it is quite general.

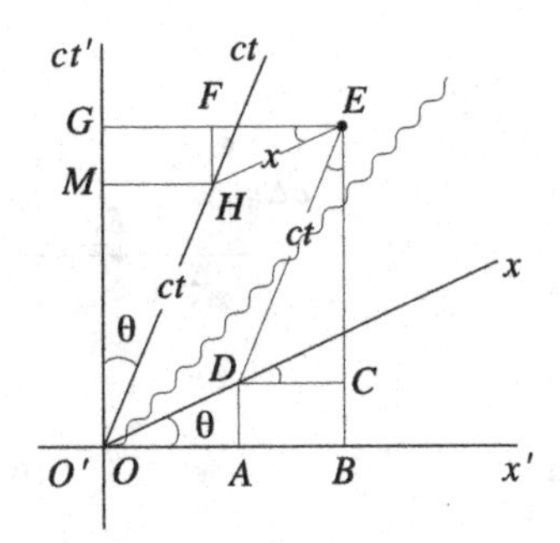

FIGURE 4.17 Deriving the Lorentz transformation from the diagram.

4.4 LORENTZ TRANSFORMATION REVISITED

I derived the Lorentz transformation (LT) in Chapter 3 from the invariance of the spacetime distance using algebraic manipulations. It is also possible to derive it via geometric methods as I'll do below.

Deriving Lorentz transformation from spacetime diagrams.

Figure 4.17 shows an event E with coordinates (x, ct) in O. My aim is to calculate its coordinates (x', ct') in O' in terms of x and ct. First note that $x' = \overline{O'A} + \overline{AB}$, and $\overline{O'A}$ is the projection of $\overline{OD}$—lying on the x-axis of O—onto the x'-axis. By rule 4 of Note 4.2.7, $\overline{O'A} = \gamma\,\overline{OD} = \gamma x$. Next, look at $\overline{AB}$. The figure shows that $\overline{AB} = \overline{CD}$. But rule 2 of Note 4.2.6 gives $\beta = \overline{CD}/\overline{CE}$, because the marked angles are all equal. Now $\overline{CE} = \overline{O'M}$ (verify this!) and $\overline{O'M}$ is the projection of $\overline{OH}$ onto the time axis of O'. Again by rule 2 of Note 4.2.6, $\overline{O'M} = \gamma\,\overline{OH} = \gamma ct$; so

$$\overline{AB} = \overline{CD} = \beta\overline{CE} = \beta\overline{O'M} = \beta\gamma ct.$$

Therefore,

$$x' = \overline{O'A} + \overline{AB} = \gamma x + \beta\gamma ct = \gamma(x + \beta ct). \tag{4.6}$$

How is ct' related to x and ct? Obviously, $ct' = \overline{O'M} + \overline{MG}$. I have already calculated $\overline{O'M}$: it is equal to γct. To find $\overline{MG}$, I use $\overline{MG} = \overline{FH} = \overline{DA} = \beta\overline{O'A}$ (the last equality follows from rule 2 again). But $\overline{O'A} = \gamma x$, as I showed above. Therefore, $\overline{MG} = \beta\gamma x$, and

$$ct' = \overline{O'M} + \overline{MG} = \gamma ct + \beta\gamma x = \gamma(\beta x + ct). \tag{4.7}$$

Equations (4.6) and (4.7) are identical to the two relations in (3.20). You can appreciate the ease and power of geometric and diagrammatic reasoning if you compare this derivation with the one done algebraically in Section 3.2.

4.4.1 Lorentz Transformation and Time Travel

Combining the graphical geometric methods with the algebraic LT, I intend to look at the possibility of time traveling to John F. Kennedy's assassination in

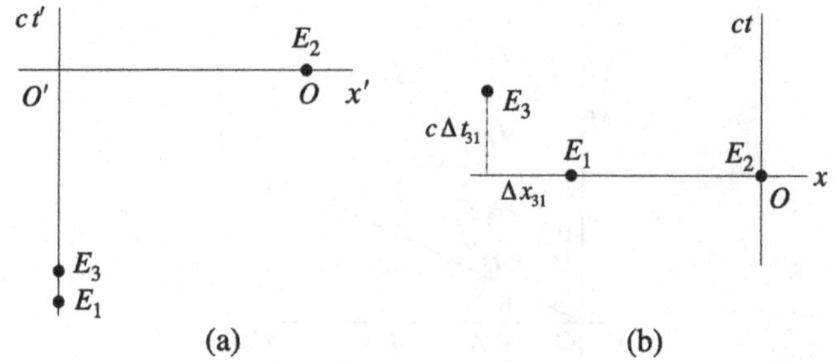

FIGURE 4.18 (a) The events E_1, E_2, and E_3 in the Earth RF. Note that the time of the origin (and thus the time of all the events on the x'-axis) is NOW, which is the year 2163. (b) The same three events as seen by the crew of Diracus II.

1963. We of course know from Note 4.3.4 that it is impossible. Nevertheless, it is worth reexamining it from the LT perspective. It is also a good exercise in calculations involving LTs.

Lorentz transformation and time travel.

The year is 2163, and the American delegation to the Intergalactic Space Federation, stationed on Earth, is submitting a proposal to use the laws of relativity to stop the assassination of President Kennedy by finding a RF for which 1961, two years before the assassination, is NOW. That way, argues ISF's project director, the crew will have some extra years to "prepare" for the event. The ISF accepts the proposal and after some initial search it decides that the Spaceship Diracus II, 205 ly away, is a good candidate.

Q: How fast and in what direction should Diracus II be moving relative to Earth?

A: Let the RF of Diracus II be O and the Earth's RF be O'. The two events of interest are E_1, "the building—in Earth year 1961—in which the shooting will take place" and E_2 the "passage of Diracus II by an outpost in Earth year 2163." These events are shown in Figure 4.18(a) in the Earth reference frame O'. E_1 has coordinates (0, −202 ly) because NOW is the year 2163. Similarly, E_2 has coordinates (205 ly, 0).

Since there are three events to deal with, it is helpful to label the Δ quantities. For example, I use Δx_{21} to denote $x_2 - x_1$, and $c\Delta t_{31}$ to denote $ct_3 - ct_1$, etc. Then, I get $\Delta x'_{21} = 205$ ly, $c\Delta t'_{21} = 202$ ly, and, because I want the two events to be simultaneous on Diracus II, $\Delta t_{21} = 0$. Thus, Equation (3.25) reduces to

$$205 \text{ ly} = \gamma \Delta x_{21}, \quad 202 \text{ ly} = \gamma\beta \Delta x_{21}. \tag{4.8}$$

These two equations lead to $202 = \beta(205)$ or $\beta = 0.9854$. Since β is positive, Diracus II must be moving *away* from Earth.

Q: According to the Diracus II crew, how far away is the building in which the assassin will be hiding?

A: I can find the spatial distance Δx_{21}—between the two events E_1 and E_2—for the Diracus II crew by using the first equation in (4.8) with $\gamma = 1/\sqrt{1-(0.9854)^2} = 5.87$:

$$\Delta x_{21} = x_2 - x_1 = \frac{205 \text{ ly}}{\gamma} = \frac{205}{5.87} = 34.9 \text{ ly}.$$

Since $x_2 = 0$ (origin of Diracus II), $x_1 = -34.9$ ly. Figure 4.18(b) shows these two events in the RF of Diracus II.

The chief physicist reports these results to the commander. The commander, knowing that nothing can move as fast as light, asks him to look into the possibility of sending a powerful laser beam to stun the assassin exactly at the time of the shooting. The chief physicist starts to calculate.

Q: According to the Diracus II RF, what is the time and space difference between the two events "assassination building in Earth year 1961" and "assassination building in Earth year 1963?" Denote the latter event by E_3 and designate their space and time difference with the subscript 31.

A: The (inverse) Lorentz transformation—i.e., the transformation with the *primed quantities* on the right-hand side—gives the answer. Since O' (the Earth RF) is moving *in the negative* direction of O (the Diracus II RF), we must use $-\beta$ in the formulas. Now, the space difference between the two events is zero according to the Earth RF, because they both occur in the same building. The Earth time interval is 2 years. It follows that

$$\Delta x_{31} = \gamma(\Delta x'_{31} - \beta c \Delta t'_{31}) = 5.87(0 - 0.9854c \times 2) = -11.57 \text{ ly}$$
$$c\Delta t_{31} = \gamma(-\beta \Delta x'_{31} + c\Delta t'_{31}) = 5.87(0 + c \times 2) = 11.74 \text{ ly}$$

or $\Delta t_{31} = 11.74$ yrs.

Q: According to the Diracus II RF, what are the coordinates of the event E_3, "assassination building in Earth year 1963"?

A: Recall that $\Delta x_{31} = x_3 - x_1$, and I have already calculated x_1 to be -34.9 ly. Therefore,

$$-11.57 = x_3 - (-34.9),$$

or $x_3 = -46.47$ ly. Similarly, $c\Delta t_{31} = ct_3 - ct_1$, or $11.74 = ct_3 - 0$ ($ct_1 = 0$ because E_1 is happening NOW for Diracus II). Thus, $ct_3 = 11.74$ ly. All these events are shown in Figure 4.18(b) in the RF of Diracus II. You may wonder why we did not use the equations in Note 3.3.1. The reason is that those equations require the two coordinate systems to have the *same origin*; O and O' do not have the same origin (see Problem 4.20).

Q: How fast should the laser beam be moving to be present in the assassination building in time?

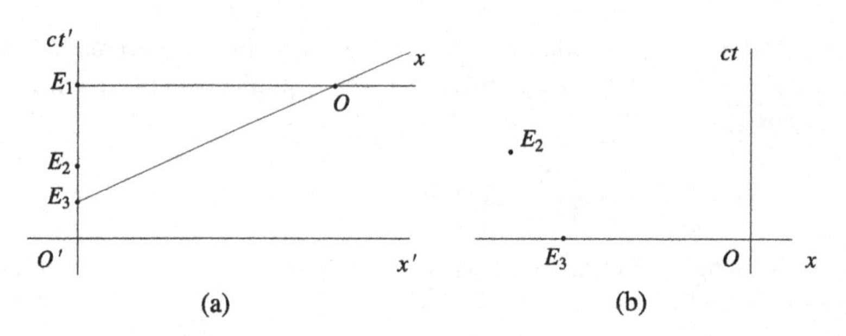

FIGURE 4.19 (a) The events as they appear to O'. (b) The relevant events as they appear to O.

A: The beam must cover a distance of 46.47 ly in 11.74 years. Therefore, $\beta = \Delta x / c\Delta t = 46.47/11.74 = 3.96$, i.e., it must move at about four times the speed of light! So, although the crew of Diracus II knows that Kennedy's assassination *will* take place 11.74 years from now, they can do nothing about it. Unfortunately, Kennedy cannot be saved!

You have already seen how an extremely simple diagrammatic approach proves the impossibility of traveling back in time (see the discussion after Note 4.3.4). The tedious algebraic proof in the following example should make you appreciate the power and elegance of geometry!

Impossibility of traveling back in time.

Example 4.4.1. Let E_1 be now on Earth, E_2 the event we want to get to, and E_3 the event in the past of E_2 to which we first go to wait for E_2. You can think of O—the origin of the RF (call it Diracus) we are seeking—as an event as well [see Figure 4.19(a)]. Use prime for the coordinates of events in O' and label the *intervals* by the events they connect. For example, $\Delta x'_{10} \equiv x'_1 - x'_0$ is the space interval between E_1 and O as measured by O' (Earth); similarly, $\Delta t_{23} \equiv x_2 - x_3$ is the time interval between E_2 and E_3 as measured by O (Diracus).

First I'll find the speed β of O relative to O' using the fact that E_3 is taking place NOW in O (so that $\Delta t_{03} = 0$). Dividing both sides of the second of the two equations

$$\begin{aligned} \Delta x'_{03} &= \gamma(\Delta x_{03} + \beta c \Delta t_{03}) = \gamma(\Delta x_{03} + 0) = \gamma \Delta x_{03} \\ c\Delta t'_{03} &= \gamma(\beta \Delta x_{03} + c\Delta t_{03}) = \gamma(\beta \Delta x_{03} + 0) = \gamma\beta\Delta x_{03} \end{aligned} \tag{4.9}$$

by the first, you get β:

$$\frac{c\Delta t'_{03}}{\Delta x'_{03}} = \frac{\gamma\beta\Delta x_{03}}{\gamma\Delta x_{03}} \quad \text{or} \quad \beta = \frac{c\Delta t'_{03}}{\Delta x'_{03}}.$$

To avoid the cluttering of indices, I use X for $\Delta x'_{03}$, T for $\Delta t'_{03}$ (which is also the time interval between E_1 and E_3, the amount of time I want to go back in the past), T_1 for $\Delta t'_{12}$, and T_2 for $\Delta t'_{23}$. Then you can express β as

$$\beta = \frac{cT}{X} \quad \text{from which you get} \quad X = \frac{cT}{\beta}. \tag{4.10}$$

Now I'll figure out how O sees the events of interest. Figure 4.19(b) shows the situation for O. Event E_3 is on Diracus's x-axis. I can calculate its distance from the first equation of (4.9):

$$\Delta x_{03} = \frac{\Delta x'_{03}}{\gamma} = \frac{X}{\gamma}.$$

The intervals between E_2 and E_3 *as measured by Diracus* are given by the inverse LT:

$$\begin{aligned} \Delta x_{23} &= \gamma(\Delta x'_{23} - \beta c \Delta t'_{23}) = \gamma(0 - \beta c \Delta t'_{23}) = -\gamma\beta c T_2 \\ c\Delta t_{23} &= \gamma(-\beta \Delta x'_{23} + c\Delta t'_{23}) = \gamma(0 + c\Delta t'_{23}) = \gamma c T_2. \end{aligned} \tag{4.11}$$

Therefore, the space interval Δx_{20} between E_2 and Diracus (which is the distance between E_3 and Diracus plus the space interval between E_2 and E_3) is

$$\Delta x_{20} = \Delta x_{30} + \Delta x_{23} = -\Delta x_{03} + \Delta x_{23} = -\frac{X}{\gamma} - \gamma\beta c T_2,$$

and the time interval Δt_{20} between E_2 and Diracus's NOW, i.e., the time that E_2 will take place in Diracus's future is just $\gamma c T_2$, because Δt_{20} is the same as Δt_{23}.

Summarizing, Diracus wants to send a probe to Earth so that it will reach Earth at the exact time that E_2 is happening. This probe has to travel a distance of $-\frac{X}{\gamma} - \gamma\beta c T_2$ in a time interval of $\gamma c T_2$. Therefore, its speed should be (ignore the negative sign of the speed)

$$\beta_{\text{probe}} = \frac{X/\gamma + \gamma\beta c T_2}{\gamma c T_2} = \beta + \frac{X}{\gamma^2 c T_2}.$$

Substitute X from Equation (4.10) and use $T = T_1 + T_2$ and $1/\gamma^2 = 1 - \beta^2$ to obtain (see Problem 4.21)

$$\beta_{\text{probe}} = \beta + \frac{cT/\beta}{\gamma^2 c T_2} = \beta + \frac{T_1 + T_2}{\beta\gamma^2 T_2} = \frac{1}{\beta} + \frac{T_1}{\beta\gamma^2 T_2}. \tag{4.12}$$

The last line shows that $\beta_{\text{probe}} > 1$ because the first term is greater than 1 and the second term is positive. ■

4.5 CURVED WORLDLINES

The straight worldlines considered so far do not describe the most general type of motion. Reference frames often change their speed, i.e., they accelerate. The worldline of an accelerating RF is a curve in the spacetime plane such as the one shown in Figure 4.20(a), where the slope changes from event to event. If the curve represents a real worldline, then the tangent to the curve at every point must have a slope that is larger than 1.

A curved worldline is the union of a lot of small straight worldlines.

Although straight worldlines are restrictive, they are fundamental in the same sense that straight lines are fundamental in Euclidean geometry: *any* worldline can be built up from small straight worldline segments. For example, Figure 4.20(b) shows a worldline consisting of five straight segments ($\overline{OA}$ through

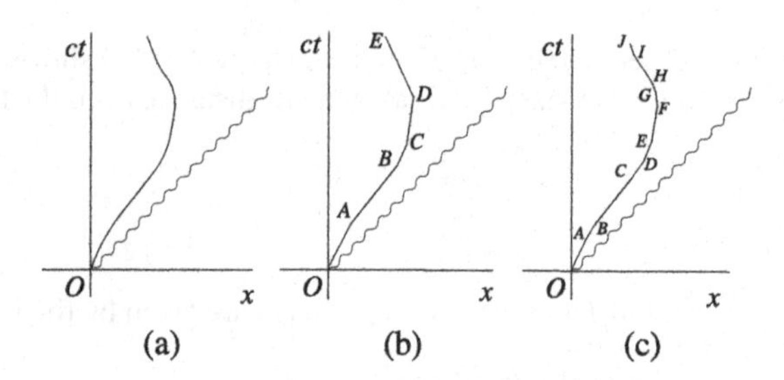

FIGURE 4.20 (a) The worldline of an accelerated frame. (b) Five *straight* worldlines approximating the curve. (c) Ten *straight* worldlines approximating the curve.

$\overline{DE}$). This collection of five straight worldlines approximates the curve of Figure 4.20(a) fairly well, but if you want a better approximation, you can increase the number of straight worldlines. Figure 4.20(c) shows a worldline consisting of ten straight segments ($\overline{OA}$ through $\overline{IJ}$), which obviously approximate the curve more accurately. For even better accuracy you can further increase the number of worldline segments. All these straight segments must have slopes that are larger than 1.

4.5.1 The Spacetime Triangle Inequality

Given any two points in Euclidean geometry, there are infinitely many curves that connect those points. These curves have different lengths, and only one—the one we call *straight*—has the shortest length. At the heart of this property lies the *triangle inequality*, which states that the sum of the lengths of any two sides of a triangle is greater than the length of the third side. If triangle inequality holds, then you can show that indeed a straight line is the shortest path. Here is how.

Consider two points P_1 and P_2 in a Euclidean plane. Draw a straight line and a curve through the points as shown in Figure 4.21. Pick a point A on the curve and form the triangle P_1AP_2. By triangle inequality, $\overline{P_1A}+\overline{AP_2}>\overline{P_1P_2}$. Now choose a point B on the curve between P_1 and A and another point C between A and P_2. Invoking the triangle inequality again, you can see that $\overline{P_1B}+\overline{BA}>\overline{P_1A}$ and $\overline{AC}+\overline{CP_2}>\overline{AP_2}$. Substitute these inequalities in the previous one to get

$$\overline{P_1B}+\overline{BA}+\overline{AC}+\overline{CP_2}>\overline{P_1P_2}.$$

Continue this process ad infinitum and conclude that the length of the curve is larger than the length of the straight line.

Is there a triangle inequality in spacetime geometry? A **spacetime triangle** consists of three mutually causally connected events connected by straight worldlines (the slopes of these lines are therefore larger than 1). Call the space-

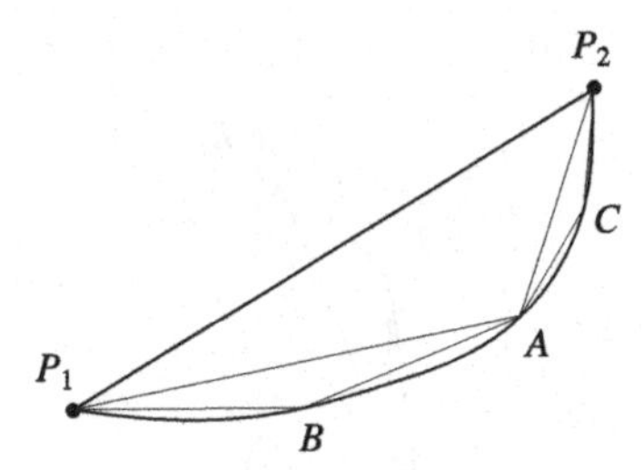

FIGURE 4.21 The triangle inequality in Euclidean geometry proves that a straight line is the shortest distance between two points.

time distance between two events at the two ends of a side of the triangle the **spacetime length** of that side.[8]

Proof of spacetime triangle inequality.

Figure 4.22 shows a spacetime triangle whose vertices are the three events E_1, E_2, and E_3. The space and time intervals between any two events are shown on the corresponding axes. For example, the space interval between E_1 and E_2 is x_{12} and the time interval between E_2 and E_3 is t_{23}. Denote the spacetime interval between any two events similarly. It is clear that $x_{13} = x_{12} + x_{23}$ and $t_{13} = t_{12} + t_{23}$. In my attempt at obtaining an inequality, I try to find s_{13} in terms of s_{12} and s_{23}:

$$\begin{aligned}
s_{13}^2 &= (ct_{13})^2 - x_{13}^2 = (ct_{12} + ct_{23})^2 - (x_{12} + x_{23})^2 \\
&= (ct_{12})^2 + (ct_{23})^2 + 2(ct_{12})(ct_{23}) - x_{12}^2 - x_{23}^2 - 2x_{12}x_{23} \\
&= s_{12}^2 + s_{23}^2 + 2[(ct_{12})(ct_{23}) - x_{12}x_{23}].
\end{aligned}$$

Since $s_{12}^2 + s_{23}^2 = (s_{12} + s_{23})^2 - 2s_{12}s_{23}$, write the last equality as

$$s_{13}^2 = (s_{12} + s_{23})^2 + 2[(ct_{12})(ct_{23}) - x_{12}x_{23} - s_{12}s_{23}]. \tag{4.13}$$

The sign of the expression in the square brackets determines the direction of the inequality. To find this sign, manipulate the (square of the) last term in the square brackets:

$$\begin{aligned}
(s_{12}s_{23})^2 &= s_{12}^2 s_{23}^2 = [(ct_{12})^2 - x_{12}^2][(ct_{23})^2 - x_{23}^2] \\
&= \underbrace{(ct_{12})^2(ct_{23})^2 + x_{12}^2 x_{23}^2}_{\text{Call this "term1"}} - \underbrace{[(ct_{12})^2 x_{23}^2 + (ct_{23})^2 x_{12}^2]}_{\text{Call this "term2"}}.
\end{aligned}$$

Rewrite term1 as

$$\text{term1} = [(ct_{12})(ct_{23}) - x_{12}x_{23}]^2 + 2(ct_{12})(ct_{23})x_{12}x_{23},$$

[8] Since all events are causally connected (or timelike), I can talk about spacetime *length* because in this case $(\Delta s)^2 \geq 0$.

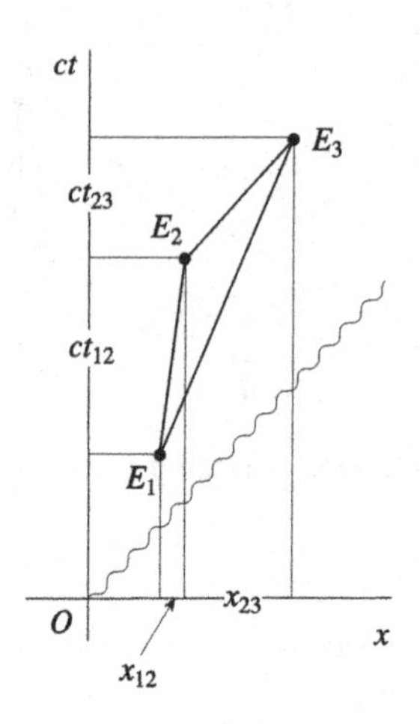

FIGURE 4.22 The three events E_1, E_2, and E_3 form a spacetime triangle.

add the second term to term2 (which you can now express as a complete square), and show that

$$[(ct_{12})(ct_{23}) - x_{12}x_{23}]^2 = (s_{12}s_{23})^2 + [(ct_{12})x_{23} - (ct_{23})x_{12}]^2,$$

or, since the second term on the right-hand side is positive,

$$[(ct_{12})(ct_{23}) - x_{12}x_{23}]^2 > (s_{12}s_{23})^2.$$

This shows that

$$(ct_{12})(ct_{23}) - x_{12}x_{23} > s_{12}s_{23},$$

i.e., that the expression in the square brackets of Equation (4.13) is positive, and therefore that $s_{13}^2 > (s_{12} + s_{23})^2$, or that $s_{13} > s_{12} + s_{23}$. This proves that

A weird inequality!

> **Note 4.5.1. (Spacetime Triangle Inequality)** *The sum of the spacetime length of any two sides of a spacetime triangle is* **less than** *the spacetime length of the third side.*

This is a surprising result, but by now you should be used to surprises in relativity! The inequality, which is quite the opposite of what you are accustomed to, is the direct result of the formula for the spacetime distance, especially the negative sign in the equation of Note 4.2.1, which is itself a direct result of the second postulate of relativity.

Given any two causally connected events in spacetime geometry, there are infinitely many worldlines that connect those events. Using the triangle inequality of Note 4.5.1 and an argument similar to the one used in the Euclidean case of Figure 4.21, you can show that out of all these worldlines the straight

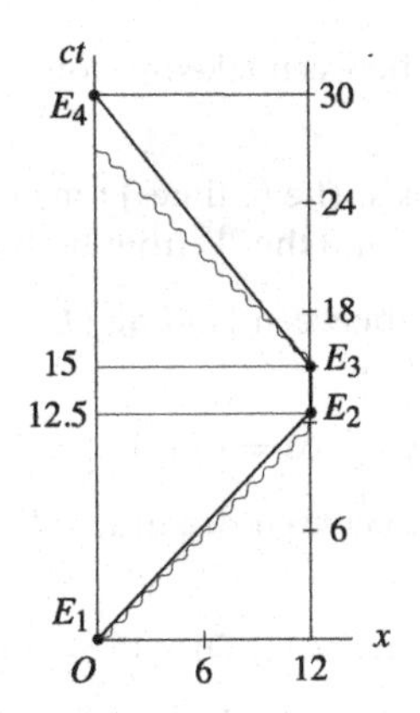

FIGURE 4.23 Sam moves on the heavy worldline while Sonya remains on Earth with the ct-axis being her worldline.

one has the *longest* spacetime length. These worldlines represent observers who travel with different speeds and accelerations, and the spacetime length becomes essentially the proper time of those observers.[9] Since straight worldlines represent inertial observers, we have the important result:

The straight worldline between two events is the *longest* spacetime distance between them!

Note 4.5.2. *The longest proper time between two events is measured by the inertial observer present at those events. Stated differently, the inertial observer ages more than any accelerated observer between two events.*

Example 4.5.3. Sam gets on a spaceship that travels to a planet of a star system 12 ly away on a worldline drawn with thick lines in Figure 4.23 as seen by observer O, Sonya. All units are in light years, and for easier reading most of the calibration of the ct-axis is made on the worldline parallel to it.

Q: What is the speed of the spaceship between E_1 and E_2? Between E_2 and E_3? Between E_3 and E_4?

A: Recall from Note 4.2.6 that β is the slope of the angle between axes of the two RFs. Since the worldline represents the time axis, β is the slope of $\overline{E_1E_2}$ relative to the ct-axis, or, more intuitively, just distance divided by time:

$$\beta = \frac{\Delta x_{21}}{c\Delta t_{21}} = \frac{12 \text{ ly}}{12.5 \text{ ly}} = 0.96 \quad \text{or} \quad v = 0.96c.$$

Similarly $\beta = 0$ for the speed between E_2 and E_3, and

$$\beta = \frac{\Delta x_{43}}{c\Delta t_{43}} = \frac{x_4 - x_3}{ct_4 - ct_3} = \frac{-12 \text{ ly}}{15 \text{ ly}} = -0.8 \quad \text{or} \quad v = -0.8c$$

for the speed between E_3 and E_4.

[9] Recall from Note 4.2.1 that the spacetime distance is c times the proper time.

Q: How long is the time interval between take-off from Earth (E_1) and landing on the planet (E_2) according to Sonya?

A: The vertical axis is Sonya's ct-axis. The (c times) time interval is shown to be 12.5 ly. Thus, $c\Delta t = 12.5$ ly; cancel the "c" and the "l" from both sides to get $\Delta t = 12.5$ y.

Q: How long is the time interval between landing (E_2) and departure (E_3) from the planet according to Sonya?

A: E_3 is at the 15 ly mark. Therefore, $c\Delta t = 15 - 12.5 = 2.5$ ly, or $\Delta t = 2.5$ y.

Q: How long is the time interval between departure (E_3) and landing on Earth (E_4) according to Sonya?

A: E_4 is at the 30 ly mark. Therefore, $c\Delta t = 30 - 15 = 15$ ly, or $\Delta t = 15$ y.

Q: How long does the entire trip take according to Sonya?

A: The time interval between E_1 and E_4 is 30 ly. Therefore, $c\Delta t = 30$ ly, or $\Delta t = 30$ y.

Q: What is Δs_{21}, the spacetime interval for the two events E_1 and E_2?

A: With E_1 and E_2 having coordinates $(0, 0)$ and $(12, 12.5)$, respectively, you get

$$\Delta s_{21} = \sqrt{(c\Delta t_{21})^2 - (\Delta x_{21})^2} = \sqrt{12.5^2 - 12^2} = 3.5 \text{ ly}.$$

Q: What is Δs_{32}, the spacetime interval for the two events E_2 and E_3?

A: E_2 has coordinates $(12, 12.5)$, and E_3 has coordinates $(12, 15)$. Therefore,

$$\Delta s_{32} = \sqrt{(c\Delta t_{32})^2 - (\Delta x_{32})^2} = \sqrt{2.5^2 - 0^2} = 2.5 \text{ ly}.$$

Q: What is Δs_{43}, the spacetime interval for the two events E_3 and E_4?

A: As in the two previous cases,

$$\Delta s_{43} = \sqrt{(c\Delta t_{43})^2 - (\Delta x_{43})^2} = \sqrt{15^2 - 12^2} = 9 \text{ ly}.$$

Q: What is Δs for the entire trip? How long did this trip take according to Sam?

A: Add the three spacetime intervals to obtain

$\Delta s = \Delta s_{21} + \Delta s_{32} + \Delta s_{43} = 3.5 + 2.5 + 9 = 15$ ly.

Since the broken worldline is that of Sam and Δs is the spacetime interval (or length) of this worldline, Δs is related to Sam's proper time via $\Delta s = c\Delta\tau$. Thus, $c\Delta\tau = 15$ ly and $\Delta\tau = 15$ y.

Q: Who measures the proper time interval between E_1 and E_4, Sam or Sonya (or both)?

A: Since they are both present at the two events, they both measure the proper time. ■

4.5.2 The Twin Paradox Revisited

I discussed the twin paradox in Section 2.4 and emphasized the *symmetry* between the twin observers: Each twin sees the other age slower. You may accept

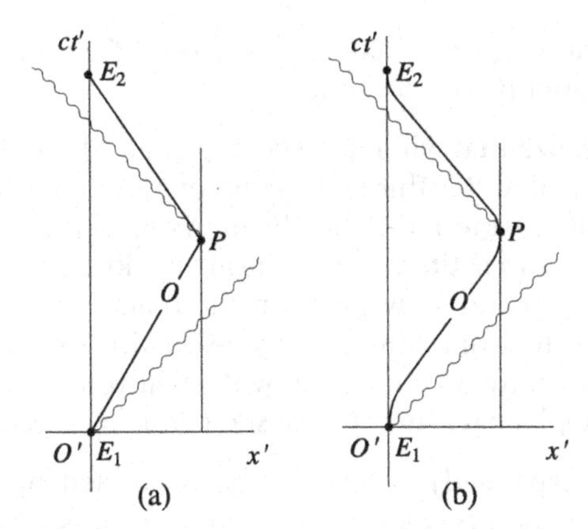

FIGURE 4.24 The twins' worldlines. (a) The accelerations are abrupt (infinite). (b) The accelerations are smooth.

this reluctantly as long as the twins are far apart and one is moving relative to the other. But what if you bring them face to face? Which one will be younger?

Suppose Sonya's rocket ship takes off (event E_1), immediately reaches a speed close to light speed, and heads to a distant planet. As soon as she reaches the planet, she abruptly turns around and heads back to Earth to land at event E_2. Sam, her twin brother, is waiting for her at the landing site. From our experience with the example of the captain and his daughter (e.g., see Example 2.1.6), we know that Sonya will be younger, but why?

Sam's worldline in Figure 4.24(a) is the vertical axis labeled ct'; Sonya's worldline is the broken line E_1PE_2; and the planet's worldline is the vertical line passing through P. The spacetime triangle inequality of Note 4.5.1 implies that Sonya's spacetime distance between E_1 and E_2 is *shorter* than Sam's, and, therefore, she will be younger.

"But," you may say "we haven't really resolved the paradox. There is still a symmetry between the two twins: Sonya looks at her brother and sees that he is moving away from her as she takes off toward the planet. On her way home, she sees Sam approaching her, exactly as Sam sees his twin sister. So, there is a complete symmetry. How do you explain the fact that Sonya stays younger?" The answer is that there is *no symmetry* between the two! Sonya is *accelerating* for parts of her journey; Sam never experiences an acceleration. In fact, Sonya experiences four kinds of acceleration: her take-off requires an acceleration until she reaches her steady speed; once she approaches the planet, she has to slow down to a stop; then she has to accelerate toward Earth until she reaches her steady speed for return; finally, she has to slow down for landing. Sonya's true worldline is shown in Figure 4.24(b), where the curvatures at the beginning, in the middle, and at the end of the journey indicate the acceleration and non-inertiality of Sonya's RF. And, as Note 4.5.2 indicates, accelerated RFs experience a shorter passage of time.

It is important to realize that while the velocity of an RF cannot be measured by instruments inside that RF (this is the consequence of the first postulate of relativity as shown after Note 1.1.2), acceleration *can* be measured. The reason is that acceleration creates fictitious forces inside the RF which are measurable. For example, when you suddenly push on the brakes of a fast moving car, things are thrown to the front. Similarly, when an airplane takes off from the ground, you feel a pressure on your back by the seat you are sitting on, which disappears when the plane reaches the constant cruising speed.

Two given (greatly separated) events can be connected by many different worldlines, just as two points can be connected by various curves in Euclidean space. To get the (spacetime) length of any of these worldlines, you'll have to add many, say N, infinitesimally separated events of that worldline connected by infinitesimal *straight* worldlines whose lengths are $\Delta s_i = c\Delta\tau_i$, with $i = 1, 2, \ldots, N$. That is, to obtain the length of a worldline connecting two events, you'll have to integrate $ds = c d\tau$ for that worldline between the two events. I illustrate this in the following example.

Example 4.5.4. Sam, Sonya, and Pat are newly born triplets. At time $t = 0$ for the three, Sam is put on a spaceship that immediately accelerates to $c\beta_0$ and begins its journey on a worldline described in Sonya's coordinate system by

$$x_s(t,\kappa) = \frac{c\beta_0\sqrt{\kappa^2+1}}{2}\left(\sqrt{\kappa^2+1}\,T - \sqrt{\kappa^2T^2+(2t-T)^2}\right), \tag{4.14}$$

where κ is a positive parameter describing the shape of the worldline (see Figure 4.25). Pat travels on another ship whose worldline is described by $x_p(t) \equiv x_s(t, 0)$.

Q: What is the equation for $x_p(t)$?

A: Substitute $\kappa = 0$ in Equation (4.14) and get

$$x_s(t,0) = \frac{c\beta_0}{2}\left(T - \sqrt{(2t-T)^2}\right) = \frac{c\beta_0}{2}\left(T - |2t-T|\right),$$

which you can write as

$$x_p(t) = \begin{cases} c\beta_0 t & \text{if } t \le T/2 \\ c\beta_0(T-t) & \text{if } t \ge T/2. \end{cases}$$

Q: When do the triplets meet again and what are Sam's and Pat's velocities when they do?

A: Note that $x_s(0,\kappa) = 0 = x_s(T,\kappa)$, independent of κ. Therefore, all three are together at $t = 0$ and meet again at $t = T$. To find Sam's speed, differentiate Equation (4.14):

$$\frac{dx_s}{dt} = -\frac{c\beta_0\sqrt{\kappa^2+1}(2t-T)}{\sqrt{\kappa^2T^2+(2t-T)^2}}. \tag{4.15}$$

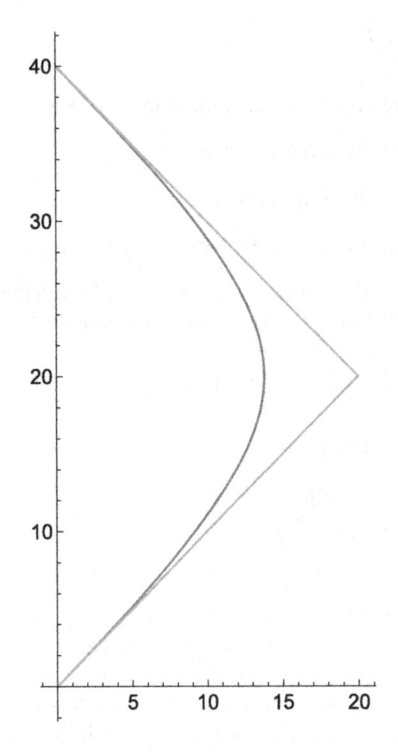

FIGURE 4.25 **The curved worldline is Sam's; the broken line is Pat's. The plots are drawn for** $\beta_0 = 0.99$, $T = 40$, **and** $\kappa = 0.5$.

Sam's initial and final velocities are given by

$$\left.\frac{dx_s}{dt}\right|_{t=0} = c\beta_0 \quad \text{and} \quad \left.\frac{dx_s}{dt}\right|_{t=T} = -c\beta_0,$$

respectively. Since these are independent of r, they are also Pat's initial and final velocities.

Q: At what time does Sam's spaceship stop to turn around? How far away is it when it does so?

A: Set the derivative in (4.15) equal to zero to obtain $t = T/2$ as the time when Sam's spaceship stops. Therefore, Sam is at a distance of

$$x_s(T/2, \kappa) = \frac{c\beta_0 T}{2}\left(\kappa^2 + 1 - \kappa\sqrt{\kappa^2 + 1}\right)$$

when his spaceship stops momentarily and starts returning.

Q: At what time does Pat's spaceship stop to turn around? How far away is it when it does so?

A: Since the zero of (4.15) is independent of κ, Pat's ship also stops at $t = T/2$, and its distance is

$$x_p(T/2) = x_s(T/2, 0) = \frac{c\beta_0 T}{2}.$$

This is just speed times time because Pat is moving at constant speed.

Q: Who goes farther from Sonya, Sam or Pat?

A: Since $\kappa^2 - \kappa\sqrt{\kappa^2 + 1} < 0$, Pat goes farther.

Q: Of the three triplets, who is the oldest? Who is the youngest?

A: You have to find the length of each spacetime path by integration. You can find all three by taking appropriate limits of Sam's spacetime length. Note that

$$d\tau_s = \frac{1}{c}\sqrt{(cdt)^2 - (dx_s)^2} = \sqrt{1 - \frac{1}{c^2}\left(\frac{dx_s}{dt}\right)^2}\, dt,$$

and therefore by Equation (4.15),

$$\tau_s(\beta_0, \kappa) = \int_0^T \sqrt{1 - \frac{\beta_0^2(\kappa^2 + 1)(2t - T)^2}{\kappa^2 T^2 + (2t - T)^2}}\, dt. \tag{4.16}$$

You can obtain Sonya's age by setting $\beta_0 = 0$. This gives $\tau_s(0, \kappa) = T$ as Sonya's age. You can obtain Pat's age by setting $\kappa = 0$. This gives $\tau_s(\beta_0, 0) = \sqrt{1 - \beta_0^2}\, T$ as Pat's age. Therefore, Pat is younger than Sonya. You can't integrate (4.16) for arbitrary κ and β_0. However, you *can* compare the integrand with the special cases of Sonya and Pat. It is obvious that the integrand is smaller than the special case of $\beta_0 = 0$. Therefore, the integral must also be smaller. So, Sam's age is less than Sonya's. I leave it for you to show that the integrand (and hence the integral itself) is larger than the special case of $\kappa = 0$ and smaller than the special case of $\beta_0 = 0$. Therefore, Sam is older than Pat and younger than Sonya. ■

4.6 PROBLEMS

4.1. When axes are perpendicular to each other, it does not matter whether you draw lines that are parallel or perpendicular to the axes to find the coordinates of a point. Why can't you draw *perpendicular* lines when axes are not at right angles to each other? Hint: If a point lies on an axis, what do you expect its "other" coordinate to be?

4.2. Consider a non-perpendicular coordinate system with axes x and y. Take any two points P_1 and P_2 with coordinates (x_1, y_1) and (x_2, y_2), respectively. Show that if

$$\overline{P_1P_2}^2 = (x_2 - x_1)^2 + (y_2 - y_1)^2,$$

then the axes must be perpendicular.

4.3. Draw the x- and ct-axes with acute angles. Draw a wavy line through the origin to represent the worldline of a light signal. Show that if the speed of this light signal is to be c, then its worldline has to make equal angles with both axes. That is, it has to be the bisector of the angle between the x- and ct-axes.

4.4. A train of rest length L_0 travels at speed β in the positive x'-direction of Sam. As the front of the train passes Sam at $t' = 0$, a light signal is sent from the front of the train to the rear.

(a) Draw a spacetime diagram showing the worldlines of the front and rear of the train and the photon in Sam's RF.
(b) When does the rear of the train pass Sam?
(c) When does the signal reach the rear of the train according to Sam?

4.5. Sonya and Sam live on two different planets one light hour apart. A space station is located half way between the two planets. The planets and the space station are all stationary relative to each other. Sonya wants to arrange a meeting with Sam at 1:00 PM on the space station. She decides to send him a light signal so that as soon as he receives it he starts to move toward the space station to get there at exactly 1:00 PM. Sonya's spaceship moves at half the speed of light while Sam's moves at $0.75c$. All times are according to the common reference frame of the planets and the space station.

(a) At what time should Sonya send the signal?
(b) At what time should she leave her planet?
(c) At what time should Sam leave his planet?

4.6. Same as the previous problem except that Sonya's spaceship moves at $0.2c$.

(a) At what time should Sonya leave her planet?
(b) At what time should she send the signal?
(c) As she tries to contact Sam, Sonya realizes that she can't contact him directly. She decides to send a radio signal to her planet telling them to contact Sam instead. What is the latest time that Sonya can contact her planet so that Sam receives the message in time to be able to make it to the meeting?

4.7. In Example 4.2.4, I showed that when two events are causally disconnected, you can always find an observer for whom the order of occurrence of the events is switched. I used algebraic method to prove the statement. Geometry makes the argument much easier! Let O' have perpendicular axes. Pick two events E_1 and E_2 that are causally disconnected. Now choose a set of axes passing through the origin of O' in such as way that the order of occurrence of E_1 and E_2 is switched. Hint: Draw parallel lines from the two events in such a way that the earlier event cuts the ct'-axis at a *later* time. From this decide what the new x-axis should be. Make sure that the angle between the x- and x'-axes is not larger than $45°$.

4.8. Figure 4.26 shows five firecrackers separated by 4 light seconds from one another in Sam's reference frame O. F_1 and F_3 occur simultaneously 4 seconds into the future; F_2, F_4, and F_5 occur 16 seconds, 8 seconds, and 20/3 seconds into the future, respectively. Sonya (reference frame O') moves at 1/3 the speed

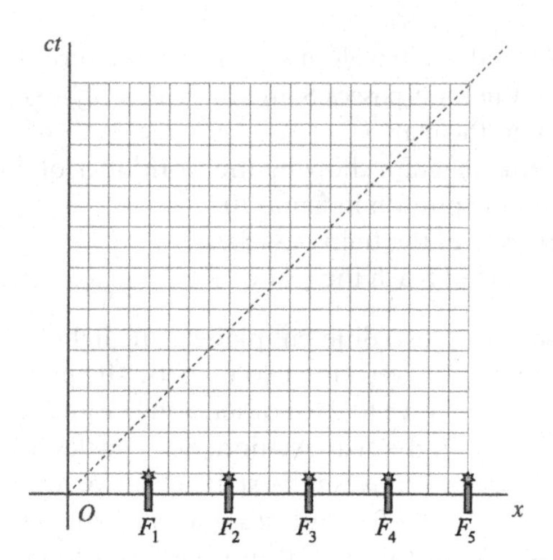

FIGURE 4.26 The distance between consecutive firecrackers is 4 light seconds. The unit on the vertical axis is also light second. The dashed line is the light worldline.

of light relative to Sam in the positive x-direction in such a way that at $t = t' = 0$ the origins of the two RFs coincide.

(a) With dots, show the events corresponding to the explosion of the firecrackers.
(b) At what times does Sam receive the light signals from the firecrackers?
(c) Draw Sonya's spacetime axes, x' and ct'.
(d) In what order do the firecrackers explode according to Sonya?
(e) In what order do the light signals from the firecrackers reach Sonya?
(f) At what times according to Sam do the light signals from the firecrackers reach Sonya?

4.9. Sonya, who is in reference frame O, sees a flash of red light at $x = 1500$ m, and after 5 μs, a flash of green light at $x = 300$ m. Use spacetime diagrams for the problem.

(a) What should Sam's speed be relative to Sonya so that he sees the two events at the same point in his RF?
(b) Which event occurs first according to Sam and what is the time interval between the two flashes?

4.10. In a galactic rocket race, Sam and Sonya drive two rockets in opposite directions toward their respective finish lines as shown in Figure 4.27 (the units are not the same on the two axes). Sam moves to the right and Sonya to the left. The drivers start their motion at $t = 0$ according to all three RFs. When they reach their finish lines—placed at 20 light minutes on either side of the referee—each driver sends a light signal to the referee relaying their arrival. The referee receives the signals from Sam and Sonya exactly 45 minutes after their departure, indicating a tie.

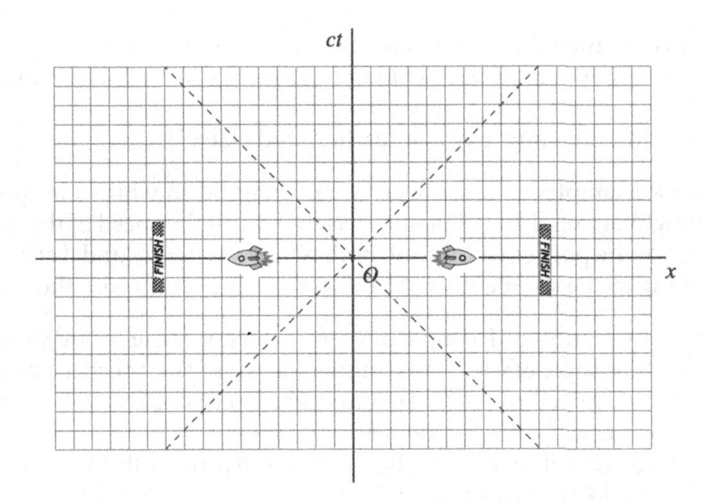

FIGURE 4.27 The diagram shows the referee's RF. The unit on the x-axis is 2 light minutes, and on the ct-axis is 5 light minutes. The dashed lines are the light worldlines.

(a) Draw the two events of the arrival of the rockets at the finish lines on the referee's RF.
(b) What is the speed of each rocket?
(c) Draw the spacetime axes of both rockets.
(d) Graphically show the coordinates of the two arrival events on the axes of all three observers.
(e) What is the order of arrivals according to Sam? Does he think he is the winner?
(f) What is the order of arrivals according to Sonya? Does she think she is the winner?

4.11. Sonya, who is in reference frame O, throws a ball with speed β_b at time t_0 (event E_0) in her positive x-direction, which reaches a point (event E) Δx from the origin at time $t_0 + \Delta t$. Sam, in RF O', with respect to whom Sonya is moving with speed β in the positive x'-direction, looks at the same two events. Use a spacetime diagram in which Sam's axes are perpendicular to derive the relativistic law of addition of velocities.

4.12. Alpha Centauri is about 4 light years away from Earth and does not move significantly relative to it, so they are both in the same RF. Assume Earth is at the origin of this RF. Event E_1 occurs at $t = 0$ on Earth. Event E_2 occurs on Alpha Centauri 3 years later. Use spacetime diagrams.

(a) What is the time difference between the two events according to an observer moving from Earth to Alpha Centauri at $0.9c$? Which event occurs first?

(b) What is the time difference between the two events according to an observer moving from Alpha Centauri to Earth at $0.9c$? Which event occurs first?

What happened to the invariance of "earlier" and "later"?

4.13. Rework Example 3.5.3 using geometric method in which the speeder's RF has orthogonal axes. Instead of numbers, use β_1 for the speed of the speeder and $\beta_2 > \beta_1$ for the policeman's speed, both relative to the ground. Let T be the time according to the speeder that the policeman catches up with the speeder.

4.14. This is a variation of Example 3.5.3 for which you are to use spacetime diagrams. Let the speeder's RF have orthogonal axes. The speeder passes the policeman with a speed of $0.99c$. Two minutes later (policeman's time), the policeman starts chasing the speeder with a speed of $0.995c$.

(a) How long does it take the policeman to catch up with the speeder according to the speeder's watch?

(b) How long does it take the policeman to catch up with the speeder according to the policeman's watch?

To answer these questions you have to find all the following on the speeder's spacetime coordinates: the time when the policeman starts the chase, the distance between the two when the chase starts, and the time it takes for the policeman to catch up with the speeder after the start of the chase.

4.15. Two clocks C_1 and C_2 separated by a fixed distance L' are placed on the x'-axis of an observer O' and synchronized according to O'.[10] Reference frame O is moving relative to O' in the positive direction of x' with speed β. Observer O_1 is at the origin of O and observer O_2 is placed strategically along the x-axis in such a way that they can read the two moving clocks *at the same time*. O_1 records the reading as 12:00 (the zero time). Draw a spacetime diagram with the O' axes perpendicular, and show the worldlines of the two clocks in the O' reference frame.

(a) What is the location of O_2 in O?

(b) What time does O_2 record?

4.16. Two photons are moving in O at a fixed distance L apart. Show that in O' with respect to which O moves with fractional speed β in the *negative* direction, the two photons are separated by

$$L' = \sqrt{\frac{1+\beta}{1-\beta}}\,L.$$

4.17. Sam and Sonya are twins. Sam is put on a spaceship O moving with relative speed β in the positive x'-direction of Sonya's reference frame O'. At time T'_0, O' sends a light signal toward O. Draw a spacetime diagram showing the relevant axes and the light signal. Use the rules of spacetime geometry.

---

[10] This is the same problem done in Example 3.4.10.

(a) Show that when Sam receives the light signal, the spaceship is at a distance of $\beta c T'_0/(1-\beta)$ from Sonya.
(b) Show that Sam receives the light signal when he is $\sqrt{(1+\beta)/(1-\beta)}\,T'_0$ older than when he left his sister.

4.18. Tomorrow is "only one day away"; it is probably not too much to ask the theory of relativity to help us get there. Utilizing the experience you gained in the case of Bruno's death, find observer O who is only 24.5 light hours away.[11]

Time traveling to future.

(a) What speed should O have? In which direction?
(b) How far away is "tomorrow" taking place from O?

4.19. Sonya (observer O) moves at $0.99c$ relative to Sam (observer O') as shown in Figure 4.9. Assume that the length of the train is 50 m. Find the coordinates of all the points marked as triangle, square, circle, oval, and stars in Figure 4.9(b). Hint: Go through Example 4.3.1 for warmup.

4.20. Use the result of Problem 3.7 to find the coordinates of E_1, E_2, and E_3 of Figure 4.18 in both coordinate systems O and O'.

4.21. Provide the missing steps leading to the final result of Equation (4.12).

4.22. Sam, Sonya, and Pat are newly born triplets. Sam and Sonya are put on two different spaceships that travel to a planet of a star system 10 ly away. Sonya lands on the planet 10.1 years later as seen by observer O, Pat. She waits 4.9 years until Sam, who is traveling slower, lands on the same planet (see Figure 4.28). After six months they both return home on the same spaceship and land on Earth 26 years after their departure according to the Earth calendar. All times and distances of the figure are given according to the Earth observers, and all units shown are in light years, and for easier reading most of the calibration of the ct-axis is made on the worldline parallel to it.
(a) What is the speed of Sonya's spaceship on her journey to the planet?
(b) What is the speed of Sam's spaceship on his journey to the planet?
(c) How old is Sonya when she meets Sam? How old is Sam?
(d) How old is Sonya when she lands back on Earth? How old is Sam? How old is Pat?

4.23. Verify that Equation (4.14) is the left half of a hyperbola with center at $\left[c\beta_0(\kappa^2+1)T/2, T/2\right]$, semi-major axis $a = c\beta_0\kappa\sqrt{\kappa^2+1}\,T/2$, and semi-minor axis $b = \kappa T/2$.

4.24. Show that the absolute value of the speed in (4.15) is always less than c for the duration of all the three trips.

4.25. In Example 4.5.4, assume that Pat makes four identical round trips between $t = 0$ and $t = T$ with the same speed β_0.

---

[11] A light hour is the distance that light travels in one hour. For comparison, Saturn is about 1.25 light hours away from Sun.

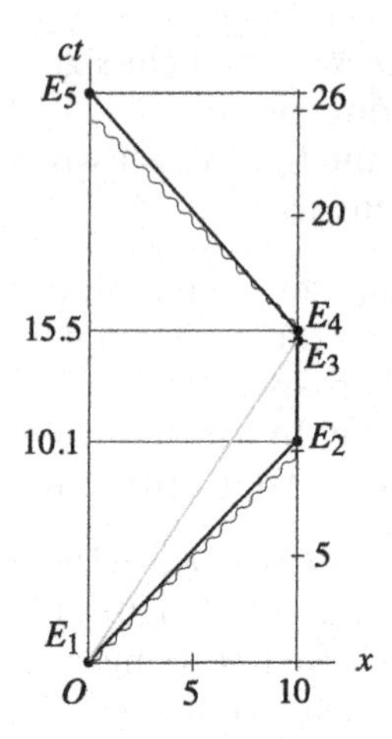

FIGURE 4.28 Sonya moves on the heavy black worldline. Sam moves on the grey worldline first and then joins Sonya to return home. Pat remains on Earth.

(a) How far does he have to go before turning around in each trip?
(b) How much does he age in each trip?
(c) Compare his total age with his age in the example. Can you give a simple geometric reason for this? Hint: The geometric reason is identical to ordinary Euclidean geometry.

4.26. In Example 4.5.4, assume that Pat travels with the initial speed of $0.999c$ to a destination 40 light years away.
(a) How old is Sonya when Pat returns? How old is Pat?
(b) Sam goes to a destination 30 light years away with the same initial speed as Pat. What is κ?
(c) How old is Sam when he returns? You'll have to numerically integrate (4.16) to find the answer.

4.27. Sonya gets on a spaceship that travels on a parabolic worldline given by

$$x(t) = c\kappa\left(t - \frac{t^2}{T}\right),$$

in Sam's coordinate system O, where κ is a positive constant. Note that $x(0) = x(T) = 0$. This means that Sonya leaves Sam at $t = 0$ and meets him again at $t = T$.
(a) What are Sonya's velocities relative to Sam at the times of her departure and return? What restriction does this put on κ?
(b) Show that during Sonya's entire trip, her speed is always less than the speed of light.
(c) When and where does Sonya stop momentarily and come back?

(d) Verify that it takes Sonya

$$T\left(\frac{\sin^{-1}\kappa + \kappa\sqrt{1-\kappa^2}}{2\kappa}\right)$$

to make her round trip. Show that as $\kappa \to 0$, this travel time reduces to Sam's time, as it should.

(e) Show that no matter how fast Sonya starts to travel, the ratio of her increase in age to Sam's can't be smaller than $\pi/4$.

4.28. Sonya gets on a spaceship that travels on a worldline given parametrically by

$$x(\theta) = cT\kappa(\sqrt{2}\cos\theta - 1), \quad ct(\theta) = cT(\sqrt{2}\sin\theta + 1), \quad -\frac{\pi}{4} \le \theta \le \frac{\pi}{4}$$

in Sam's coordinate system O, where κ is a positive constant less than one. Note that $x(-\pi/4) = ct(-\pi/4) = 0$ and $x(\pi/4) = 0, ct(\pi/4) = 2cT$. This means that Sonya leaves Sam at $t = 0$ and meets him again at $t = 2T$.

(a) Show that during Sonya's entire trip, her speed is always less than the speed of light.

(b) What are Sonya's velocities relative to Sam at the times of her departure and return?

(c) When and where does Sonya stop momentarily and come back?

(d) Verify that it takes Sonya

$$\sqrt{2}T\int_{-\pi/4}^{\pi/4}\sqrt{1-(\kappa^2+1)\sin^2\theta}\,d\theta$$

to make her round trip. Show that as $\kappa \to 0$, this travel time reduces to Sam's time, as it should.

(e) Numerically integrate the integral in (d) for various κ and plot the ratio of Sam's round-trip time to Sonya's. What is the limit of this ratio as $\kappa \to 1$?

CHAPTER 5

Spacetime Momentum

In Chapters 3 and 4, I introduced the (straight) spacetime distance between two events and the relevant transformations (the Lorentz transformations) of coordinates that leave this distance unchanged. I also showed that, except for a factor of c, the spacetime distance is simply the proper time interval between the two events. The invariance of $\Delta s = c\Delta\tau$ implies that *all observers* calculate exactly the same value for the proper time interval between two events. This time interval is measured *directly* by the clock of an observer moving with *uniform* velocity from one event to another.

If an observer moves on a curved worldline, you can still calculate the proper time of that observer by integration. That is what I did in Example 4.5.4. Integration gives you the length of a worldline. In dynamics of particles, you also need differentiation. In this chapter I use the infinitesimal proper time (or spacetime distance) to define some important dynamic quantities.

5.1 SPACETIME VELOCITY

The first quantity I want to introduce is velocity, and in order to understand the spacetime version of it, I'll first look at the ordinary velocity. An object moves in a Euclidean plane from a point P_1 to an infinitesimally close point P_2 on a straight line as shown in Figure 5.1. The displacement $\Delta\mathbf{r}$, which could be a directed line element (at some given time) of a curve on which the object moves, has components Δx and Δy. If I divide $\Delta\mathbf{r}$ by Δt, the time it takes the object to go from P_1 to P_2, I'll get the *instantaneous* velocity of the object at a given time:

$$\mathbf{v} = \frac{\Delta\mathbf{r}}{\Delta t} \quad \text{or} \quad \mathbf{v} = \frac{d\mathbf{r}}{dt}.$$

Being a vector, $\mathbf{v}$ has components

$$v_x = \frac{\Delta x}{\Delta t} \quad \text{or} \quad v_x = \frac{dx}{dt}$$
$$v_y = \frac{\Delta y}{\Delta t} \quad \text{or} \quad v_y = \frac{dy}{dt}. \tag{5.1}$$

Special Relativity. DOI:10.1016/B978-0-12-810411-8.00005-5

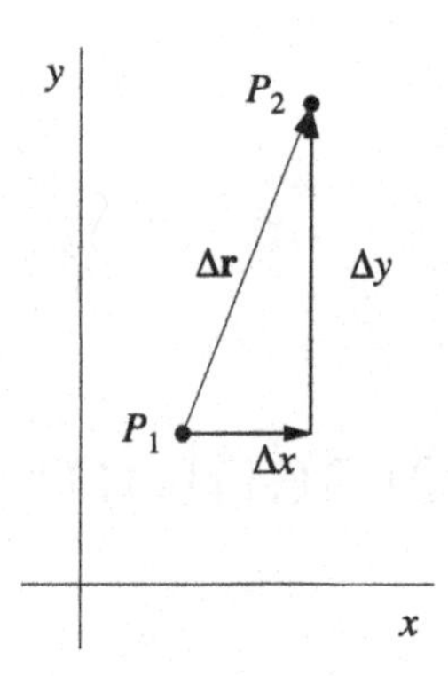

FIGURE 5.1 A general path can be "resolved" into an x path and a y path.

How do I apply the discussion above to relativity? Imagine an object moving in space while carrying a clock with it. The points of space through which the object passes, plus the instants at which it passes through those points, constitute a series of events. The object is of course present at all those events, and its clock measures the proper time. Not all observers have access to this clock, and therefore, they cannot read the proper time directly. However, each observer can measure the distance the object travels in a particular time interval in his/her own RF and, using Note 4.2.1, can calculate the proper time interval for that distance. The important point is that once the calculation is done, all observers get the same number. Thus,

Note 5.1.1. *Every moving object has a unique and universal time attached to it, its proper time, measured by the clock being carried by the object.*

It is therefore natural to define the (time-dependent) physical quantities associated with that object in terms of this universal time.[1] One such quantity is the **spacetime velocity**. How do I calculate the spacetime velocity?

Sonya looks at a bullet moving with a speed v_b *relative to her* and measures its displacement Δx_b—between two events E_1 and E_2—and the time Δt_b it takes the bullet to cover the distance Δx_b (see Figure 5.2). These are the analogues of Δx and Δy of the coordinate velocity of Figure 5.1. However, instead of dividing them by Δt, which is universal only in non-relativistic situations, she divides them by the next best thing: the bullet's proper time. To do this, Sonya calculates the spacetime distance Δs_b—from Δt_b and Δx_b—of the bullet and, from that $\Delta \tau_b$, the proper time between E_1 and E_2. This yields two components: the space component $\Delta x_b/\Delta \tau_b$ and the time component $c\Delta t_b/\Delta \tau_b$. It is

Spacetime velocity.

[1] I have to emphasize that although each object has a universal time attached to it, this time is different for different objects. It is not universal in the sense of Newtonian physics.

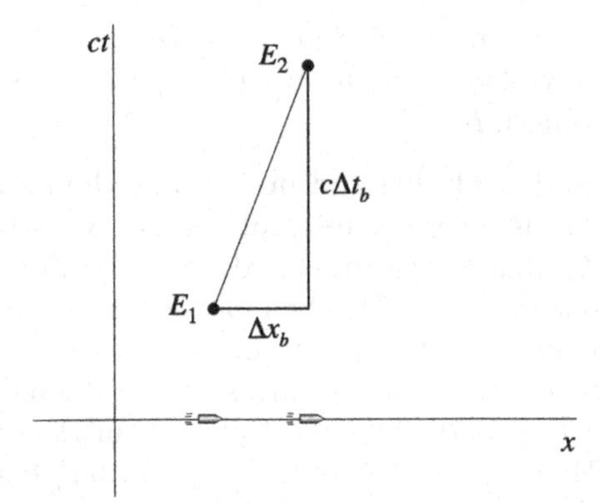

FIGURE 5.2 The displacement and flight time of the bullet according to Sonya. Note that the bullet is moving along the x-axis only.

actually more convenient (and more common) to divide these components by c and obtain dimensionless components. I denote the space part by u_{bx} and the time part by u_{bt} (b for "bullet," x for space, and t for time). Then

$$u_{bx} = \frac{\Delta x_b}{c\Delta\tau_b} = \frac{\Delta x_b}{\Delta s_b} = \frac{\Delta x_b}{\sqrt{(c\Delta t_b)^2 - (\Delta x_b)^2}}$$
$$= \frac{\Delta x_b}{c\Delta t_b\sqrt{1 - (\Delta x_b/c\Delta t_b)^2}} = \frac{\beta_b}{\sqrt{1-\beta_b^2}} = \gamma_b\beta_b,$$

where $\beta_b = v_b/c = \Delta x_b/(c\Delta t_b)$ is the fractional speed of the bullet as measured by Sonya and $\gamma_b = 1/\sqrt{1-\beta_b^2}$. Similarly, I let you show that $u_{bt} = \gamma_b$. I also let you verify the following important property of the spacetime velocity:

$$u_{bt}^2 - u_{bx}^2 = 1. \tag{5.2}$$

Since the right-hand side is a constant, it must be the same for all observers, as you can verify by calculating the components of the spacetime velocity in any other RF and showing directly that they too satisfy (5.2). Therefore, you can think of $u_{bt}^2 - u_{bx}^2$ as the **invariant length** of spacetime velocity in the two-dimensional spacetime geometry.

Invariant length of spacetime velocity.

Note 5.1.2. *The spacetime velocity u_b of an object (a bullet) moving with fractional speed β_b has components u_{bx} and u_{bt} satisfying the following relations:*

$$u_{bx} = \gamma_b\beta_b, \quad u_{bt} = \gamma_b; \qquad u_{bt}^2 - u_{bx}^2 = 1. \tag{5.3}$$

The "bullet" represents *any* moving object of interest, except light. Why not light? Because $\Delta s = 0$ for light, and I can't put it in the denominator of the definition of u_{bx} or u_{bt}. You should also not be surprised that the components of the spacetime velocity are larger than 1 (in fact $u_{bt} > 1$ for any nonzero speed). This does not violate relativity because spacetime velocity is *not* the actual velocity of an object; β_b is.

Sam watches Sonya and the bullet and notices that the bullet moves with a speed v'_b *relative to him*. He measures its displacement $\Delta x'_b$—between the same two events E_1 and E_2 that Sonya measured—and the time $\Delta t'_b$ it takes the bullet to cover the distance $\Delta x'_b$. He then calculates the spacetime distance Δs_b of the bullet between E_1 and E_2, and he gets exactly what Sonya gets, of course. The spacetime velocity that Sam measures for the bullet has the space part $\Delta x'_b/\Delta s_b$ and the time part $c\Delta t'_b/\Delta s_b$. Denote Sam's space part by u'_{bx} and the time part by u'_{bt}. Then, $u'_{bx} = \gamma'_b \beta'_b$ and $u'_{bt} = \gamma'_b$, with β'_b being the fractional speed of the bullet relative to Sam.

As Equation (5.3) and the corresponding one for Sam show, the *space* component of the spacetime velocity is related directly to the ordinary velocity of the bullet. Indeed, when the bullet moves with speeds considerably smaller than light speed, γ_b will be indistinguishable from 1 and u_{bx} from β_b, the classical (fractional) velocity. The time component has no classical analogue.

How are the components of the spacetime velocity of the bullet as measured by Sam related to those measured by Sonya? If you divide both sides of Equation (3.25)—with the index b inserted on all Δ-quantities—by Δs_b, you get the primed spacetime velocity on the left-hand side and the unprimed spacetime velocity on the right-hand side. Therefore,

Components of spacetime velocity are related via a Lorentz transformation.

$$\begin{aligned} u'_{bx} &= \gamma(u_{bx} + \beta u_{bt}) \\ u'_{bt} &= \gamma(\beta u_{bx} + u_{bt}), \end{aligned} \tag{5.4}$$

i.e., the spacetime velocity transforms like the spacetime coordinates: via a LT.

Since the components of the spacetime velocity are given in terms of the ordinary velocity, Equation (5.4) is a relation between the ordinary velocity of the bullet as measured by Sam and the ordinary velocity of the bullet as measured by Sonya (as well as the relative speed of the two observers). It follows that the relativistic law of addition of velocities should be somehow hidden in Equation (5.4). In the following example, I show how to uncover the relativistic law of addition of velocities from the first equation in (5.4). You should try to get the relativistic law of addition of velocities from the second equation (see Problem 5.2).

Lorentz transformation of spacetime velocity leads to the relativistic law of addition of velocities.

Example 5.1.3. To connect the two velocities, all I need to do is to write the first equation in (5.4) in terms of β'_b and β_b. This yields

$$\gamma'_b \beta'_b = \gamma(\gamma_b \beta_b + \beta \gamma_b) = \gamma \gamma_b (\beta_b + \beta)$$

or

$$\frac{\beta'_b}{\sqrt{1-\beta'^2_b}} = \frac{1}{\sqrt{1-\beta^2}} \frac{1}{\sqrt{1-\beta^2_b}} (\beta_b + \beta).$$

Squaring both sides yields

$$\frac{\beta_b'^2}{1-\beta_b'^2} = \frac{(\beta_b+\beta)^2}{(1-\beta^2)(1-\beta_b^2)},$$

which, after a little algebra, you should be able to simplify to:

$$\beta_b'^2 \underbrace{[(1-\beta^2)(1-\beta_b^2)+(\beta_b+\beta)^2]}_{=[1+\beta^2\beta_b^2+2\beta_b\beta]=(1+\beta\beta_b)^2} = (\beta_b+\beta)^2,$$

which immediately yields

$$\beta_b' = \frac{\beta_b+\beta}{1+\beta\beta_b}.$$

This is identical to Equation (3.26). ■

Example 5.1.4. Sonya is on a train moving at $0.9c$ in the positive direction of Sam's axis. She sees a bullet dashing by with a speed of $0.95c$ in the forward direction.

Q: What are the components of the bullet's spacetime velocity according to Sonya?

A: Equation (5.3) gives these components:

$$u_{bx} = \frac{\beta_b}{\sqrt{1-\beta_b^2}} = \frac{0.95}{\sqrt{1-(0.95)^2}} = 3.042434922$$

$$u_{bt} = \frac{1}{\sqrt{1-\beta_b^2}} = \frac{1}{\sqrt{1-(0.95)^2}} = 3.202563076.$$

These two components satisfy

$$u_{bt}^2 - u_{bx}^2 = (3.202563076)^2 - (3.042434922)^2 = 1.000000001,$$

verifying the last equation of (5.3).

Q: What are the components of the bullet's spacetime velocity according to Sam?

A: Equation (5.4) gives these components in terms of Sonya's. In that equation, γ is given in terms of the *relative speed of the two observers*:

$$\gamma = \frac{1}{\sqrt{1-(v/c)^2}} = \frac{1}{\sqrt{1-(0.9)^2}} = 2.294157339,$$

yielding

$$u_{bx}' = 2.294157339(3.042434922 + 0.9 \times 3.202563076) = 13.59228963$$
$$u_{bt}' = 2.294157339(0.9 \times 3.042434922 + 3.202563076) = 13.62902555.$$

These two components satisfy

$$u_{bt}'^2 - u_{bx}'^2 = (13.62902555)^2 - (13.59228963)^2 = 1.000000057,$$

again verifying the last equation of (5.3) and illustrating that $u_{bt}^2 - u_{bx}^2$ is frame-independent.

Q: What is the speed of the bullet relative to Sam?

A: You don't have to start all over again! Just take the ratio u'_{bx}/u'_{bt}:

$$\beta'_b = \frac{u'_{bx}}{u'_{bt}} = \frac{13.59228963}{13.62902555} = 0.9973,$$

which you can easily verify to be the same as what you get when you use the formula for the relativistic law of addition of velocities. ■

5.2 SPACETIME MOMENTUM

Spacetime velocities are important in their own right. However, their true significance lies in their use for building other quantities. One such quantity is the **spacetime momentum**, which is naturally defined as the product of mass and the spacetime velocity. Since the spacetime velocity has two components in plane spacetime geometry, so does the spacetime momentum. Thus, I define the space component (denoted by p_{bx}) and the time component (denoted by p_{bt}) of the spacetime momentum (of a bullet) as

Spacetime momentum.

$$p_{bx} = mcu_{bx} = \gamma_b m v_b = \frac{mv_b}{\sqrt{1-\beta_b^2}},$$
$$p_{bt} = mcu_{bt} = \gamma_b mc = \frac{mc}{\sqrt{1-\beta_b^2}}, \tag{5.5}$$

where m is the mass of the bullet and I have introduced a factor of c to give p_{bx} and p_{bt} the right dimension.

How are these two components related to the classical properties of the bullet? To answer this question, I'll take the limit of low velocities and compare the results to the corresponding Newtonian quantities (see Note 2.3.4). The first equation in (5.5) becomes $p_{bx} \approx mv_b$ when β_b is small. Therefore, it is natural to identify $\gamma_b m v_b$ as the **relativistic momentum** of the bullet.

Relativistic momentum.

5.2.1 Relativistic Energy

The identification of p_{bt} is a little harder, because its comparison with the corresponding Newtonian quantity requires some algebra. If I completely neglect β_b and set γ_b equal to 1, I obtain $p_{bt} = mc$, which cannot be identified with any Newtonian quantity.[2] So, I am going to do the next best thing and approximate $1/\sqrt{1-\beta_b^2}$ using Equation (2.22). Then, I obtain

Identification of p_{bt} with a known classical quantity.

$$p_{bt} = \frac{mc}{\sqrt{1-\beta_b^2}} \approx mc\left(1+\tfrac{1}{2}\beta_b^2\right) = mc + \tfrac{1}{2}m\frac{v_b^2}{c}.$$

[2] You can interpret the result as the momentum of the bullet when it moves with light speed, but that is prohibited by relativity.

Now I multiply both sides by c to get

$$p_{bt}c \approx mc^2 + \tfrac{1}{2}mv_b^2.$$

This tells me that in the Newtonian limit, $p_{bt}c - mc^2$ is the kinetic energy. I therefore extrapolate to relativity, call $p_{bt}c - mc^2$ the *relativistic kinetic energy*, and write

$$p_{bt}c = mc^2 + \text{relativistic KE} \quad \text{or} \quad p_{bt}c = E_b,$$

where E_b is simply the **relativistic energy** of the bullet:

Relativistic energy.

$$E_b = p_{bt}c = mc^2 u_{bt} = \gamma_b mc^2 = \frac{mc^2}{\sqrt{1-\beta_b^2}}. \tag{5.6}$$

From this equation and the first equation in (5.5),—ignoring the subscript x—you can get

$$\beta_b = \frac{p_b c}{E_b} \quad \text{or} \quad v_b = \beta_b c = \frac{p_b c^2}{E_b}, \tag{5.7}$$

which holds for *any* particle. If you apply it to a photon, you get $E = pc$, which I'll derive later using a different technique.

The spacetime momenta of an object as measured by two observers are related by a LT. You can see this immediately if you multiply both sides of (5.4) by mc. However, it is better to change this result slightly: use Equation (5.6) to replace p_{bt} with E_b and change p_{bx} to p_b, because, with p_{bt} removed, the x subscript is superfluous. So, multiply both sides of (5.4) by mc and keep these changes in mind to get

Spacetime momenta of an object in two different RFs are related by a LT.

$$\begin{aligned} p_b' &= \gamma(p_b + \beta E_b/c) \\ E_b' &= \gamma(\beta p_b c + E_b). \end{aligned} \tag{5.8}$$

These equations connect the energy and momentum of an object as measured by an observer O to those measured by O', relative to whom O is moving with speed $v = \beta c$.

Two special cases of Equation (5.6) are important: when $\beta_b = 0$ and when $\beta_b = 1$. In the first case the denominator is 1 and you get $E_b = mc^2$. Remove the subscript b and write the most famous equation in physics:

The most famous equation in physics.

$$E = mc^2, \tag{5.9}$$

showing the equivalence of mass and energy. It implies that it is in principle possible to convert mass into energy and vice versa. Because of the enormity of c, a tiny amount of mass can turn into a tremendous amount of energy. In a nuclear power plant uranium mass is turned into energy through the nuclear process of **fission**, in which the large uranium nucleus fragments into two smaller **daughter** nuclei whose total mass is *smaller* than the uranium mass. The remaining mass of the uranium is partially converted into the total kinetic energy of the daughter nuclei, which is ultimately converted into electrical energy.

Example 5.2.1. To see the enormity of the energy hidden in even a small mass, suppose that one kilogram of uranium is converted into usable electric output. The energy produced is then

$$E = mc^2 = 1 \times (3 \times 10^8)^2 = 9 \times 10^{16} \text{ J}.$$

A large power plant feeding a small city produces about one *gigawatt* of power, or 10^9 Joules per second. If this power plant had only one kilogram of uranium that it could convert entirely into energy, the production of electricity could continue for

$$\frac{9 \times 10^{16} \text{ J}}{10^9 \text{ J/s}} = 9 \times 10^7 \text{ s},$$

or almost three years! ■

Stars shine because of the thermonuclear process of **fusion**, in which, most often, four protons fuse to form a helium nucleus (plus other particles). If you add the masses of the initial protons, you'll get a number that is larger than the total masses of the final products. The *missing mass* is converted into the kinetic energies of the final particles.

Example 5.2.2. In the **proton-proton cycle** occurring in the core of stars like the Sun, 4 protons fuse to give a helium nucleus, two positrons (positive electrons), two neutrinos, and two photons.[3] A proton has a mass of 1.6726×10^{-27} kg; a helium nucleus has a mass of 6.6447×10^{-27} kg; a positron has a mass of 9.1094×10^{-31} kg; and neutrinos are essentially massless, as are photons (see Note 5.2.3 below).

Q: What is the total kinetic energy (KE) of the final products?

A: Denote by Δm the difference between the initial and the final total masses. Then, the total KE is Δmc^2. But

$$\Delta m = 4(1.6726 \times 10^{-27}) - [6.6447 \times 10^{-27} + 2(9.1094 \times 10^{-31})]$$
$$= 4.3878 \times 10^{-29} \text{ kg},$$

and therefore,

$$KE_{\text{tot}} = \Delta mc^2 = 4.3878 \times 10^{-29} \times (3 \times 10^8)^2 = 3.949 \times 10^{-12} \text{ J},$$

or 24.68 MeV. ■

5.2.2 No Mass at Light Speed

The second special case of Equation (5.6), $\beta_b = 1$, i.e., the case of a photon, makes the denominator of Equation (5.6) equal to zero, giving rise to an infinite energy. This is unacceptable *unless the numerator also vanishes*, i.e., $m = 0$. The reverse argument is also true: if $m = 0$ (and the particle has some energy), then the denominator must also vanish, i.e., $v = c$. Therefore,

[3] The process takes place in several fusion steps, with each step fusing two particles and releasing some energy. The sum of these energies is slightly different from what I obtain below.

Note 5.2.3. *Only massless particles such as photons can travel at the speed of light. Moreover, if a particle moves at light speed, it is necessarily massless. No massive particle can attain light speed, and no massless particle can have any speed but the speed of light.*

A consequence of the formula for the relativistic energy is that it gets progressively harder to increase the speed of an object as this speed gets closer and closer to light speed. The following example illustrates this point.

Example 5.2.4. In this example, I demonstrate how difficult it is to speed up massive objects further when they are already moving close to light speed. Suppose that a vehicle of mass 1000 kg is moving at $0.9999c$ and I want to speed it up to $0.99999c$.

Energy consumption for approaching light speed.

Q: How much energy do I need?

A: According to Equation (5.6), the initial energy of the vehicle is (ignore the subscript b)

$$E_i = \frac{mc^2}{\sqrt{1-\beta^2}} = \frac{1000 \times (3 \times 10^8)^2}{\sqrt{1-0.9999^2}} = 6.36 \times 10^{21} \text{ J}$$

and the final energy is

$$E_f = \frac{1000 \times (3 \times 10^8)^2}{\sqrt{1-0.99999^2}} = 2.01 \times 10^{22} \text{ J}.$$

So, the energy needed is the difference between these two energies, or 1.38×10^{22} J.

Now suppose that I have already reached a speed of $0.999999c$ and I want to speed it up to $0.9999999c$. Then the energy that I need is the difference between

$$E_i = \frac{1000 \times (3 \times 10^8)^2}{\sqrt{1-0.999999^2}} = 6.36 \times 10^{22} \text{ J}$$

and

$$E_f = \frac{1000 \times (3 \times 10^8)^2}{\sqrt{1-0.9999999^2}} = 2.01 \times 10^{23} \text{ J}$$

or 1.38×10^{23} J, which is ten times larger than the previous energy difference. Thus, the closer I get to the speed of light, the harder it gets to increase my speed.

It is instructive to compare these answers with the classical predictions. In the first case the vehicle has an initial energy of

$$E_i = \tfrac{1}{2}mv^2 = \tfrac{1}{2}(1000)(0.9999 \times 3 \times 10^8)^2 = 4.4991 \times 10^{19} \text{ J}$$

and a final energy of

$$E_f = \tfrac{1}{2}(1000)(0.99999 \times 3 \times 10^8)^2 = 4.49991 \times 10^{19} \text{ J}.$$

Thus the energy needed is 8.1×10^{15} J. In the second case, the energy needed is the difference between

$$E_i = \tfrac{1}{2}(1000)(0.999999 \times 3 \times 10^8)^2 = 4.499991 \times 10^{19} \text{ J}$$

and

$$E_f = \tfrac{1}{2}(1000)(0.9999999 \times 3 \times 10^8)^2 = 4.4999991 \times 10^{19}\ \text{J},$$

or 8.1×10^{13} J, which is only one percent of the previous energy difference! From a classical point of view, the closer you get to light speed, the easier it gets to speed up. In fact, classical physics places no limit on speeds: Not only can you reach light speed, but also surpass it. This conclusion is of course completely wrong as classical physics fails drastically when objects move with speed comparable to light speed. ■

5.2.3 Invariant Length of Spacetime Momentum

Spacetime momentum is defined directly in terms of the spacetime velocity. Since the latter satisfies the invariant "length" equation [the last equation in (5.3)], the former—being just mc times the spacetime velocity—also satisfies a similar equation: $p_{bt}^2 - p_{bx}^2 = m^2c^2$. In terms of energy (and removing the subscript x), this becomes $(E_b/c)^2 - p_b^2 = m^2c^2$. Multiplying both sides of this equation by c^2 and removing the subscript b as well yields

Relation between energy, momentum, and mass.

$$E^2 - p^2c^2 = m^2c^4. \tag{5.10}$$

Once again, this holds in all RFs, and you can think of it as the **invariant length** of the spacetime momentum.

An interesting consequence of Equation (5.10) is that it allows the possibility of setting $m = 0$. When I do so, I get

$$E^2 - p^2c^2 = 0 \quad \text{or} \quad E^2 = p^2c^2 \quad \text{or} \quad E = |p|c \tag{5.11}$$

for massless particles such as photons [see the comment right after Equation (5.7)]. I put the absolute value sign around p because E is always positive, but p could be positive or negative. However, often I just write $E = pc$, and introduce a negative sign if p points in the negative direction.

Example 5.2.5. If you combine the energy-momentum Lorentz transformations (5.8), the relation between energy and momentum of a photon (5.11), and the Planck-Einstein quantization rule for EM waves, $E = hc/\lambda$, you get an interesting result.

Derivation of relativistic Doppler formula from energy and momentum of light.

Sonya (observer O) is moving with speed β toward Sam (observer O') in his positive direction as shown in Figure 5.3(a). She emits a beam of light, whose energy and momentum she measures to be E and p with $E = pc$. Sam receives this signal and measures its energy and momentum to be E' and p', related to E and p via LT of Equation (5.8):

$$\begin{aligned} p' &= \gamma(p + \beta E/c) = \gamma(p + \beta p) = \frac{p + \beta p}{\sqrt{1-\beta^2}} = \sqrt{\frac{1+\beta}{1-\beta}}\, p \\ E' &= \gamma(\beta pc + E) = \gamma(\beta E + E) = \sqrt{\frac{1+\beta}{1-\beta}}\, E. \end{aligned} \tag{5.12}$$

Some interesting results come out of this equation. First note that, because $E = pc$, $E' = p'c$ as well. This is what you should expect, as the energy of a photon is its momentum

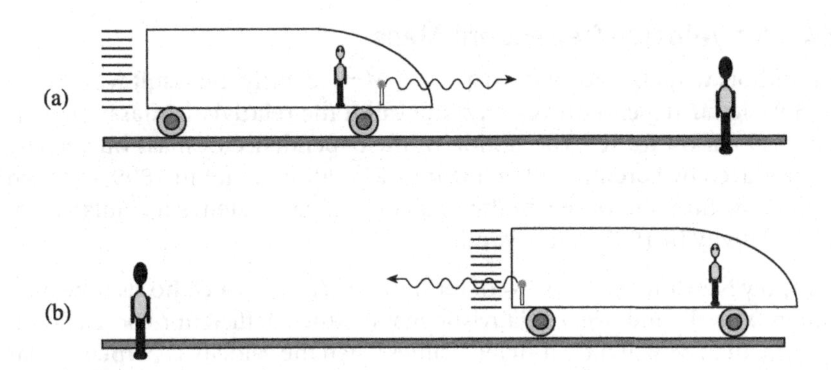

FIGURE 5.3 The light emitted by Sonya has a different energy and momentum than that received by Sam.

times its speed in *all* RFs. The second result is that $p' \neq p$. Thus, although a photon moves with the same speed with respect to all observers, its momentum is not the same. The third result is obtained by using the Planck-Einstein relation on both sides of the second equation:

$$\frac{hc}{\lambda'} = \sqrt{\frac{1+\beta}{1-\beta}}\frac{hc}{\lambda} \quad \text{or} \quad \lambda' = \sqrt{\frac{1-\beta}{1+\beta}}\lambda. \tag{5.13}$$

The wavelength that Sam receives is smaller than that emitted by Sonya. This is the relativistic Doppler formula, which I obtained in Section 4.3.3 using a completely different approach.

As Sonya passes Sam and moves away from him, the wavelength of the photon detected by Sam should increase. Does Equation (5.12) give the correct answer? The photons that Sam receives are moving in the *negative* direction [see Figure 5.3(b)]. Therefore, I have to introduce a negative sign for p, and write (5.12) as

$$p' = \gamma(-p + \beta E/c) = \gamma(-p + \beta p) = -\frac{p - \beta p}{\sqrt{1-\beta^2}} = -\sqrt{\frac{1-\beta}{1+\beta}}\,p$$

$$E' = \gamma(-\beta pc + E) = \gamma(-\beta E + E) = \frac{E(1-\beta)}{\sqrt{1-\beta^2}} = \sqrt{\frac{1-\beta}{1+\beta}}\,E.$$

The second equation now yields

$$\frac{hc}{\lambda'} = \sqrt{\frac{1-\beta}{1+\beta}}\frac{hc}{\lambda} \quad \text{or} \quad \lambda' = \sqrt{\frac{1+\beta}{1-\beta}}\lambda, \tag{5.14}$$

showing an *increase* in the wavelength. You can use this formula for *both* approach and recession by assigning a positive value to β when the source and the detector are receding from each other, and a negative value when they are approaching one another. ■

5.2.4 No Velocity-Dependent Mass

Some authors write Equation (5.6) as $E_b = Mc^2$, identify the result with Equation (5.9), let M depend on velocity, and call it the **relativistic mass**! There is a historical reason for it.[4] The notion of the dependence of mass on velocity was introduced by Lorentz (of Lorentz transformation fame) in 1899 and then developed by him and others in the years preceding Einstein's formulation of special relativity in 1905 and beyond.

If you apply Newton's second law of motion $\mathbf{F} = d\mathbf{p}/dt$, which holds in relativity theory as well, and *define* relativistic mass as the coefficient of acceleration in $\mathbf{F} = M\mathbf{a}$, then M will have different values when the velocity is perpendicular to the force (the transverse mass) than when it is in the direction of the force (longitudinal mass).[5] This already should be an indication of the trouble with the notion of "relativistic mass." Nevertheless, these are the very expressions with which Lorentz introduced the two masses. Together with the "relativistic mass" m_r in the relation $\mathbf{p} = m_r\mathbf{v}$, these masses formed the basis of the language physicists used at the beginning of the twentieth century. Making the trouble even more lasting, it was decided to call the "relativistic mass" simply "mass" and to denote it by m, while the normal mass m was nicknamed "rest mass" and denoted by m_0.

The real trouble with the notion of relativistic mass comes in when we associate a mass of E/c^2 to photons when they experience a gravitational force. When the correct theory of gravity, the general theory of relativity, is applied to photons, it is found that a photon would have different masses for different inclinations relative to the gravitating body. If it falls directly toward the center, its mass is E/c^2, but if it moves perpendicular to the radial direction, its mass is $2E/c^2$!

Einstein advises against velocity-dependent mass!

Equation (5.10) clearly identifies mass (on the right-hand side of the equation) as an *invariant* quantity, which is the same for *all observers*. Only if the particle is at rest relative to an observer (so that $\mathbf{p} = 0$) does E, which is now called the rest energy and usually denoted by E_0, equal mc^2. Einstein, who at the beginning of relativity theory talked about a "relativistic mass," in a letter to Lincoln Barnett—an American journalist—dated 19 June 1948, writes, "It is not good to introduce the concept of the mass $M = m/\sqrt{1 - v^2/c^2}$ of a moving body for which no clear definition can be given. It is better to introduce no other mass concept than the 'rest mass' m. Instead of introducing M it is better to mention the expression for the momentum and energy of a body in motion."[6]

[4] See [Okun 89] for a detailed discussion of the notion of the "relativistic mass" and why it should be avoided.

[5] I have elaborated on this in Section 6.8.3.

[6] The letter is reprinted in [Okun 89] p. 32.

Every couple of years, the Particle Data Group publishes the *Particle Physics Booklet*. In it, you can find a huge amount of information about elementary particles, cosmology, astrophysics, latest values of the fundamental constants of physics, and a lot more. One important piece of information included in the Booklet is the *masses* of elementary particles. For example, the July 2014 edition of the Booklet lists the mass of the electron as $9.10938291 \times 10^{-31}$ kg, and that of the proton as $1.672621777 \times 10^{-27}$ kg. If mass were velocity-dependent, the authors of the Booklet would not bother listing *any* mass, because particles constantly move with unspecified speeds. Furthermore, the precision with which these masses are reported should tell you that mass is indeed a *unique* property of a particle.

Note 5.2.6. *Any object has a unique mass (zero included). If the object is moving with momentum p and energy E, then this unique mass is given by* $m = \sqrt{E^2 - p^2c^2}/c^2$. *All observers in the universe obtain the exact same mass for that object regardless of its energy and momentum.*

5.3 CONSERVATION OF MOMENTUM

Conservation laws are the most universal laws of physics. With due alterations, they have survived all the dramatic changes that took place with the advent of quantum mechanics and relativity theory. While momentum conservation was discovered early on alongside the laws of motion, energy conservation required more time because of the variety of forms in which it shows itself. It was not until the middle of the nineteenth century that its formulation appeared as the first law of thermodynamics.

By the first postulate of relativity theory, the discussion above implies that energy and momentum conservation holds for *all* observers. This universality has some consequences that I investigate now. Conservation laws are best studied in collisions. To appreciate the relativistic conservation laws and their consequences, I'll first look at the simpler case of the classical laws. I'll assume that all collisions, both classical and relativistic, take place in one dimension.

5.3.1 The Classical Case

Two masses m_1 and m_2 are moving with velocities v_1 and v_2, respectively, as shown in Figure 5.4. Since $v_1 > v_2$ a collision takes place and two new masses M_1 and M_2 are produced, which move with velocities V_1 and V_2. All of this happens on a table top in Sonya's laboratory in the reference frame O. Conservation of momentum implies that

$$m_1v_1 + m_2v_2 = M_1V_1 + M_2V_2.$$

The same process is observed by Sam who sees Sonya and her lab move in the positive direction with velocity v. Therefore everything that moves in Sonya's

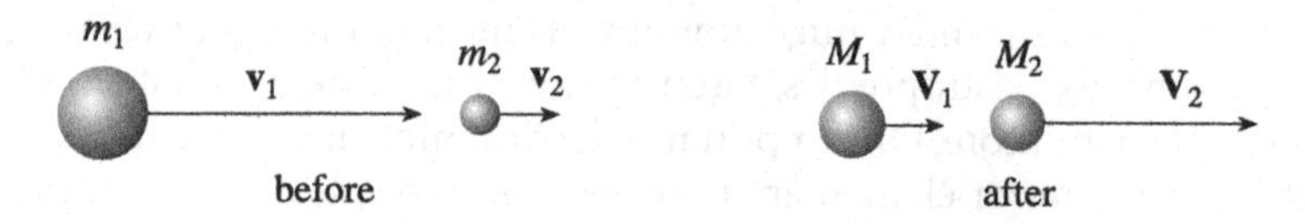

FIGURE 5.4 The conservation of momentum in a collision.

lab appears to move faster by v; for example, m_1 moves with a speed $v_1 + v$ relative to Sam.[7] For him, the conservation of momentum becomes

$$m_1(v_1 + v) + m_2(v_2 + v) = M_1(V_1 + v) + M_2(V_2 + v),$$

or

$$m_1v_1 + m_2v_2 + (m_1 + m_2)v = M_1V_1 + M_2V_2 + (M_1 + M_2)v.$$

Sonya's conservation law equates the first two terms on the left-hand side to the first two terms on the right-hand side. It follows that

$$(m_1 + m_2)v = (M_1 + M_2)v \quad \text{or} \quad m_1 + m_2 = M_1 + M_2,$$

i.e., the total mass does not change in a collision. I just showed the following important result:

Conservation of mass in classical reactions.

> **Note 5.3.1.** *In any classical process (be it a collision or a chemical reaction) in which masses of the participating objects change, in addition to the conservation of classical momentum, the initial total mass must equal the final total mass. This is restated by saying that* ***the total mass is conserved*** *in a classical reaction.*

5.3.2 The Relativistic Case

Now I look at the process of Figure 5.4 from a relativistic viewpoint. I can go through the same argument as above, but use relativistic expression for momentum and the *relativistic* law of addition of velocities to come up with a relativistic version of Note 5.3.1. This turns out to be rather cumbersome (nevertheless a good exercise for you to do and I've asked you to do it in Problem 5.7). The easier way is to use the LTs for momentum and energy as given in Equation (5.8).

Call the relativistic momenta before collision p_1 and p_2, the *relativistic* energies before collision E_1 and E_2, the relativistic momenta after collision P_1 and P_2, and the relativistic energies $\mathcal{E}_1$, $\mathcal{E}_2$. Then Sonya's conservation of momentum becomes $p_1 + p_2 = P_1 + P_2$. For Sam, the conservation of momentum yields

---

[7] The law of addition of velocities holds in a *classical* process.

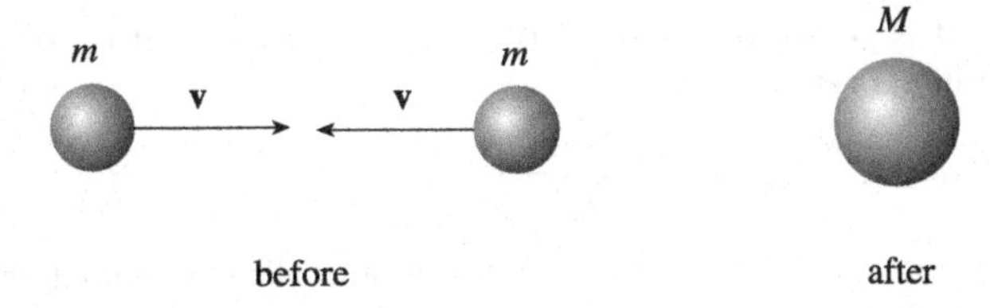

FIGURE 5.5 The relativistic conservation of momentum in this collision implies the non-conservation of mass.

$p'_1 + p'_2 = P'_1 + P'_2$. The primed quantities are related to the unprimed quantities via Equation (5.8) (with b replaced by 1 or 2):

$$p'_1 = \gamma(p_1 + \beta E_1/c), \quad p'_2 = \gamma(p_2 + \beta E_2/c),$$
$$P'_1 = \gamma(P_1 + \beta \mathcal{E}_1/c), \quad P'_2 = \gamma(P_2 + \beta \mathcal{E}_2/c).$$

Substitute these in Sam's conservation law

$$\gamma(p_1 + \beta E_1/c) + \gamma(p_2 + \beta E_2/c) = \gamma(P_1 + \beta \mathcal{E}_1/c) + \gamma(P_2 + \beta \mathcal{E}_2/c)$$

and divide all terms by γ to get

$$p_1 + p_2 + \beta(E_1/c + E_2/c) = P_1 + P_2 + \beta(\mathcal{E}_1/c + \mathcal{E}_2/c).$$

Sonya's conservation law equates the first two terms on the left-hand side to the first two terms on the right-hand side. It follows that

Conservation of relativistic momentum and energy.

$$\beta(E_1/c + E_2/c) = \beta(\mathcal{E}_1/c + \mathcal{E}_2/c) \quad \text{or} \quad E_1 + E_2 = \mathcal{E}_1 + \mathcal{E}_2$$

i.e., the total *relativistic energy* does not change in a collision.

Note 5.3.2. *In any relativistic process (be it a collision or a nuclear reaction) in which masses of the participating objects change, in addition to the conservation of relativistic momentum, the initial total energy must equal the final total energy; i.e.,* ***the total energy is conserved*** *in a relativistic reaction.*

This does not say anything about the conservation of mass. In fact, the following example shows that *mass is not always conserved in a relativistic reaction.*

Example 5.3.3. Figure 5.5 shows two equal masses moving in opposite directions with equal speed. They collide and coalesce to form a single mass M. Both classically and relativistically the two momenta are equal but opposite in direction, giving a total momentum of zero. Therefore, after the collision the momentum of M is zero.

Example of non-conservation of mass.

Q: What is the mass M in terms of the initial mass m?

A: Classically, the mass is conserved, so $M = 2m$. Relativistically, it is the energy that is conserved. The total energy before the collision is

$$E_{\text{tot}} = E_1 + E_2 = \gamma mc^2 + \gamma mc^2 = 2\gamma mc^2.$$

After collision, M is at rest, but its relativistic energy is *not* zero; it is Mc^2. Equating these two quantities, I get

$$Mc^2 = 2\gamma mc^2 \quad \text{or} \quad M = 2m\gamma = \frac{2m}{\sqrt{1-(v/c)^2}}.$$

Therefore, mass is not generally conserved in relativistic collisions. For v much smaller than light speed, the denominator is almost 1 and $M \approx 2m$, regaining the classical conservation of mass.

As a numerical example, let two 1-kg masses approach each other with a speed of $0.9c$, then the mass formed at the end of their collision is

$$M = \frac{2m}{\sqrt{1-(v/c)^2}} = \frac{2}{\sqrt{1-(0.9)^2}} = 4.588 \text{ kg}.$$

So 2 kg of mass has turned into 4.588 kg. Where has the extra 2.588 kg come from? It is the kinetic energy of the colliding particles that has transformed into mass. To see this, note that each particle has a KE of

$$KE = \frac{mc^2}{\sqrt{1-(v/c)^2}} - mc^2 = \frac{9 \times 10^{16}}{\sqrt{1-(0.9)^2}} - 9 \times 10^{16} = 1.165 \times 10^{17} \text{ J}$$

and the two of them carry twice this much KE or 2.33×10^{17} J. On the other hand, the energy "hidden" in the extra mass of 2.588 kg is

$$2.588 \times (3 \times 10^8)^2 = 2.33 \times 10^{17} \text{ J},$$

equal to the total KE of the two particles! This process of the transformation of KE into mass is the underlying principle of particle creation in accelerators like the Large Hadron Collider, which produced the first ever Higgs boson in July 2012. ■

Example 5.3.3 demonstrates a problem with classical physics. I concentrated on the classical conservation of momentum without paying attention to the energy conservation. Before collision, each particle has a KE of $\frac{1}{2}mv^2$. After collision, there is no KE! What happened to the energy? Is it not conserved?

In these situations, the explanation is that the initial KE turns into other forms of energy: the two initial particles may produce a sound as they stick together; they also get warmer. So, I can say that the initial KE turned into the energy of the sound wave produced and the heat that warmed up the final larger particle. I can even give such collisions—the ones that do not conserve the mechanical KE—names and call them **inelastic**. But there is a fundamental incompleteness about such a treatment.

Problem with classical collisions!

It is of course true that when two pieces of clay—initially in motion—stick (soundlessly) together, the initial KE is transferred to the atoms and molecules of the larger piece, raising its temperature. But, what if the particles are fundamental and have no internal structures? I may argue that the final piece *must* have an internal structure and that it is impossible for two fundamental particles to coalesce into another fundamental particle. However, these would be "explanations" that I have to conjure up simply to support my argument.

Relativity has no problem explaining such collisions. Whether I collide two pieces of mud or two fundamental particles, the equivalence of mass and energy can beautifully explain the disappearance of the initial KE. Even if the final product is a composite when two fundamental particles collide, relativity assigns an equivalent mass to the binding energy of the final particle. If this binding energy is small—as in the case of an electron and a proton "colliding" to form a hydrogen atom with the binding energy of 13.6 eV—its contribution to the mass of the composite is negligible. However, in nuclear processes, the binding energy of the composite particle is so large that it indeed contributes to its mass, and relativistic energy and momentum conservations can accurately explain the outcome of such collisions, as I demonstrated in Example 5.3.3.

Relativity has no problem with collisions!

Example 5.3.4. Relativity is "contemptuous" of the spectacular phenomena of classical physics! Take two 0.5-kg pieces of clay each moving at the *classically enormous* rate of 2000 m/s toward the other. They stick together after collision to form a larger piece of clay. The relativistic mass of the final piece is given by the dull result of 1.000000000022 kg, very close to the classical mass of 1 kg.

Classically, in which mass is conserved, the mass of the final piece of clay is *exactly* 1 kg. The total KE of the initial pieces has turned into heat. What kind of temperature are we talking about? I let you calculate the temperature (see Problem 5.15). You'll find that the final clay evaporates at a temperature of over 40,000 K! So, while relativity regards the mass as negligibly close to its classical limit, classical physics finds both the KE of the initial particles and the heat inside the final one enormous! I am not implying that relativity can ignore this kind of extreme temperature, but simply that the energies associated with such temperatures are small compared to the energy stored in a mass given by $E = mc^2$. ■

5.4 PROBLEMS

5.1. Show that $u_{bt} = \gamma_b$, and that $u_{bt}^2 - u_{bx}^2 = 1$. Show that for any other observer O', $u_{bt}'^2 - u_{bx}'^2 = 1$ as well.

5.2. Derive the relativistic law of addition of velocities from the second equation in (5.4).

5.3. I defined spacetime velocity by differentiating x and ct with respect to $s = c\tau$. Now define **spacetime acceleration** a_b by differentiating spacetime velocity with respect to s:

Spacetime acceleration.

$$a_{bt} = \frac{du_{bt}}{ds}, \qquad a_{bx} = \frac{du_{bx}}{ds}.$$

(a) Show that

$$a_{bt}u_{bt} - a_{bx}u_{bx} = 0. \tag{5.15}$$

[Hint: Use Equation (5.2).] You'll see later that it is possible to define a "dot product" in two-dimensional relativity by *subtracting* the product of

Dot product in spacetime plane.

the two components of the vectors. More specifically, if $\mathbf{A} = (A_x, A_t)$ and $\mathbf{B} = (B_x, B_t)$ are two vectors, then the "dot product" is defined as

$$\mathbf{A} \cdot \mathbf{B} = A_t B_t - A_x B_x. \tag{5.16}$$

Equation (5.15), therefore, says that spacetime acceleration is orthogonal to spacetime velocity.

(b) Verify that spacetime acceleration transforms via LTs. That is, if O measures the components of the spacetime acceleration of a particle to be (a_{bx}, a_{bt}) and O' measures them to be (a'_{bx}, a'_{bt}), then

$$a'_{bx} = \gamma(a_{bx} + \beta a_{bt}), \qquad a'_{bt} = \gamma(\beta a_{bx} + a_{bt}),$$

where β is the speed of O relative to O'.

5.4. Let $y = f(x)$ describe a curve in the xy-plane on which a particle moves.

(a) Verify that for the velocity vector of the particle to be perpendicular to the position vector of that particle, $f(x)$ must satisfy the following differential equation:

$$f(x)\frac{df}{dx} + x = 0.$$

(b) Solve this simple equation and show that the solution is

$$f(x) = \pm\sqrt{C - x^2},$$

where C is the constant of integration. This is a circle of radius $\sqrt{C}$.

(c) Now let $x = g(t)$ describe the worldline of a particle in the spacetime plane. Refer to Equation (5.16) for the definition of the dot product in the spacetime plane, and show that for the spacetime velocity vector of the particle to be perpendicular to the "position" vector (x, ct) of that particle, $g(t)$ must satisfy the following differential equation:

$$g(t)\frac{dg}{dt} - c^2 t = 0.$$

(d) Solve this simple equation and show that the solution is

$$g(t) = \pm\sqrt{C + c^2 t^2},$$

where C is the constant of integration. What kind of a curve is this worldline?

5.5. Use Equations (5.7) and (5.8) to derive the relativistic law of addition of velocities.

5.6. Derive Equation (5.10) directly from the definition of energy and momentum in Equations (5.6) and (5.5). Now use (5.8) to show that Equation (5.10) holds in all RFs.

5.7. Using the definition of relativistic momentum and the relativistic law of addition of velocities, show that if relativistic momentum is conserved in all RFs, then relativistic energy must also be conserved.

5.8. A particle of mass m moving at speed v collides with another particle of the same mass at rest. They stick together and move with speed V. What is V in terms of v? What is the mass of the final combined particle?

5.9. Electrons in projection TV sets are accelerated through a potential difference of 50 kV.

(a) Find the speed of the electrons using the *relativistic* form of KE (relativistic energy minus rest energy) assuming that the electrons start from rest.
(b) Find the speed of the electrons using the *classical* form of KE.
(c) Does the difference in speed have to be taken into account when designing the TV set?

5.10. Two identical particles of mass m approaching each other with velocity β collide, and as a result of their collision a particle of mass M and a photon are produced.

(a) Can M remain stationary as in Example 5.3.3?
(b) If the answer to (a) is no, then what are the energies of M and the photon?
(c) What are the momenta of M and the photon?

5.11. Two particles of masses m_1 and m_2 are moving in opposite directions with the same relativistic momentum p.

(a) Find their speeds v_1 and v_2, in terms of m_1, m_2, and p.
(b) Find their energies E_1 and E_2, in terms of m_1, m_2, and p.
(c) The two particles collide and form a particle of mass M at rest. Find the mass of the final particle.
(d) Verify that the extra mass (times c^2) is equal to the total initial kinetic energy of the two particles.

5.12. A particle of mass M at rest decays into two particles of masses m_1 and m_2.

(a) What is the sum of the momenta of the two particles?
(b) Find their energies E_1 and E_2, in terms of M, m_1, and m_2.
(c) Verify that the common momentum of the particles can be expressed as

$$p = \frac{\sqrt{(M^2 - m_1^2)^2 + (M^2 - m_2^2)^2 - M^4 - 2m_1^2 m_2^2}}{2M}\, c.$$

Note the symmetry of the expression in m_1 and m_2.

(d) Show that if the masses are equal, then $E_1 = E_2$ and that these are independent of the common mass of the particles.
(e) Without going through the previous parts, show that if a particle of mass M at rest decays into two identical particles, the energies of the two par-

ticles are equal and this energy is the same regardless of their mass. In particular, whether M decays into two massive or massless identical particles, the particles carry the same amount of energy. What is that energy?

5.13. The neutral pion π^0 has a mass of 2.41×10^{-28} kg. It decays into two photons 98.8% of the times.

(a) What is the energy of each photon in the rest frame of the pion?
(b) What is the wavelength of each photon? Remember that $\lambda = hc/E$ where h is the Planck constant.
(c) Assume that the pion is moving at $0.9c$ in the lab frame in the same direction as one of the photons. What are the energies and wavelengths of each photon in the lab?

5.14. The negative pion π^- has a mass of 2.49×10^{-28} kg. It decays into a muon and a neutrino 99.988% of the times. The muon has a mass of 1.89×10^{-28} kg and the neutrino is essentially massless.

(a) What are the energies of the muon and the neutrino in the pion's rest frame?
(b) What is the momentum of the neutrino?
(c) What is the speed of the muon?
(d) Assume that the pion is moving at $0.9c$ in the lab frame in the same direction as the neutrino. What are the energies and momenta of the muon and the neutrino in the lab?

5.15. This problem continues the discussion of Example 5.3.4. Two 0.5-kg pieces of clay, each moving at the rate of 2000 m/s toward the other, collide.

(a) What is the total KE distributed among the molecules of the end piece?
(b) Each kilogram of clay has about 4 moles, or about 2.4×10^{24} molecules. What is the average KE that each molecule receives?
(c) Use $KE_{\text{avg}} = \frac{3}{2}k_B T$, where $k_B = 1.38 \times 10^{-23}$ J/K is the Boltzmann constant, to find the temperature of the final piece of clay.

CHAPTER 6

Relativity in Four Dimensions

All discussions of relativity have so far been confined to a two-dimensional spacetime. While these discussions led to the essential features of special relativity, the real physics takes place in three dimensions, which with the addition of time, become a four-dimensional spacetime. The main task of this chapter is to generalize the results obtained previously to the full four dimensions.

6.1 FOUR-VECTORS

The starting point of the machinery of special relativity was the discovery of the spacetime distance Δs. As all distances should be, it is an *invariant* quantity in the two-dimensional spacetime. Later, we saw more invariant quantities such as the spacetime velocity and spacetime momentum. The common feature of all these quantities was that they had two components, the time component and the x (or space) component. Let **a** represent any one of these quantities. You can think of **a** as a two-dimensional vector, or a **two-vector**, or a **2-vector**. Let a_0 denote the time component of this 2-vector and a_1 its space component. Then the invariant quantity is $a_0^2 - a_1^2$, the square of the time component minus the square of the space component.

Now recall that the two-dimensional spacetime was obtained by removing two of the three space dimensions. This resulted in the reduction of the space components of Δs to one. Restoring the removed space components leads to the four-dimensional vector $\Delta \mathbf{s}$:[1]

$$\Delta \mathbf{s} = (c\Delta t, \Delta x, \Delta y, \Delta z) \equiv (\Delta s_0, \Delta s_1, \Delta s_2, \Delta s_3)$$

for which the invariant length becomes

$$(\Delta s)^2 = (\Delta s_0)^2 - (\Delta s_1)^2 - (\Delta s_2)^2 - (\Delta s_3)^2.$$

The fact that all space components appear with a negative sign is the result of their complete equivalence: In deriving Equation (3.13), we could have used

[1] I am changing my practice here and putting the time component as the first component to be consistent with my subscripting: 0 comes before 1!

Special Relativity. DOI:10.1016/B978-0-12-810411-8.00006-7

y, or z, instead of x. The addition of the y and z components changes all the spacetime quantities into four-dimensional vectors. For example, the spacetime velocity and momentum become

$$\mathbf{u} = (u_t, u_x, u_y, u_z) \equiv (u_0, u_1, u_2, u_3)$$
$$\mathbf{p} = (p_t, p_x, p_y, p_z) \equiv (p_0, p_1, p_2, p_3).$$

Dot product determines the geometry of a space.

A general "space" of any dimension is useless unless it admits some kind of geometry. A geometry is defined by a "straight" line as well as angles between those straight lines. For example, the surface of a sphere admits a geometry in which straight lines are the great circles of the sphere and angles are defined as usual. The straight lines of any geometry are determined—as a differential equation in the coordinate variables—by how the distance between two infinitesimally close points are calculated and how the angle between two such distances are determined in terms of the coordinates of the two points. It turns out that both of these are outcomes of the notion of dot products. In ordinary space, the dot product determines both the length of a displacement (between two points) and the angle between two such displacements.

Four-vectors introduced.

What is the geometry of the four-dimensional spacetime? Start with a generic four-dimensional spacetime vector of the form $\mathbf{a} = (a_0, a_1, a_2, a_3)$, called a **four-vector** or a **4-vector**. It is also denoted by $(a_0, \vec{a})$ where $\vec{a} \equiv (a_1, a_2, a_3)$ is the *space part* (or the *3-vector part*) of the 4-vector. The invariant length of $\mathbf{a}$ is[2]

$$a_0^2 - a_1^2 - a_2^2 - a_3^2 \equiv a_0^2 - \vec{a} \cdot \vec{a} = a_0^2 - |\vec{a}|^2. \tag{6.1}$$

From invariant length to four-dimensional dot product.

Is it possible to get a dot product out of the invariant length? Consider two 4-vectors and two observers O and O', relative to whom the two vectors are $\mathbf{a}$ and $\mathbf{b}$ and $\mathbf{a}'$ and $\mathbf{b}'$, respectively. Calculate the invariant length of their sum in the O coordinate systems to get

$$\begin{aligned}(\text{inv. length})^2 &= (a_0 + b_0)^2 - (\vec{a} + \vec{b}) \cdot (\vec{a} + \vec{b}) \\ &= a_0^2 + b_0^2 + 2a_0 b_0 - \vec{a} \cdot \vec{a} - \vec{b} \cdot \vec{b} - 2\vec{a} \cdot \vec{b} \\ &= (a_0^2 - \vec{a} \cdot \vec{a}) + (b_0^2 - \vec{b} \cdot \vec{b}) + 2(a_0 b_0 - \vec{a} \cdot \vec{b}),\end{aligned}$$

and in the O' coordinate systems to get

$$\begin{aligned}(\text{inv. length})^2 &= (a'_0 + b'_0)^2 - (\vec{a}' + \vec{b}') \cdot (\vec{a}' + \vec{b}') \\ &= a'^2_0 + b'^2_0 + 2a'_0 b'_0 - \vec{a}' \cdot \vec{a}' - \vec{b}' \cdot \vec{b}' - 2\vec{a}' \cdot \vec{b}' \\ &= (a'^2_0 - \vec{a}' \cdot \vec{a}') + (b'^2_0 - \vec{b}' \cdot \vec{b}') + 2(a'_0 b'_0 - \vec{a}' \cdot \vec{b}').\end{aligned}$$

---

[2] I have to warn you from the beginning that there are different definitions of invariant length of a four-vector. Some regard time component as the fourth component. Some use positive signs for the space and negative sign for the time in Equation (6.1). Unfortunately, the physics and math communities have not come up with a unified notation, and you need to pay attention to this fact when you read the literature on relativity.

The equality of the left-hand side and the invariance of the length of the two vectors imply that $a'_0 b'_0 - \vec{a}' \cdot \vec{b}' = a_0 b_0 - \vec{a} \cdot \vec{b}$. This is the four-dimensional dot product we are after:

Four-dimensional dot product.

Note 6.1.1. *For any two four-vectors* **a** *and* **b**, *the four-dimensional dot product,*

$\mathbf{a} \bullet \mathbf{b} \equiv a_0 b_0 - \vec{a} \cdot \vec{b}$,

is an invariant quantity, i.e., independent of observers. In particular, the invariant length of any 4-vector **v** *is* $\mathbf{v} \bullet \mathbf{v}$, *which is also denoted by* $\mathbf{v}^2 : \mathbf{v}^2 \equiv \mathbf{v} \bullet \mathbf{v}$.

Denote a 4-vector as a column vector, i.e., as a 4×1 matrix. Then the dot product can be written in matrix form as

$$\mathbf{a} \bullet \mathbf{b} = \begin{pmatrix} a_0 & a_1 & a_2 & a_3 \end{pmatrix} \begin{pmatrix} 1 & 0 & 0 & 0 \\ 0 & -1 & 0 & 0 \\ 0 & 0 & -1 & 0 \\ 0 & 0 & 0 & -1 \end{pmatrix} \begin{pmatrix} b_0 \\ b_1 \\ b_2 \\ b_3 \end{pmatrix} \tag{6.2}$$

or

$$\mathsf{a} \bullet \mathsf{b} = \tilde{\mathsf{a}} \eta \mathsf{b} \qquad \text{where} \quad \eta = \begin{pmatrix} 1 & 0 & 0 & 0 \\ 0 & -1 & 0 & 0 \\ 0 & 0 & -1 & 0 \\ 0 & 0 & 0 & -1 \end{pmatrix}, \tag{6.3}$$

and $\tilde{\mathsf{a}}$ and b are the row and column vectors in Equation (6.2). Here $\sim$ denotes the transpose of a matrix.

6.2 LORENTZ TRANSFORMATION IN 4D

The language of matrices is a powerful technique to study the geometry of spacetime, in particular the LTs. Recall from Chapter 3 that LTs were defined as those *linear* transformations which left the spacetime distance $(\Delta s)^2$ invariant. Since linear transformations are nicely represented by matrices, the condition of invariance can be translated into a relation among matrices.

Let Λ denote the matrix that transforms the 4-vectors **a** and **b** in O to the corresponding 4-vectors in O'. In matrix notation this means $\mathsf{a}' = \Lambda \mathsf{a}$ and $\mathsf{b}' = \Lambda \mathsf{b}$. The invariance of the four-dimensional dot product means that $\tilde{\mathsf{a}}' \eta \mathsf{b}' = \tilde{\mathsf{a}} \eta \mathsf{b}$, or that

$$\widetilde{\Lambda \mathsf{a}} \eta \Lambda \mathsf{b} = \tilde{\mathsf{a}} \tilde{\Lambda} \eta \Lambda \mathsf{b} = \tilde{\mathsf{a}} \eta \mathsf{b}.$$

This equality must hold for arbitrary a and b. Therefore,

Four-dimensional Lorentz transformation.

$$\tilde{\Lambda} \eta \Lambda = \eta. \tag{6.4}$$

This is the defining property of the four-dimensional LT. You are already familiar with the two-dimensional LT. Does this definition reduce to that familiar case?

Confine yourself to two dimensions. Then vectors are of the form $\mathbf{a} = (a_0, a_1)$, $\mathbf{b} = (b_0, b_1)$, the inner product is of the form $\mathbf{a} \bullet \mathbf{b} \equiv a_0 b_0 - a_1 b_1$, the matrix η reduces to

$$\eta = \begin{pmatrix} 1 & 0 \\ 0 & -1 \end{pmatrix},$$

and the LTs become 2×2 matrices.

Let $\Lambda = \begin{pmatrix} a_{11} & a_{12} \\ a_{21} & a_{22} \end{pmatrix}$ be a two-dimensional LT that acts on 2-vectors in O to give the corresponding 2-vectors in O'. Then Λ must satisfy Equation (6.4) or

$$\begin{pmatrix} a_{11} & a_{21} \\ a_{12} & a_{22} \end{pmatrix} \begin{pmatrix} 1 & 0 \\ 0 & -1 \end{pmatrix} \begin{pmatrix} a_{11} & a_{12} \\ a_{21} & a_{22} \end{pmatrix} = \begin{pmatrix} 1 & 0 \\ 0 & -1 \end{pmatrix}, \tag{6.5}$$

which (you should show) is equivalent to the following three equations:

$$a_{11}^2 - a_{21}^2 = 1, \qquad a_{11}a_{12} - a_{21}a_{22} = 0, \qquad a_{12}^2 - a_{22}^2 = -1. \tag{6.6}$$

These are three equations in four unknowns. Therefore, you can solve all of them in terms of one, say a_{11}. When you do, you get (see Problem 6.1)

$$a_{22}^2 = a_{11}^2, \qquad a_{12}^2 = a_{21}^2, \qquad a_{12}^2 = a_{11}^2 - 1. \tag{6.7}$$

To determine a_{11}, consider the 2-vector $(c\Delta t, \Delta x)$, the difference between the time and position of two events in O. This 2-vector is represented by $(c\Delta t', \Delta x')$ in O', and, by the definition of the Lorentz transformations,

$$\begin{pmatrix} c\Delta t' \\ \Delta x' \end{pmatrix} = \begin{pmatrix} a_{11} & a_{12} \\ a_{21} & a_{22} \end{pmatrix} \begin{pmatrix} c\Delta t \\ \Delta x \end{pmatrix}. \tag{6.8}$$

Now suppose that $\Delta x = 0$, i.e., that the two events occur at the same location. Then O is measuring the *proper time*, so that $\Delta t = \Delta\tau$. From Equation (6.8), you also have $c\Delta t' = a_{11} c\Delta t$ or $\Delta t' = a_{11}\Delta\tau$. Comparison with Equation (2.4) yields

$$a_{11} = \frac{1}{\sqrt{1-(v/c)^2}} = \frac{1}{\sqrt{1-\beta^2}} \equiv \gamma. \tag{6.9}$$

The rest of the matrix elements can now be found. The first equation in (6.7) gives $a_{22} = \pm\gamma$. To choose the correct sign for a_{22}, note that if O and O' are not moving relative to one another, the coordinates do not change. Therefore Λ must be the unit matrix. So, $a_{22} = 1$ when $v = 0$. This can happen only if $a_{22} = +\gamma$. The second equation in (6.7) now gives $a_{12} = a_{21}$, and the third equation yields

$$a_{12}^2 = \gamma^2 - 1 = \frac{1}{1-\beta^2} - 1 = \frac{\beta^2}{1-\beta^2} = \beta^2\gamma^2 \Rightarrow a_{12} = \pm\beta\gamma.$$

The ambiguity in the sign comes from the choice we have for the direction of motion. We absorb this choice of sign in β, and write

$$\Lambda = \begin{pmatrix} \gamma & \gamma\beta \\ \gamma\beta & \gamma \end{pmatrix}. \tag{6.10}$$

Although the matrix was derived for a particular 2-vector, it applies to the transformation of *all* 2-vectors. For the important case of spacetime "position" vector (ct, x), this yields

$$\begin{aligned} ct' &= \gamma\,(ct + \beta x) \\ x' &= \gamma(x + \beta ct), \end{aligned} \tag{6.11}$$

which is the LT you encountered in Chapter 3.

How do you generalize this to four-dimensions? Let observer O move in an arbitrary direction relative to O'. Take advantage of the freedom of choice of the coordinate axes and let both x and x' coincide with the direction of motion of O. Now recall that lengths perpendicular to the direction of motion do not change. Therefore, Equation (6.11) generalizes to

$$\begin{aligned} ct' &= \gamma\,(ct + \beta x) \\ x' &= \gamma(x + \beta ct) \\ y' &= y \\ z' &= z, \end{aligned} \tag{6.12}$$

or in matrix form,

$$\begin{pmatrix} ct' \\ x' \\ y' \\ z' \end{pmatrix} = \begin{pmatrix} \gamma & \gamma\beta & 0 & 0 \\ \gamma\beta & \gamma & 0 & 0 \\ 0 & 0 & 1 & 0 \\ 0 & 0 & 0 & 1 \end{pmatrix} \begin{pmatrix} ct \\ x \\ y \\ z \end{pmatrix}. \tag{6.13}$$

These equations are not general enough. In many situations, you cannot assume that the direction of motion is along the x-axis. While you can always choose the direction of relative motion of *two* observers to be their common x-axis, when you have three reference frames moving in arbitrary directions, this is impossible. How do you generalize Equation (6.12)? Appendix B shows how, and if you are comfortable with matrices, you *should read* that appendix for insight, practice with matrices, and seeing the power of mathematics in obtaining physical inferences. I reproduce the result of Appendix B here. The matrix generalizing (6.13) to arbitrary direction of motion is

$$\Lambda = \begin{pmatrix} \gamma & \gamma\beta_x & \gamma\beta_y & \gamma\beta_z \\ \gamma\beta_x & 1+\hat{\beta}_x^2(\gamma-1) & \hat{\beta}_x\hat{\beta}_y(\gamma-1) & \hat{\beta}_x\hat{\beta}_z(\gamma-1) \\ \gamma\beta_y & \hat{\beta}_x\hat{\beta}_y(\gamma-1) & 1+\hat{\beta}_y^2(\gamma-1) & \hat{\beta}_y\hat{\beta}_z(\gamma-1) \\ \gamma\beta_z & \hat{\beta}_x\hat{\beta}_z(\gamma-1) & \hat{\beta}_y\hat{\beta}_z(\gamma-1) & 1+\hat{\beta}_z^2(\gamma-1) \end{pmatrix} = \begin{pmatrix} \gamma & \gamma\tilde{\vec{\beta}} \\ \gamma\vec{\beta} & \overleftrightarrow{\Lambda} \end{pmatrix}, \tag{6.14}$$

where the last matrix is the block form of Λ in which

$$\vec{\beta} = \begin{pmatrix} \beta_x \\ \beta_y \\ \beta_z \end{pmatrix} \quad \text{and} \quad \overleftrightarrow{\Lambda} = \begin{pmatrix} 1 & 0 & 0 \\ 0 & 1 & 0 \\ 0 & 0 & 1 \end{pmatrix} + (\gamma - 1) \begin{pmatrix} \hat{\beta}_x^2 & \hat{\beta}_x\hat{\beta}_y & \hat{\beta}_x\hat{\beta}_z \\ \hat{\beta}_x\hat{\beta}_y & \hat{\beta}_y^2 & \hat{\beta}_y\hat{\beta}_z \\ \hat{\beta}_x\hat{\beta}_z & \hat{\beta}_y\hat{\beta}_z & \hat{\beta}_z^2 \end{pmatrix}. \tag{6.15}$$

Applying (6.14) to the spacetime coordinates yields

$$\begin{aligned} ct' &= \gamma(ct + \vec{\beta}\cdot\vec{r}) \\ \vec{r}\,' &= \vec{r} + \gamma\vec{\beta}ct + (\gamma - 1)\hat{\beta}\hat{\beta}\cdot\vec{r} \\ &= \vec{r} + \gamma\vec{\beta}\left(ct + \frac{\gamma}{\gamma+1}\vec{\beta}\cdot\vec{r}\right), \end{aligned} \tag{6.16}$$

where $\vec{r}\,' = (x', y', z')$, $\vec{r} = (x, y, z)$, and $\vec{\beta} = \vec{v}/c$ is the velocity of O relative to O'. You should show (see Problem 6.2) that if $\vec{v}$ is along the x-axis, then (6.14) reduces to (6.13) and (6.16) to (6.12).

The second equation in (6.16) is sometimes expressed in terms of components parallel to $\vec{\beta}$ and perpendicular to it. Write $\vec{r}\,' = \vec{r}\,'_{\parallel} + \vec{r}\,'_{\perp}$ and $\vec{r} = \vec{r}_{\parallel} + \vec{r}_{\perp}$ and show that (6.16) leads to

$$\begin{aligned} ct' &= \gamma\left(ct + |\vec{\beta}|\,|\vec{r}_{\parallel}|\right) \\ \vec{r}\,'_{\parallel} &= \gamma(\vec{r}_{\parallel} + \vec{\beta}ct) \\ \vec{r}\,'_{\perp} &= \vec{r}_{\perp}. \end{aligned} \tag{6.17}$$

This is consistent with the fact that lengths perpendicular to the direction of motion don't change.

The speed of light is 1!

As you have undoubtedly noticed, the speed of light c enters in almost all formulas of relativity, and its presence sometimes makes formulas unnecessarily complicated. It is therefore common and convenient to set $c = 1$. If desired, you can easily restore the factors of c at the end by a simple dimensional analysis.

Note 6.2.1. (Convention) *From now on, whenever convenient and unambiguous, c is set equal to unity. If need be, the factors of c can be restored by dimensional analysis.*

6.3 GENERAL VELOCITY ADDITION FORMULA

Assume that $\vec{r}\,'$ and $\vec{r}$ are position 3-vectors of a bullet. So subscript them with a b. Then divide the differential of the second equation of (6.16) by the differential of the first to obtain the general relativistic law of addition of velocities:

$$\vec{\beta}'_b = \frac{\vec{\beta}_b + \gamma\vec{\beta} + (\gamma - 1)\hat{\beta}\hat{\beta}\cdot\vec{\beta}_b}{\gamma(1+\vec{\beta}\cdot\vec{\beta}_b)} = \frac{\vec{\beta}_b + \gamma\vec{\beta}\left(1 + \dfrac{\gamma}{\gamma+1}\vec{\beta}\cdot\vec{\beta}_b\right)}{\gamma(1+\vec{\beta}\cdot\vec{\beta}_b)} \tag{6.18}$$

where $c\vec{\beta}'_b = \vec{v}'_b = d\vec{r}'_b/dt'$ and $c\vec{\beta}_b = \vec{v}_b = d\vec{r}_b/dt$. Although it should be obvious, it is worth noting that the three vectors $\vec{\beta}$, $\vec{\beta}_b$, and $\vec{\beta}'_b$ all lie in the same plane.[3] Using the composition of two Lorentz transformations, you should be able to derive (6.18) as well. I urge you to look at Problem 6.5 and go through the derivation as a great exercise in relativity and matrix manipulation.

You should verify that (6.18) reduces to the non-relativistic law when all velocities are small and to (3.26) when all velocities are in the same direction. Furthermore, if "b" is a photon (so that $|\vec{\beta}_b| = 1$), then $|\vec{\beta}'_b| = 1$ as well (see Problem 6.13).

Example 6.3.1. TILT OF AXES
Sam, observer O, moves in the positive y-direction of Sonya, observer O', with 3-velocity $\vec{\alpha}$. Sonya in turn moves in the positive x-direction of Sally, observer O'', with 3-velocity $\vec{\beta}$. A firecracker located at point P with coordinates (x, y) in Sam's xy-plane explodes at time t according to Sam. What are the coordinates of P according to Sally, and when does the event occur according to her?

This is just the composition of two LTs: apply the $\vec{\alpha}$ transformation to go from O to O' and $\vec{\beta}$ transformation to go from O' to O''. So, start with[4]

$$\vec{\alpha} = \alpha\hat{\mathbf{e}}_y, \qquad \vec{\beta} = \beta\hat{\mathbf{e}}_x, \qquad \vec{r} = x\hat{\mathbf{e}}_x + y\hat{\mathbf{e}}_y$$

and use (6.16) to find t' and $\vec{r}\,'$. With $c = 1$, you should show that

$$\begin{aligned} t' &= \gamma_\alpha(t + \alpha y) \\ \vec{r}\,' &= x\hat{\mathbf{e}}_x + \left[y + \alpha\gamma_\alpha\left(t + \frac{\gamma_\alpha}{\gamma_\alpha + 1}\alpha y\right)\right]\hat{\mathbf{e}}_y, \end{aligned} \tag{6.19}$$

and when you use $\gamma_\alpha^2\alpha^2 = \gamma_\alpha^2 - 1$ (with a similar expression for β), that

$$\begin{aligned} t'' &= \gamma_\beta(t' + \vec{\beta}\cdot\vec{r}\,') = \gamma_\beta[\gamma_\alpha(t + \alpha y) + \beta x] \\ \vec{r}\,'' &= \vec{r}\,' + \gamma_\beta\vec{\beta}\left(t' + \frac{\gamma_\beta}{\gamma_\beta + 1}\beta x\right) \\ &= \gamma_\beta\left[x + \beta\gamma_\alpha(t + \alpha y)\right]\hat{\mathbf{e}}_x + \gamma_\alpha(y + \alpha t)\,\hat{\mathbf{e}}_y. \end{aligned} \tag{6.20}$$

Now consider the special case where the firecracker is on the x-axis (i.e., when $y = 0$):

$$\begin{aligned} t'' &= \gamma_\beta(\gamma_\alpha t + \beta x) \\ \vec{r}\,'' &= \gamma_\beta\,(x + \beta\gamma_\alpha t)\,\hat{\mathbf{e}}_x + \alpha\gamma_\alpha t\hat{\mathbf{e}}_y. \end{aligned} \tag{6.21}$$

---

[3] If it is not obvious, then convince yourself that it is true. Hint: Two vectors always form a plane, which includes their sum.

[4] Here I am using α for $|\vec{\alpha}|$ and β for $|\vec{\beta}|$.

What does Sally see when she looks at P at the same time that her origin passes O? Obviously this is when $t'' = 0$. So, the first equation gives $\gamma_\alpha t = -\beta x$, which upon substitution in the second equation yields

$$\vec{r}'' = \gamma_\beta \left(x - \beta^2 x\right) \hat{\mathbf{e}}_x - \alpha\beta x \hat{\mathbf{e}}_y = \frac{x}{\gamma_\beta} \hat{\mathbf{e}}_x - \alpha\beta x \hat{\mathbf{e}}_y.$$

So, at the same time that Sally sees Sam's origin pass her origin, P is at $(x/\gamma_\beta, -\alpha\beta x)$. Thus, Sally concludes that Sam's x-axis has tilted by an angle ϕ given by

$$\tan\phi = -\frac{\alpha\beta}{\gamma_\beta}. \tag{6.22}$$

In particular, if Sam has a ruler lined up with his x-axis, it appears tilted to Sally.

Such an effect does not exist in non-relativistic physics because α and β are extremely small. For example, with $v_\alpha = v_\beta = 1000$ m/s (a very large speed by any human standard), you get

$$\alpha = \beta = \frac{1000}{3 \times 10^8} = 3.33 \times 10^{-6}$$

and

$$\tan\phi = -\frac{(3.33 \times 10^{-6})^2}{1} = -1.1 \times 10^{-11},$$

corresponding to $\phi = -2.3 \times 10^{-6}$ arcsecond!

If P is rigidly attached to the x-axis (i.e., it does not move relative to Sam), then from (6.21), the velocity of P relative to Sally is

$$\frac{d\vec{r}''}{dt''} = \frac{\gamma_\beta\,(0 + \beta\gamma_\alpha dt)\,\hat{\mathbf{e}}_x + \alpha\gamma_\alpha dt \hat{\mathbf{e}}_y}{\gamma_\beta(\gamma_\alpha dt + 0)} = \beta\hat{\mathbf{e}}_x + \frac{\alpha}{\gamma_\beta}\hat{\mathbf{e}}_y. \tag{6.23}$$

Since this is independent of the position of P, the entire x-axis moves as a rigid rod with speed $\sqrt{\beta^2 + (\alpha/\gamma_\beta)^2}$ at an angle ϑ with respect to the positive x''-axis where $\tan\vartheta = \alpha/(\beta\gamma_\beta)$. You should verify that the velocity (6.23) is consistent with the relativistic law of addition of velocities (6.18). ■

Example 6.3.2. A spaceship moves with velocity $\vec{\beta}_s$—whose magnitude cannot be changed—due north with respect to an intergalactic medium which moves eastward with speed $\vec{\beta}$ relative to Earth. What is the magnitude and direction of the velocity of the spaceship $\vec{\beta}'_s$ relative to Earth?

This is a straightforward example of the law of addition of velocities (6.18). Let $\hat{\mathbf{e}}_x$ and $\hat{\mathbf{e}}_y$ be unit vectors in the eastward and northward directions, respectively. Then $\vec{\beta}_s \cdot \vec{\beta} = 0$ and (6.18) yields

$$\vec{\beta}'_s = \frac{\vec{\beta}_s + \gamma\vec{\beta}}{\gamma} = \frac{\beta_s\hat{\mathbf{e}}_y + \gamma\beta\hat{\mathbf{e}}_x}{\gamma}.$$

Therefore,

$$|\vec{\beta}'_s|^2 = \frac{\beta_s^2 + \gamma^2\beta^2}{\gamma^2} = \frac{\beta_s^2 + \gamma^2 - 1}{\gamma^2} = \frac{\gamma^2 - \gamma_s^{-2}}{\gamma^2} = \frac{\gamma^2\gamma_s^2 - 1}{\gamma^2\gamma_s^2},$$

and

$$|\vec{\beta}_s'| = \frac{\sqrt{\gamma^2\gamma_s^2 - 1}}{\gamma\gamma_s}.$$

Let θ be the angle formed by $\vec{\beta}_s'$ and the y-axis. Then

$$\cos\theta = \frac{\beta_s/\gamma}{|\vec{\beta}_s'|} = \frac{\beta_s/\gamma}{\sqrt{\gamma^2\gamma_s^2 - 1}/\gamma\gamma_s} = \frac{\beta_s\gamma_s}{\sqrt{\gamma^2\gamma_s^2 - 1}},$$

with θ being east of north.

In which direction should the spaceship move relative to the medium so that it maintains a northward motion relative to Earth? What is the magnitude of that northward velocity? Let $\vec{\beta}_s = \beta_{sx}\hat{\mathbf{e}}_x + \beta_{sy}\hat{\mathbf{e}}_y$. Then (6.18) yields

$$\vec{\beta}_s' = \frac{\beta_{sx}\hat{\mathbf{e}}_x + \beta_{sy}\hat{\mathbf{e}}_y + \hat{\mathbf{e}}_x[\gamma\beta + (\gamma - 1)\beta_{sx}]}{\gamma(1 + \beta\beta_{sx})}. \tag{6.24}$$

For the x-component to be zero, we must have

$$\beta_{sx} + \gamma\beta + (\gamma - 1)\beta_{sx} = 0 \quad \text{or} \quad \beta_{sx} = -\beta,$$

which is the same as the non-relativistic result. Thus,

$$\vec{\beta}_s = -\beta\hat{\mathbf{e}}_x + \sqrt{\beta_s^2 - \beta^2}\,\hat{\mathbf{e}}_y,$$

and $\cos\theta = \beta/|\vec{\beta}_s|$, with θ west of north. For this to make sense, $|\vec{\beta}_s|$ must be larger than β, otherwise there is no solution. To find the magnitude, note that $\beta_{sx}' = 0$; then use (6.24) to get

$$|\vec{\beta}_s'| = \beta_{sy}' = \frac{\beta_{sy}}{\gamma(1 + \beta\beta_{sx})} = \frac{\sqrt{\beta_s^2 - \beta^2}}{\gamma(1 - \beta^2)} = \gamma\sqrt{\beta_s^2 - \beta^2}.$$

Note that if $\beta_s \to 1$ then $|\vec{\beta}_s'| \to 1$ as well. ■

6.3.1 Aberration

Equation (6.18) has an interesting consequence when "b" becomes a *point* source of light. With "b" as light, $|\vec{\beta}_b| = |\vec{c}\,| = 1$ and, as a result of Problem 6.13, $|\vec{c}\,'| = 1$ as well. Now take the dot product of (6.18) with the unit vector $\hat{\beta}$ to get

$$\vec{c}\,' \cdot \hat{\beta} = \frac{\vec{c} \cdot \hat{\beta} + \hat{\beta} \cdot \hat{\beta}\left[|\vec{\beta}|\gamma + (\gamma - 1)\hat{\beta} \cdot \vec{c}\right]}{\gamma(1 + \vec{\beta} \cdot \vec{c})} = \frac{|\vec{\beta}| + \hat{\beta} \cdot \vec{c}}{1 + |\vec{\beta}|\hat{\beta} \cdot \vec{c}},$$

or, using φ and φ' for the angles between the light beams and the direction of motion,

Aberration formulas.

$$\cos\varphi' = \frac{|\vec{\beta}| + \cos\varphi}{1 + |\vec{\beta}|\cos\varphi}$$
$$\sin\varphi' = \frac{\sin\varphi}{\gamma\left(1 + |\vec{\beta}|\cos\varphi\right)}, \tag{6.25}$$

where the second equation is a direct consequence of the first. The equations in (6.25) are called the **aberration formulas**. Remember that $\vec{\beta}$ is the velocity of O relative to O', and it is important to keep in mind that

Note 6.3.3. *φ and φ' are the angles formed in the two reference frames by the* ***direction of motion*** *of a* ***single*** *point source and a* ***single*** *ray of light coming from it. The rays of light in the two frames and the direction of motion are all in the same plane.*

The last statement means that the change from φ to φ' occurs in the plane formed by the direction of motion and the ray of light.

You can also use the trigonometric identity

$$\tan \tfrac{1}{2}\phi = \frac{\sin\phi}{1+\cos\phi}$$

to write the aberration formulas in the form

$$\tan \tfrac{1}{2}\varphi' = \sqrt{\frac{1-|\vec{\beta}|}{1+|\vec{\beta}|}}\tan \tfrac{1}{2}\varphi. \tag{6.26}$$

This suggests a magnification of the angle, which is different when the source is approaching ($\pi/2 > \varphi > 0$) than when it is receding ($\pi > \varphi > \pi/2$). On approach, φ' is smaller than φ because $\sqrt{(1-|\vec{\beta}|)/(1+|\vec{\beta}|)} < 1$. On recession, let $\eta \equiv \pi - \varphi$ and $\eta' \equiv \pi - \varphi'$. Then Equation (6.26) gives

$$\tan\left(\frac{\pi}{2}-\frac{\eta'}{2}\right) = \sqrt{\frac{1-|\vec{\beta}|}{1+|\vec{\beta}|}}\tan\left(\frac{\pi}{2}-\frac{\eta}{2}\right)$$

or

$$\tan \tfrac{1}{2}\eta' = \sqrt{\frac{1+|\vec{\beta}|}{1-|\vec{\beta}|}}\tan \tfrac{1}{2}\eta, \tag{6.27}$$

indicating that $\eta' > \eta$. I'll elaborate on this notion of magnification in Chapter 7.

Suppose that the observer O carries an EM radiation source that radiates uniformly in all directions. The angles of each ray of this source in the two RFs are related by the aberration formulas. If β is very close to 1, then (6.25) implies that $\sin\varphi' \to 0$, *regardless of* φ. This means that each ray makes an angle $\varphi' \approx 0$ with the direction of the motion of O relative to O'. Thus,

Note 6.3.4. *An ultra-relativistic source of electromagnetic radiation has (almost) only forward radiation.*

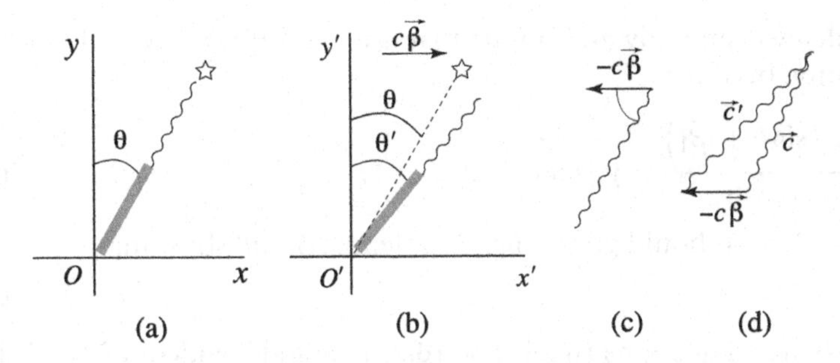

FIGURE 6.1 (a) The line of sight to a star if Earth were not moving. (b) The telescope has to be tilted *forward* to see the star. Note that $\vec{\beta}$ is the velocity of O' (the Earth) relative to O (the star). (c) The angles φ and φ' are the angles between the light ray and the direction of motion of its point source (see Note 6.3.3). (d) The non-relativistic law of addition of velocities for light. Both $\vec{c}$ and $\vec{c}'$ point downward.

Aberration of Starlight

As you look at a star, you direct your sight in a specific direction. You may think that this direction should change due to the motion of Earth around the Sun. And it does! However, this change is so small that for all practical purposes you can ignore it. In fact, you should be able to calculate the *maximum* change in angle per second by taking the ratio of the speed of the Earth as it moves around the Sun (about 30 km/s) and the distance to the *closest* star (about 4 light years). This is about 1.6×10^{-7} arcsecond per second! Thus, if you look at a star for an hour through your telescope, you need to change its tilting *at most* by 0.0006 arcsecond in that one hour. So, the direction of sight of a star is, for all practical purposes, fixed. This lack of **parallax** was the main reason that the Greeks did not accept the heliocentric model even though Aristarchus was led to that model through his estimates of the Earth-Moon and Earth-Sun distances.

Parallax is too small to detect.

Although parallax is practically unobservable, the direction in which you should *point your telescope* to look at the star does depend on the motion of the Earth around the Sun! This effect, known as the **aberration of starlight**, can be calculated by considering two cases: one in which you look at the star from a frame in which the star is not moving (or equivalently, Earth is not moving around the Sun), call it O, and the other in which the motion of Earth is taken into account, call it O'. Figure 6.1(a) shows the starlight coming down along a line making angle θ with the vertical. Therefore, the angle φ of Equation (6.25) is $\varphi = \pi/2 - \theta$. Substituting this in the ratio of the second equation in (6.25) to the first yields

Aberration of starlight.

$$\tan\varphi' = \frac{\sin\varphi}{\gamma\left(|\vec{\beta}| + \cos\varphi\right)} = \frac{\cos\theta}{\gamma\left(|\vec{\beta}| + \sin\theta\right)}.$$

With θ' defined similarly and noting that $\tan(\pi/2-\theta')=\cot\theta'=1/\tan\theta'$, the last equation becomes

$$\tan\theta' = \frac{\gamma\left(\sin\theta + |\vec{\beta}|\right)}{\cos\theta} = \gamma\left(\tan\theta + |\vec{\beta}|\csc\theta\right). \tag{6.28}$$

For $|\vec{\beta}| << 1$, you should go through Problem 6.11 and show that

$$\theta' = \theta + |\vec{\beta}|\cos\theta. \tag{6.29}$$

Unlike parallax, Equations (6.28) and (6.29) are independent of the distance of the star. They depend only on the direction of sight of a star.

From (6.29) you note that $\theta' > \theta$, implying that the telescope should be tilted *forward* (in the direction of Earth's motion). Therefore, the observation of a given star requires different tilt angles depending on the position of Earth on its orbit around the Sun. In particular, the tilting is in the *opposite* direction six months after any given initial observation! This is how aberration was discovered in the 18th century.

For $\beta \approx 0.0001$, corresponding to the Earth's speed around the Sun, (6.29) gives

$$\theta' - \theta \approx 0.0001\cos\theta \text{ rad} \quad \text{or} \quad \theta' - \theta \approx 20.6\cos\theta \text{ arcsec}. \tag{6.30}$$

For small θ, corresponding to stars almost directly above, this is over 30,000 times larger than the parallax of the closest star, Alpha Centauri. In fact if this were due to the parallax, Alpha Centauri would be inside the solar system! James Bradley, the British astronomer, was the first to explain the apparent motion of the star Epsilon Draconis—which was at first thought to be due to parallax—by deriving the small-β approximation of (6.28) using the non-relativistic law of addition of velocities as shown in Figure 6.1(d) and Problem 6.10.

6.4 GENERAL DOPPLER EFFECT

The treatment of the Doppler effect in Chapter 4 is good only if the source of the EM wave is moving directly toward or away from the observer. Now that we have the general LT (6.16), we can find the Doppler effect for a general motion of the source.

An EM source is fixed at the origin of a coordinate system O moving with velocity $\vec{\beta}$ relative to an observer O'. At some time t_1, the source emits a wave front, constituting an event E_1 whose coordinates in O are $(t_1, \vec{0})$. A little later it emits the next wave front at event E_2 with coordinates $(t_2, \vec{0})$ in O, where $t_2 = t_1 + T$, T being the period of the EM wave in O. From (6.16)—with $c=1$—these two events have coordinates

$$t_1' = \gamma(t_1 + \vec{\beta}\cdot\vec{0}) = \gamma t_1, \qquad t_2' = \gamma t_2$$

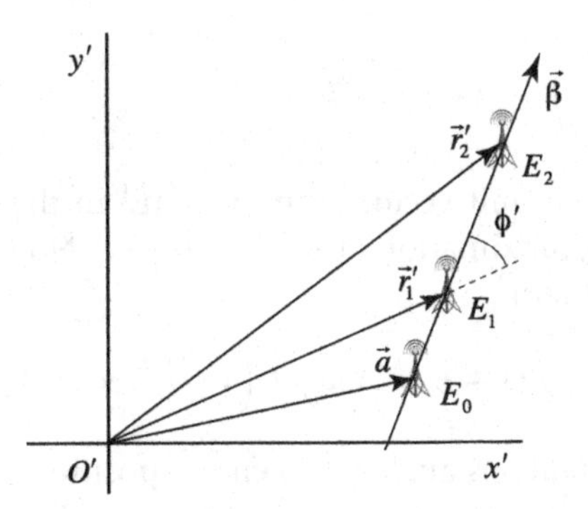

FIGURE 6.2 Events E_0, E_1, and E_2, occur at $t' = 0$, $t' = t_1'$, and $t' = t_2'$, respectively. The vector connecting E_1 to E_2 is $\vec{\beta}(t_2' - t_1') = \gamma\vec{\beta}T$.

$$\vec{r}\,'_1 = \vec{a} + \gamma\vec{\beta}\left(t_1 + \frac{\gamma}{\gamma+1}\vec{\beta}\cdot\vec{0}\right) = \vec{a} + \gamma\vec{\beta}t_1, \qquad \vec{r}\,'_2 = \vec{a} + \gamma\vec{\beta}t_2 \tag{6.31}$$

with respect to the observer O'. Here $\vec{a}$ is the position of the source relative to O' at $t' = t = 0$. Its presence is necessary, because if $\vec{a}$ were $\vec{0}$, then $\vec{r}\,'_1$ and $\vec{r}\,'_2$ would be in the direction of $\vec{\beta}$ and, as Figure 6.2 shows, the source would be passing through the origin of O' and its motion would be either directly away or toward, the special cases we already considered in Chapter 4.

Rewrite the second line of Equation (6.31) as

$$\vec{r}\,'_2 = \vec{r}\,'_1 - \gamma\vec{\beta}t_1 + \gamma\vec{\beta}t_2 = \vec{r}\,'_1 + \gamma\vec{\beta}T. \tag{6.32}$$

Observer O' *receives* signal number one $|\vec{r}\,'_1|$ (in reality $|\vec{r}\,'_1|/c$, but $c = 1$) after t_1' and signal number two $|\vec{r}\,'_2|$ after t_2'. Hence, to O', the period of oscillation is

$$\begin{aligned} T' &= t_2' + |\vec{r}\,'_2| - (t_1' + |\vec{r}\,'_1|) \\ &= \gamma(t_2 - t_1) + |\vec{r}\,'_2| - |\vec{r}\,'_1| \equiv \gamma T + r_2' - r_1', \end{aligned} \tag{6.33}$$

where $r_1' \equiv |\vec{r}\,'_1|$ and $r_2' \equiv |\vec{r}\,'_2|$.

The only thing left to do is to find $r_2' - r_1'$. First note that $\gamma\beta T$ is small compared to r_1'.[5] Next take the square root of the dot product of either side of (6.32) with itself to get

$$\begin{aligned} r_2' &= \left[r_1'^2 + (\gamma\beta T)^2 + 2\gamma T\vec{\beta}\cdot\vec{r}\,'_1\right]^{1/2} \\ &= r_1'\left[1 + \left(\frac{\gamma\beta T}{r_1'}\right)^2 + 2\frac{\gamma T\vec{\beta}\cdot\vec{r}\,'_1}{r_1'^2}\right]^{1/2} \end{aligned}$$

[5] For example, for $T \approx 10^{-14}$, corresponding to visible light and $\beta = 0.99$, $\gamma\beta T \approx 20$ μm.

$$\approx r_1' \left(1 + 2\frac{\gamma T \vec{\beta} \cdot \hat{r}'_1}{r_1'}\right)^{1/2} \approx r_1' + \gamma T \vec{\beta} \cdot \hat{r}'_1$$

where $\hat{r}'_1 \equiv \vec{r}'_1 / r_1'$ is the unit vector along $\vec{r}'_1$, and in the last step you have to use the binomial approximation $(1+\epsilon)^n = 1 + n\epsilon$. Now substitute this last result in (6.33) and obtain

$$T' = \gamma T + \gamma T \vec{\beta} \cdot \hat{r}'_1 = \gamma T (1 + \vec{\beta} \cdot \hat{r}'_1) = \gamma T \left(1 + |\vec{\beta}| \cos\phi'\right), \tag{6.34}$$

where ϕ' is the instantaneous angle between the position vector of the source and its direction of motion (see Figure 6.2). Since $\lambda = cT$, you get

$$\lambda' = \gamma \left(1 + |\vec{\beta}| \cos\phi'\right) \lambda. \tag{6.35}$$

Transverse Doppler effect.

Equation (6.35) agrees with the special cases of $\phi' = 0$ (the source moving away from the observer) and $\phi' = \pi$ (the source approaching the observer). When $\phi' = \pi/2$, (6.35) gives the **transverse Doppler effect**, $\lambda' = \gamma\lambda$. This is a strictly relativistic effect, resulting—as (6.34) shows—solely from time dilation. There is no transverse Doppler effect in classical physics, because there is no net approach or recession in that case.

Relativistic beaming.

Equation (6.25) indicates that a source radiating isotropically at rest will have more radiation in its forward direction than backward when it moves. In terms of photons, it means that more photons are emitted forward than backward. Equation (6.35) states that the photons moving in the forward direction of an approaching source have shorter wavelengths and, therefore, higher energies.[6] This double enhancement of energy means that most of the *energy flux* of the approaching source is concentrated in its forward direction. This is the phenomenon of **relativistic beaming**.

Relativistic jets.

Active galactic nuclei (AGN) are compact astronomical objects at the center of many distant galaxies. They emit EM radiation of all wavelengths. One of the main characteristics of AGN is their emission of **relativistic jets**, which can be hundreds of thousands of light years long. Another feature is the extraordinary luminosities of these jets. The source of these jets is believed to be the accretion disk of supermassive black holes, which are believed to occupy the center of most large galaxies. Although the symmetry of the source implies a more or less isotropic luminosity, relativistic beaming biases the intensity in the direction of motion of the source. The jets moving toward us indicate very high speeds (the γ factor for some of these jets has been estimated to be as high as 100) of the sources.

---

[6] Recall that the energy of a photon of wavelength λ is $E = hc/\lambda$.

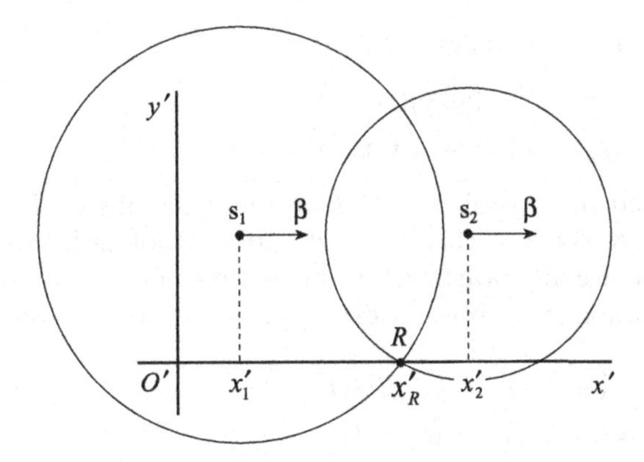

FIGURE 6.3 The frame of the two satellites S_1 and S_2 is moving with speed β to the right relative to the Earth frame.

6.5 GLOBAL POSITIONING SYSTEM

The global positioning system (GPS) uses a constellation of satellites to locate the position of a "receiver" on Earth. Each satellite carries an atomic clock, which keeps time to an accuracy of a fraction of a nanosecond per year. Nevertheless, these clocks are corrected from the ground several times per day for even more accuracy. The operation of the GPS is crucially dependent on relativity theory (both special and general). The details of this operation are very complicated, but the basic principle is relatively simple.

Three satellites moving with the same velocity relative to the Earth emit microwave signals simultaneously—according to their clocks—when they are at known locations in their rest frame. If an observer on Earth receives these signals simultaneously—according to Earth clocks[7]—they can calculate their position relative to the satellites by finding the intersection of the three spheres formed by the wave fronts of the signal.

To understand the basic principles, take away one dimension and concentrate on a one-dimensional Earth, and instead of three satellites take two, S_1 and S_2 as in Figure 6.3. At time $t = t' = 0$, when the origins of O and O' coincide, the two satellites are located at x_1 and x_2 in O and send signals simultaneously (as measured in their own frame). A receiver R picks up these signals simultaneously at time t'_R. These pieces of information are sufficient to find x'_R.

Let L be the separation between the two satellites in O. Ignoring the z-coordinate, the emission of signals, call them events E_1 and E_2, have spacetime coordinates $(x_1, y, 0)$ and $(x_1 + L, y, 0)$, respectively, in O. Lorentz trans-

[7] The *reception* of the two signals is simultaneous. The signals are not emitted simultaneously according to the Earth observer.

formation gives the coordinates of E_1 and E_2 in O':

$$
\begin{aligned}
x'_1 &= \gamma x_1, & y'_1 &= y, & t'_1 &= \gamma\beta x_1 = \beta x'_1 \\
x'_2 &= \gamma(x_1 + L), & y'_2 &= y, & t'_2 &= \gamma\beta(x_1 + L) = \beta x'_2.
\end{aligned}
\tag{6.36}
$$

To find x'_R, insert the coordinates of R in the equations of the two circles, noting that the center of each circle is the location of each satellite and its radius is the distance the wave travels from the time of its emission (t'_1 or t'_2) to the time of reception (t'_R in both cases). With $y'_R = 0$, you get two equations

$$
\begin{aligned}
\left(x'_R - x'_1\right)^2 + y^2 &= \left(t'_R - t'_1\right)^2 = \left(t'_R - \beta x'_1\right)^2 \\
\left(x'_R - x'_2\right)^2 + y^2 &= \left(t'_R - ct'_2\right)^2 = \left(t'_R - \beta x'_2\right)^2.
\end{aligned}
\tag{6.37}
$$

Solving each equation for ct'_R and setting them equal gives

$$
\beta\left(x'_2 - x'_1\right) + \sqrt{\left(x'_R - x'_2\right)^2 + y^2} - \sqrt{\left(x'_R - x'_1\right)^2 + y^2} = 0
$$

or

$$
\beta\gamma L + \sqrt{\left(x'_R - x'_2\right)^2 + y^2} = \sqrt{\left(x'_R - x'_1\right)^2 + y^2}. \tag{6.38}
$$

To ease the algebraic manipulation, let $\xi = x'_2 - x'_R$. Then, $x'_R - x'_1 = \gamma L - \xi$. Substitute these in the equation above, square both sides, and simplify to get

$$
\beta^2\gamma L + 2\beta\sqrt{\xi^2 + y^2} = \gamma L - 2\xi,
$$

or

$$
2\beta\sqrt{\xi^2 + y^2} = \gamma L(1 - \beta^2) - 2\xi = \frac{L}{\gamma} - 2\xi.
$$

Squaring both sides again and simplifying yields

$$
\xi^2 - \gamma L\xi + \frac{L^2}{4} - \beta^2\gamma^2 y^2 = 0,
$$

whose solution is

$$
\xi = \frac{1}{2}\gamma L \pm \gamma\beta\sqrt{\frac{L^2}{4} + y^2}.
$$

For $\beta \to 1$, the plus sign gives

$$
\xi = x'_2 - x'_R > \gamma L = x'_2 - x'_1 \quad \text{or} \quad x'_1 > x'_R,
$$

which is not acceptable because x'_R should lie between x'_1 and x'_2. Therefore,

$$
x'_R = x'_2 - \frac{1}{2}\gamma L + \gamma\beta\sqrt{\frac{L^2}{4} + y^2} \tag{6.39}
$$

or equivalently,

$$\begin{aligned} x'_R &= x'_1 + \frac{1}{2}\gamma L + \gamma\beta\sqrt{\frac{L^2}{4} + y^2} \\ &= \gamma x_1 + \frac{1}{2}\gamma L + \gamma\beta\sqrt{\frac{L^2}{4} + y^2}. \end{aligned} \tag{6.40}$$

The receiver is closer to S_2 because $t'_2 > t'_1$, i.e., according to O', S_2 sends its signal later than S_1, so R should be closer to S_2 to receive the two signals simultaneously.

Example 6.5.1. The number of satellites in a GPS constellation is increased once in a while for improvement in accuracy. For now, suppose that there are 30 equally spaced satellites on a circular orbit corresponding to a period of 12 hours. The second law of motion

$$\frac{v^2}{r} = \frac{GM_\oplus}{r^2}, \quad M_\oplus = \text{mass of Earth},$$

with $v = 2\pi r/T$ yields Kepler's third law:

$$T^2 = \frac{4\pi^2}{GM_\oplus} r^3.$$

Inserting $T = 12 \times 3600$, $G = 6.67 \times 10^{-11}$, and $M_\oplus = 6 \times 10^{24}$ kg yields $r = 2.66 \times 10^7$ m, a separation of

$$L = \frac{2\pi r}{30} = \frac{2\pi \times 2.66 \times 10^7}{30} \approx 5.6 \times 10^6 \text{ m},$$

a height of

$$y = r - R_\oplus = 2.66 \times 10^7 - 6.4 \times 10^6 = 2.03 \times 10^7 \text{ m},$$

and a speed of

$$v = \frac{2\pi r}{T} = \frac{2\pi \times 2.66 \times 10^7}{12 \times 3600} = 3875 \text{ m/s} \quad \beta = 1.3 \times 10^{-5}, \quad \gamma \approx 1.$$

Substituting these values in Equation (6.40) gives

$$\begin{aligned} x'_R &= x'_1 + \frac{1}{2}L + 1.3 \times 10^{-5}\sqrt{(2.8 \times 10^6)^2 + (2.03 \times 10^7)^2} \\ &\approx x'_1 + \frac{1}{2}L + 266 \text{ m}. \end{aligned}$$

Thus, the receiver is 266 m to the right of the midpoint between the two satellites. ■

With satellites moving so much slower than light, as illustrated in the preceding example, one might expect non-relativistic dynamics to be applicable. In fact, Problem 6.12 shows that despite the presence of γ in Equations (6.39) and (6.40), for the parameters of satellites currently in operation, classical mechanics gives the same results. So, why the claim that relativity is needed in the operation of GPS? The answer is in the synchronization of the atomic clocks on board the satellites.[8]

If relativistic time dilation is not taken into account in the synchronization of the satellite clocks, an error Δt is introduced in the measurement of t_1' and t_2' of Equation (6.37). This is carried over to (6.38), resulting is an extra term of the order of $\delta = c\Delta t$ on its right-hand side. Propagating this term, Equations (6.39) and (6.40) will acquire an extra term of about the same magnitude. Therefore, x_R' will have an error of about $c\Delta t$. If we require an accuracy of one meter in x_R', then Δt should be less than $1/3 \times 10^8$ s or about 3 nanoseconds. Taking only special relativistic effects into account, the *fractional* difference between proper time and coordinate time is $\frac{1}{2}\beta^2$ [see Equation (2.26)] or about 8.5×10^{-11}, using the speed of Example 6.5.1. This may seem very small, but once the clocks run for only 40 seconds, the difference between the proper time and the coordinate time will be

$$\tfrac{1}{2}\beta^2 \Delta t = 8.5 \times 10^{-11} \times 40 = 3.4 \times 10^{-9} \text{ sec.}$$

Relativity (both special and general) is crucial in the workings of the GPS.

This means that every 40 seconds the error in location measurement will grow by more than one meter. This translates into more than 1.5 meters in one minute, or 90 meters in one hour. So, over the course of one day, without relativistic synchronization, the GPS locators will be off by more than 2 kilometers! Including general relativistic effects magnifies the error by a factor of about three. Therefore, relativity is crucial in the workings of the GPS.

6.6 FOUR-VELOCITY

Once the generalization to four spacetime dimensions is made and the "position vector" becomes $\mathbf{r} = (ct, x, y, z) \equiv (ct, \vec{r})$, the spacetime velocity turns into the **four-velocity**. The 4-velocity usually applies to a moving *object*, which up to now has been exemplified as a bullet carrying a subscript "*b*." The sparing discussion of moving objects in different RFs, having their own relative speed, has not caused any cluttering of notation so far. But since from now on we will be discussing dynamics in different RFs, we have to avoid the subscript. Therefore, we stick to the following notation:

Important notation to keep in mind in the sequel!

Note 6.6.1. (Notation) *The 3-velocity of an object is denoted by $\vec{v}$ and its fractional velocity by $\vec{\alpha}$ with the corresponding gamma factor γ_α. The fractional relative 3-velocity of two observers is, as before, denoted by $\vec{\beta}$ with the corresponding gamma factor γ.*

---

[8] Although I am considering only special relativistic effects, the effects of general relativity cannot be ignored. In fact, the latter are more (about twice more) important than the former. However, they are of the same type, namely time dilation. So, multiplying the result of the following analysis by a factor (of about 3) incorporates both corrections.

We shall avoid using 3-velocities (non-fractional) for the relative motion of two observers.

With this convention Equation (6.18) becomes

$$\vec{\alpha}' = \frac{\vec{\alpha} + \gamma\vec{\beta} + (\gamma - 1)\hat{\beta}\hat{\beta}\cdot\vec{\alpha}}{\gamma(1+\vec{\beta}\cdot\vec{\alpha})} = \frac{\vec{\alpha} + \gamma\vec{\beta}\left(1 + \frac{\gamma}{\gamma+1}\vec{\beta}\cdot\vec{\alpha}\right)}{\gamma(1+\vec{\beta}\cdot\vec{\alpha})} \tag{6.41}$$

and the 4-velocity can be conveniently defined as

$$\begin{aligned}\mathbf{u} &\equiv (u_0, \vec{u}) \equiv \frac{d\mathbf{r}}{ds} = \frac{1}{c}\frac{d\mathbf{r}}{d\tau} = \left(\frac{dt}{d\tau}, \frac{1}{c}\frac{d\vec{r}}{d\tau}\right) \\ &= \gamma_\alpha\,(1, \vec{\alpha}) \equiv \gamma_\alpha\,(1, \vec{v}/c) \equiv \gamma_\alpha\left(1, \frac{\dot{x}}{c}, \frac{\dot{y}}{c}, \frac{\dot{z}}{c}\right),\end{aligned} \tag{6.42}$$

where a dot represents differentiation with respect to the *coordinate* time t, and you have to use $dt = \gamma_\alpha d\tau$ [see Equation (2.4)] to get to the final result.

As in the two-dimensional case, the four-velocity is of unit length:

4-velocity has unit length.

$$\mathbf{u}\bullet\mathbf{u} = u_0^2 - |\vec{u}|^2 = \gamma_\alpha^2\,(1 - \vec{\alpha}\cdot\vec{\alpha}) = 1. \tag{6.43}$$

The four-velocity of an object in the object's rest frame is $(1, 0, 0, 0)$, i.e., it is a unit vector in the *time direction*.

If $\mathbf{u} = d\mathbf{r}/ds$ in the reference frame O, then it is $\mathbf{u}' = d\mathbf{r}'/ds'$ in the RF O'. The usefulness of the 4-velocity lies in the fact that $ds = ds'$. In particular, it transforms as a 4-vector, i.e., as in Equation (6.16). In fact, writing the latter as differentials

$$\begin{aligned}dt' &= \gamma(dt + \vec{\beta}\cdot d\vec{r}) \\ d\vec{r}\,' &= d\vec{r} + \gamma\vec{\beta}\left(dt + \frac{\gamma}{\gamma+1}\vec{\beta}\cdot d\vec{r}\right),\end{aligned} \tag{6.44}$$

and dividing both sides by the invariant ds, you obtain

$$\begin{aligned}u_0' &= \gamma(u_0 + \vec{\beta}\cdot\vec{u}) \\ \vec{u}\,' &= \vec{u} + \gamma\vec{\beta}\left(u_0 + \frac{\gamma}{\gamma+1}\vec{\beta}\cdot\vec{u}\right).\end{aligned} \tag{6.45}$$

The first equation simply expresses the gamma factor in O' in terms of quantities in O:

$$\gamma_{\alpha'} = \gamma\gamma_\alpha(1 + \vec{\beta}\cdot\vec{\alpha}). \tag{6.46}$$

To appreciate the power of 4-vector calculation, try to find the same formula using Equation (6.41) (see Problem 6.13).

6.7 FOUR-MOMENTUM

4-momentum defined.

The (kinematic) 4-velocity leads to the (dynamic) 4-momentum: Just multiply **u** by mc—the c is to give dimension to the 4-velocity. In a reference frame in which an object of mass m moves with velocity $\vec{v} = c\vec{\alpha}$, the **4-momentum p** is given by

$$\mathbf{p} \equiv (p_0, p_1, p_2, p_3) \equiv (p_0, \vec{p}) = mc\mathbf{u} = (mcu_0, mc\vec{u}) = (\gamma_\alpha mc, \gamma_\alpha m\vec{v}). \qquad (6.47)$$

Relativistic momentum.

Relativistic energy.

The space part of the 4-momentum $\vec{p} = \gamma_\alpha m\vec{v}$ gives ordinary Newtonian momentum when $|\vec{v}| << c$, because in that limit, $\gamma \approx 1$. Therefore, we call $\vec{p}$ the **relativistic momentum**. What about p_0? As in Section 5.2.1, you can show that p_0c is the **relativistic energy** E. So, $\mathbf{p} \equiv (p_0, \vec{p}) = (E/c, \vec{p})$, with

$$\vec{p} = mc\vec{u} = \gamma_\alpha m\vec{v} = \frac{mc\vec{\alpha}}{\sqrt{1-\alpha^2}} = \frac{m\vec{v}}{\sqrt{1-(v/c)^2}}, \qquad \alpha \equiv |\vec{\alpha}|$$

$$E = p_0c = mcu_0 = \gamma_\alpha mc^2 = \frac{mc^2}{\sqrt{1-\alpha^2}} = \frac{mc^2}{\sqrt{1-(v/c)^2}}. \qquad (6.48)$$

An important special case of this is the 4-momentum **p** of a particle in its rest frame in which $\vec{p} = \vec{0}$ and $\gamma = 1$. Then (6.47) yields

$$\mathbf{p} = (mc, \vec{0}) = (mc, 0, 0, 0). \qquad (6.49)$$

Hence, the rest energy is $E_0 = mc^2$, which you encountered in Chapter 5 as well.

The invariance of the length of a 4-vector tells you that $\mathbf{p} \bullet \mathbf{p}$ is a quantity that is independent of observers. In fact

$$\mathbf{p} \bullet \mathbf{p} = (mc)^2 \mathbf{u} \bullet \mathbf{u} = m^2c^2 \quad \text{or} \quad E^2 - |\vec{p}|^2c^2 = m^2c^4, \qquad (6.50)$$

which is the 4-dimensional version of Equation (5.10). The case of a massless particle such as a photon becomes

$$E^2 - |\vec{p}|^2c^2 = 0 \quad \text{or} \quad E = |\vec{p}|c. \qquad (6.51)$$

Note that from Equation (6.48),

$$\vec{p}/E = \vec{v}/c^2 \quad \text{for } any \text{ particle.} \qquad (6.52)$$

With this relation, you should convince yourself that (6.50) and (6.51) imply that a particle is massless *if and only if* it moves at light speed. This is another proof of the statement in Note 5.2.3.

Example 6.7.1. A particle has 4-momentum **p** relative to an observer O' whose 4-velocity is $\mathbf{u}'$. In the rest frame of this observer $\mathbf{u}' = (1, 0, 0, 0)$, and if $\mathbf{p} = (E'/c, \vec{p}')$ in this frame, then

$$\mathbf{p} \bullet \mathbf{u}' = E'/c.$$

Now consider another observer O with respect to whom the 4-momentum of the particle is $\mathbf{p} = (E/c, \vec{p})$ and the 4-velocity of O' is $\mathbf{u}' = (\gamma, \gamma\vec{\beta})$.[9] In the frame of O,

$$\mathbf{p} \bullet \mathbf{u}' = \gamma E/c - \gamma \vec{p} \cdot \vec{\beta}.$$

The invariance of the dot product now gives

$$E' = \gamma(E - c\vec{p} \cdot \vec{\beta}). \tag{6.53}$$

In the special case in which the particle is at rest with respect to O, $\vec{p} = 0$ and $E = mc^2$. This leads to

$$E' = \gamma mc^2 = \frac{mc^2}{\sqrt{1-\beta^2}},$$

which is the expected expression for the relativistic energy of a particle moving with velocity $\vec{v}$ relative to O'. ■

The 4-momentum was obtained from the 4-velocity by multiplication by a constant. Since the 4-velocity transforms like a 4-vector as in (6.45), the 4-momentum should do the same. All you have to do is multiply both sides of (6.45) by mc and note that $E = p_0 c$ and $E' = p'_0 c$. Then, with $c = 1$, you get

$$\begin{aligned} E' &= \gamma(E + \vec{\beta} \cdot \vec{p}) \\ \vec{p}\,' &= \vec{p} + \gamma\vec{\beta}E + (\gamma - 1)\hat{\beta}\hat{\beta} \cdot \vec{p} \\ &= \vec{p} + \gamma\vec{\beta}\left(E + \frac{\gamma}{\gamma+1}\vec{\beta} \cdot \vec{p}\right). \end{aligned} \tag{6.54}$$

The first equation was also derived in Example 6.7.1.

6.8 SECOND LAW OF MOTION

In Newtonian mechanics, we *define* force as the rate of change of momentum with time. The reason that this definition has survived is because it agrees with nature. This may be surprising to you, but it is a practice that has been going on ever since Galileo. In fact, before Galileo, acceleration was defined as the rate of change of velocity with distance! Galileo found it more convenient to *define* it as the rate of change of velocity with time.

In relativity, we encounter the situation anew. As you'll see, we'll have different options for defining acceleration and force. The ultimate decision maker is nature. Whatever agrees with experiments and observations will be the definition we'll have to choose.

6.8.1 Four-Acceleration

If you define the **four-acceleration** as the rate of change of the four-velocity with respect to proper time, then by differentiating (6.43) with respect to τ,

---

[9] Because O' is an *observer*, no α subscript is needed.

you can show that the inner product of the 4-velocity and the 4-acceleration of any object is zero. Let's summarize these two properties of the 4-velocity for future reference:

4-acceleration is perpendicular to 4-velocity.

$$\mathbf{u}\bullet\mathbf{u}=1, \qquad \mathbf{u}\bullet\mathbf{a}=0. \tag{6.55}$$

Example 6.8.1. A particle is moving in the two-dimensional spacetime of an inertial frame on a path given parametrically as

$$t(\sigma)=b\sinh(\sigma), \quad x(\sigma)=b\cosh(\sigma),$$

where σ is a dimensionless parameter. The differential of the particle's proper time is[10]

$$(d\tau)^2=(dt)^2-(dx)^2=b^2\cosh^2(\sigma)\,(d\sigma)^2-b^2\sinh^2(\sigma)\,(d\sigma)^2$$
$$=b^2\,(d\sigma)^2 \Rightarrow d\sigma=\frac{1}{b}d\tau$$

and $\sigma=\tau/b$. Thus, as a function of the proper time, the path becomes

$$t(\tau)=b\sinh(\tau/b), \quad x(\tau)=b\cosh(\tau/b).$$

The components of the (dimensionless) 4-velocity are

$$u_0=\frac{dt}{d\tau}=\cosh(\tau/b), \quad u_1=\frac{dx}{d\tau}=\sinh(\tau/b),$$

which satisfy $u_0^2-u_1^2=1$ as they should.

The acceleration of the particle has components

$$a_0=\frac{du_0}{d\tau}=\frac{1}{b}\sinh(\tau/b), \quad a_1=\frac{du_1}{d\tau}=\frac{1}{b}\cosh(\tau/b).$$

It is easily verified that $\mathbf{u}\bullet\mathbf{a}=0$ and that

$$\mathbf{a}\bullet\mathbf{a}=a_0^2-a_1^2=-\left(\frac{1}{b}\right)^2.$$

So, the particle has a uniform acceleration of $1/b$. The negative sign in the last equation is due to the fact that the magnitude of the acceleration has to be defined as $-\mathbf{a}\bullet\mathbf{a}=\vec{a}^{\,2}-a_0^2$, with the space part appearing as positive (so that when a_0 is absent, you get back the Newtonian acceleration). ■

It is convenient to have the components of the 4-acceleration given entirely in terms of quantities in a single frame. The zeroth component is calculated as follows:

$$a_0=\frac{du_0}{d\tau}=\gamma_\alpha\frac{du_0}{dt}=\gamma_\alpha\frac{d\gamma_\alpha}{dt}=\gamma_\alpha\frac{d}{dt}(1-\vec{\alpha}\cdot\vec{\alpha})^{-1/2}$$
$$=\gamma_\alpha\left[-\tfrac{1}{2}(-2\vec{\alpha}\cdot\dot{\vec{\alpha}})(1-\vec{\alpha}\cdot\vec{\alpha})^{-3/2}\right]=\gamma_\alpha^4\vec{\alpha}\cdot\dot{\vec{\alpha}},$$

where $\dot{\vec{\alpha}}\equiv d\vec{\alpha}/dt$. You should calculate the space component of the 4-acceleration, and put them together to get

$$a_0=\gamma_\alpha^4\vec{\alpha}\cdot\dot{\vec{\alpha}}$$
$$\vec{a}=\gamma_\alpha^2\left[\gamma_\alpha^2(\vec{\alpha}\cdot\dot{\vec{\alpha}})\vec{\alpha}+\dot{\vec{\alpha}}\right]. \tag{6.56}$$

[10] Remember $c=1$.

You should also show that $\mathbf{a} \bullet \mathbf{u} = 0$ (see Problem 6.19).

The second term in the square bracket of (6.56) is the classical acceleration, and that is what you get when $|\vec{\alpha}| \to 0$, i.e., in the Newtonian limit. However, the presence of the first term of the square bracket should be a reminder of how nontrivial the generalization from classical to relativistic dynamics can be.

The components of the rate of change of 4-velocity with respect to the *coordinate time* are used in the dynamics of relativistic particles:

$$\dot{u}_0 \equiv \frac{du_0}{dt} = \frac{du_0}{\gamma_\alpha d\tau} = \frac{1}{\gamma_\alpha} a_0 = \gamma_\alpha^3 \vec{\alpha} \cdot \dot{\vec{\alpha}}.$$

Similarly

$$\dot{\vec{u}} \equiv \frac{d\vec{u}}{dt} = \frac{d\vec{u}}{\gamma_\alpha d\tau} = \frac{1}{\gamma_\alpha} \vec{a} = \gamma_\alpha \left[\gamma_\alpha^2 (\vec{\alpha} \cdot \dot{\vec{\alpha}}) \vec{\alpha} + \dot{\vec{\alpha}} \right].$$

Collect these in a single equation:

4-acceleration and components of derivatives of 4-velocity with respect to *coordinate time*.

$$\begin{aligned} \dot{u}_0 &= \gamma_\alpha^3 \vec{\alpha} \cdot \dot{\vec{\alpha}}, & a_0 &= \gamma_\alpha \dot{u}_0 \\ \dot{\vec{u}} &= \gamma_\alpha^3 (\vec{\alpha} \cdot \dot{\vec{\alpha}}) \vec{\alpha} + \gamma_\alpha \dot{\vec{\alpha}}, & \vec{a} &= \gamma_\alpha \dot{\vec{u}}. \end{aligned} \tag{6.57}$$

The last equation in (6.57) is another reminder of the severe restrictions relativity imposes in defining the analogues of Newtonian quantities. You would think that if you differentiate the space part $\vec{u}$ of the 4-velocity with respect to the time of a particular RF you should get the space part $\vec{a}$ of the 4-acceleration. But that is not the case! You need a factor of γ_α also. If you dot the first equation on the second line of (6.57) with $\vec{\alpha}$ and do a little algebra, you get

$$\dot{u}_0 = \vec{\alpha} \cdot \dot{\vec{u}}. \tag{6.58}$$

Convince yourself that this is true!

Since 4-acceleration is a 4-vector, it transforms like all other 4-vectors:

$$\begin{aligned} a_0' &= \gamma (a_0 + \vec{\beta} \cdot \vec{a}) \\ \vec{a}\,' &= \vec{a} + \gamma \vec{\beta} a_0 + (\gamma - 1) \hat{\beta} \hat{\beta} \cdot \vec{a} \\ &= \vec{a} + \gamma \vec{\beta} \left(a_0 + \frac{\gamma}{\gamma + 1} \vec{\beta} \cdot \vec{a} \right). \end{aligned} \tag{6.59}$$

From these, (6.46), (6.57), and (6.58), you can obtain the transformation rules for $\dot{u}_0$ and $\dot{\vec{u}}$. For instance,

$$\dot{u}_0' = \frac{a_0'}{\gamma_{\alpha'}} = \frac{\gamma (a_0 + \vec{\beta} \cdot \vec{a})}{\gamma \gamma_\alpha (1 + \vec{\beta} \cdot \vec{\alpha})} = \frac{\dot{u}_0 + \vec{\beta} \cdot \dot{\vec{u}}}{1 + \vec{\beta} \cdot \vec{\alpha}} = \frac{\vec{\alpha} \cdot \dot{\vec{u}} + \vec{\beta} \cdot \dot{\vec{u}}}{1 + \vec{\beta} \cdot \vec{\alpha}}.$$

You should obtain the transformation rule for $\dot{\vec{u}}$ and put that and the above into a single equation:

$$\dot{u}_0' = \frac{\vec{\alpha} \cdot \dot{\vec{u}} + \vec{\beta} \cdot \dot{\vec{u}}}{1 + \vec{\beta} \cdot \vec{\alpha}}$$

$$\dot{\vec{u}}' = \frac{\dot{\vec{u}} + \gamma\vec{\beta}\vec{\alpha}\cdot\dot{\vec{u}} + (\gamma-1)\hat{\beta}\hat{\beta}\cdot\dot{\vec{u}}}{\gamma(1+\vec{\beta}\cdot\vec{\alpha})} = \frac{\dot{\vec{u}} + \gamma\vec{\beta}\left(\vec{\alpha}\cdot\dot{\vec{u}} + \frac{\gamma}{\gamma+1}\vec{\beta}\cdot\dot{\vec{u}}\right)}{\gamma(1+\vec{\beta}\cdot\vec{\alpha})}. \tag{6.60}$$

6.8.2 Four-Force

Newtonian mechanics defines force as the rate of change of momentum. Generalize this to relativity and define the **four-force** as

$$\mathbf{f} = \frac{d\mathbf{p}}{d\tau} = m\frac{d\mathbf{u}}{d\tau} = m\mathbf{a} = (ma_0, m\vec{a}), \tag{6.61}$$

where τ is the proper time of the moving object with mass m, four-velocity $\mathbf{u}$, and four-momentum $\mathbf{p}$. Let's explore the meaning of the components of $\mathbf{f}$. Write the 4-force as $\mathbf{f} \equiv (f_0, \vec{f})$ and use (6.57) and (6.58) to obtain

$$\begin{aligned} f_0 &\equiv ma_0 = m\gamma_\alpha\dot{u}_0 = m\gamma_\alpha\dot{\vec{u}}\cdot\vec{\alpha} = \gamma_\alpha\dot{\vec{p}}\cdot\vec{\alpha} \\ \vec{f} &\equiv m\vec{a} = m\gamma_\alpha\dot{\vec{u}} = \gamma_\alpha\dot{\vec{p}}, \end{aligned} \tag{6.62}$$

where $\vec{p} = m\vec{u}$ is the *relativistic* 3-momentum.

The second equation of (6.62) connects the 3-vector part of the 4-force and the Newtonian force, which is *defined* as the rate of change of three-momentum:

$$\vec{f} = \gamma_\alpha\dot{\vec{p}} \Rightarrow \dot{\vec{p}} = \frac{\vec{f}}{\gamma_\alpha} \equiv \vec{F}.$$

Therefore, in a particular inertial frame, Newton's second law holds:

$$\frac{d\vec{p}}{dt} = \dot{\vec{p}} = m\dot{\vec{u}} = \vec{F}. \tag{6.63}$$

In terms of this force, the 4-force $\mathbf{f}$ can be written as

$$\mathbf{f} = \gamma_\alpha(\vec{\alpha}\cdot\vec{F}, \vec{F}). \tag{6.64}$$

We have two 3-force definitions now: the one in (6.62) and the one in (6.63). Both of these lead to the nonrelativistic second law of motion when $\gamma_\alpha \to 1$. Which one is the right one? "To gamma or not to gamma, that's the question." Note 6.8.3 below shows that the 3-force ought to be defined as $\vec{F}$ and not as $\vec{f}$. And that's what I'll stick to from now on.

What about the zeroth component of f? What physical interpretation does it contain? Write it out and find out!

$$f_0 = \frac{dp_0}{d\tau} = \gamma_\alpha\frac{dp_0}{dt}.$$

The left-hand side of this equation is $\gamma_\alpha \vec{\alpha} \cdot \vec{F}$ by (6.64) and the derivative on the right-hand side is dE/dt by (6.48)—remember that $c = 1$. Hence, the zeroth component of (6.61) translates into

$$\frac{dE}{dt} = \vec{\alpha} \cdot \vec{F}, \tag{6.65}$$

which is identical to the classical equation if you recall that $E = mc^2 + KE$. You can also obtain this equation simply by multiplying (6.58) by m and using (6.63) and (6.48). Problem 6.23 shows a third way of obtaining (6.65).

Multiplying both sides of the second equation in (6.60) by m and using (6.63) immediately yields the transformation rule for the relativistic 3-force:

$$\vec{F}' = \frac{\vec{F} + \gamma\vec{\beta}\vec{\alpha} \cdot \vec{F} + (\gamma - 1)\hat{\beta}\hat{\beta} \cdot \vec{F}}{\gamma(1 + \vec{\beta} \cdot \vec{\alpha})} = \frac{\vec{F} + \gamma\vec{\beta}\left(\vec{\alpha} \cdot \vec{F} + \frac{\gamma}{\gamma + 1}\vec{\beta} \cdot \vec{F}\right)}{\gamma(1 + \vec{\beta} \cdot \vec{\alpha})}. \tag{6.66}$$

The components of the relativistic 3-force parallel and perpendicular to the direction of motion transform as

$$\begin{aligned} \vec{F}'_\parallel &= \vec{F}_\parallel + \frac{\vec{\beta}\vec{\alpha} \cdot \vec{F}_\perp}{1 + \vec{\beta} \cdot \vec{\alpha}} \\ \vec{F}'_\perp &= \frac{\vec{F}_\perp}{\gamma(1 + \vec{\beta} \cdot \vec{\alpha})}. \end{aligned} \tag{6.67}$$

In particular, if the force is parallel to the direction of motion, it does not change. Therefore, in one dimension, force does not change when you go from one RF to another (see also Problem 6.25).

Example 6.8.2. Let a constant force act on a particle of mass m in some inertial frame. What is the speed of the particle at time t if it starts from rest?

Equation (6.63) can be trivially integrated to give $\vec{p} = \vec{F}t$. Since the force is constant and the initial velocity is zero, the motion takes place in one dimension. So, you can ignore the vector sign and write

$$m\gamma_\alpha \alpha = Ft \quad \text{or} \quad \frac{\alpha}{\sqrt{1 - \alpha^2}} = \frac{Ft}{m}.$$

Square both sides, solve for α, and at the end restore the factors of c to obtain

$$\alpha = \frac{Ft/m}{\sqrt{1 + (Ft/m)^2}} \quad \text{or} \quad v = \frac{Ft/m}{\sqrt{1 + (Ft/mc)^2}}. \tag{6.68}$$

Note that for large t (i.e., when $Ft >> mc$), $\alpha \approx 1$ or $v \approx c$. However, the particle can never attain the speed of light no matter how long you wait. On the other hand, if $Ft << mc$, then the denominator is 1 and $v = (F/m)t$, which is the Newtonian speed of a particle moving with constant acceleration.

It is instructive to consider a particle having a constant acceleration of 10 m/s^2 (approximately Earth's gravitational acceleration). How long does it take to attain a speed of 0.999c? Over 21 years! (See Problem 6.22.) On the other hand, Newtonian mechanics requires under one year to achieve the same speed! ■

Sonya places a charge q motionless in a static electric field $\vec{E}$ in her coordinate system O and measures its force to be $\vec{F} = q\vec{E}$. Sam, sitting in the reference frame O', sees Sonya move with velocity $\vec{\beta}$ relative to him. Equation (6.66) gives the force that Sam measures as

$$\vec{F}' = \frac{\vec{F} + \frac{\gamma^2}{\gamma+1}\vec{\beta}(\vec{\beta}\cdot\vec{F})}{\gamma} = \frac{\vec{F}}{\gamma} + \frac{\gamma}{\gamma+1}\vec{\beta}(\vec{\beta}\cdot\vec{F}) \tag{6.69}$$

because $\vec{\alpha} = 0$. Use the vector identity (called the "bac cab rule" for obvious reasons)

$$\vec{a} \times (\vec{b} \times \vec{c}) = \vec{b}(\vec{a}\cdot\vec{c}) - \vec{c}(\vec{a}\cdot\vec{b}) \tag{6.70}$$

to express (6.69) as

$$\begin{aligned} \vec{F}' &= \frac{\vec{F}}{\gamma} + \frac{\gamma}{\gamma+1}\left(\vec{\beta}\times(\vec{\beta}\times\vec{F}) + \beta^2\vec{F}\right) \\ &= \frac{\vec{F}}{\gamma}\left(1 + \frac{\gamma^2\beta^2}{\gamma+1}\right) + \frac{\gamma}{\gamma+1}\vec{\beta}\times(\vec{\beta}\times\vec{F}) \\ &= \vec{F} + \frac{\gamma}{\gamma+1}\vec{\beta}\times(\vec{\beta}\times\vec{F}). \end{aligned} \tag{6.71}$$

Now, manipulate the second term of the last line:

$$\begin{aligned} \text{2nd term} &= \gamma\vec{\beta}\times(\vec{\beta}\times\vec{F}) - \frac{\gamma^2}{\gamma+1}\vec{\beta}\times(\vec{\beta}\times\vec{F}) \\ &= \gamma\vec{\beta}\times(\vec{\beta}\times\vec{F}) - \underbrace{\frac{\gamma^2\beta^2}{\gamma+1}}_{=\gamma-1}\underbrace{\hat{\beta}\times(\hat{\beta}\times\vec{F})}_{\hat{\beta}(\hat{\beta}\cdot\vec{F})-\vec{F}}, \end{aligned} \tag{6.72}$$

where $\hat{\beta}$ is a unit vector in the direction of $\vec{\beta}$. Put this result back in (6.71) to obtain

$$\vec{F}' = \gamma\vec{F} - (\gamma-1)\hat{\beta}(\hat{\beta}\cdot\vec{F}) + \gamma\vec{\beta}\times(\vec{\beta}\times\vec{F}). \tag{6.73}$$

Why did I go through so much trouble to write $\vec{F}'$ in this odd-looking form? For two reasons: (1) Sam's force has acquired a velocity-dependent part, which should be expected because the stationary sources of Sonya's field are now moving relative to Sam and should produce a magnetic field for him, contained in the last term of (6.73). (2) As you'll see later (Section 11.1), electric and magnetic fields transform into one another in a fashion similar to the transformation of time and space. In fact, a *static* electric field $\vec{E} = \vec{F}/q$ for Sonya transforms into an electric field $\vec{E}'$ for Sam given by the first two terms of (6.73) divided by q, and creates a magnetic field for him contained in the last term (divided by q):

$$\begin{aligned} \vec{E}' &= \gamma\vec{E} - (\gamma-1)\hat{\beta}(\hat{\beta}\cdot\vec{E}) \\ \vec{B}' &= \gamma(\vec{\beta}\times\vec{E}). \end{aligned} \tag{6.74}$$

Therefore, the static Coulomb force of Sonya transforms into the dynamic **Lorentz force**:

Lorentz force law.

$$\vec{F}' = q\left(\vec{E}' + \vec{\beta} \times \vec{B}'\right). \tag{6.75}$$

There is also a third reason for writing (6.73) as I did, and it has to do with why the 3-force $\vec{F}$ of Equation (6.63) was defined as $m\dot{\vec{u}}$ and not as $\vec{f} \equiv m\gamma_\alpha\dot{\vec{u}}$, the space component of the 4-force. After all, both of them would reduce to the Newtonian force in the limit of small velocity. Equation (6.75) is the reason!

Note 6.8.3. *The transformation of the field gives rise to the Lorentz force law only if the 3-force is defined as in Equation (6.63).*

6.8.3 Weirdness of Velocity-Dependent Mass

At the beginning of relativity, mass was assumed to be velocity-dependent and whatever multiplied acceleration in the second law of motion was considered mass. And this led to some strange behaviors! Substitute from (6.57) in (6.63) to obtain

$$\vec{F} = m\dot{\vec{u}} = m\left[\gamma_\alpha^3(\vec{\alpha}\cdot\dot{\vec{\alpha}})\vec{\alpha} + \gamma_\alpha\dot{\vec{\alpha}}\right].$$

Now resolve the acceleration $\dot{\vec{\alpha}}$ on the right-hand side and the force $\vec{F}$ on the left into their components along and perpendicular to velocity $\vec{\alpha}$. Then the equation above becomes

$$\begin{aligned}\vec{F}_\parallel + \vec{F}_\perp &= m\left\{\gamma_\alpha^3\left[\vec{\alpha}\cdot(\dot{\vec{\alpha}}_\parallel + \dot{\vec{\alpha}}_\perp)\right]\vec{\alpha} + \gamma_\alpha(\dot{\vec{\alpha}}_\parallel + \dot{\vec{\alpha}}_\perp)\right\}\\ &= m\gamma_\alpha^3\underbrace{(\vec{\alpha}\cdot\dot{\vec{\alpha}}_\parallel)\vec{\alpha}}_{=|\vec{\alpha}|^2\dot{\vec{\alpha}}_\parallel} + m\gamma_\alpha^3\underbrace{(\vec{\alpha}\cdot\dot{\vec{\alpha}}_\perp)}_{=0}\vec{\alpha} + m\gamma_\alpha\dot{\vec{\alpha}}_\parallel + m\gamma_\alpha\dot{\vec{\alpha}}_\perp\\ &= m\gamma_\alpha\dot{\vec{\alpha}}_\parallel(1+\gamma_\alpha^2|\vec{\alpha}|^2) + m\gamma_\alpha\dot{\vec{\alpha}}_\perp = m\gamma_\alpha^3\dot{\vec{\alpha}}_\parallel + m\gamma_\alpha\dot{\vec{\alpha}}_\perp,\end{aligned}$$

or

$$\begin{aligned}\vec{F}_\parallel &= m\gamma_\alpha^3\dot{\vec{\alpha}}_\parallel\\ \vec{F}_\perp &= m\gamma_\alpha\dot{\vec{\alpha}}_\perp.\end{aligned} \tag{6.76}$$

Now conduct two experiments. In the first experiment, let a one-kilogram particle move at $0.867c$ along your x-axis. At the moment that the particle reaches the origin, apply on it a force of 8 N along your x-axis. Equation (6.76) implies that $\dot{\vec{\alpha}}_\perp = 0$, and therefore, $\dot{\vec{\alpha}} = \dot{\vec{\alpha}}_\parallel$. It also implies that

$$8 = 1 \times \left(\frac{1}{\sqrt{1-0.867^2}}\right)^3 |\dot{\vec{\alpha}}| \Rightarrow |\dot{\vec{\alpha}}| = 1\ \text{m/s}^2.$$

Speed-dependent mass creates a logical fallacy!

In the second experiment, let the same particle move with the same speed along the same direction as before. But now, when the particle reaches the origin, apply on it a force of 8 N *along your y-axis*. In this case, $\dot{\vec{\alpha}}_{\parallel} = 0$, and therefore, $\dot{\vec{\alpha}} = \dot{\vec{\alpha}}_{\perp}$. Then

$$8 = 1 \times \left(\frac{1}{\sqrt{1 - 0.867^2}}\right) |\dot{\vec{\alpha}}| \Rightarrow |\dot{\vec{\alpha}}| = 4 \text{ m/s}^2.$$

So, acceleration is not proportional to the force, but depends on the direction in which you apply the force!

Another sign of strangeness comes from defining mass as the coefficient of acceleration in the second law of motion. Then, Equation (6.76) says that when a particle is accelerated in the direction of motion, its mass is $m\gamma_\alpha^3$, and if it is accelerated perpendicular to the direction of motion, its mass changes to $m\gamma_\alpha$!

Neither acceleration nor mass should depend on the direction of motion! Therefore, the idea of a velocity-dependent mass has to be abandoned. Once again, this time with feeling:[11]

> **Note 6.8.4.** *In relativity, as in classical physics, the mass of any object is unique and independent of its speed.*

6.9 PROBLEMS

6.1. Multiply the matrices in Equation (6.5) to obtain the three equations of (6.6). Solve these equations to find all matrix elements in terms of a_{11}.

6.2. Show that if $\vec{\beta}$ is along the x-axis, then (6.14) reduces to (6.13) and (6.16) to (6.12).

6.3. Derive Equation (6.17).

6.4. Derive Equation (6.18).

6.5. Observer O moves relative to observer O' with velocity $\vec{\beta}_1$. Observer O' moves relative to observer O'' with velocity $\vec{\beta}_2$. Therefore, observer O moves relative to observer O'' with some velocity $\vec{\beta}$. Using matrices—as given in Equations (6.14) and (6.15)—find the general relativistic law of addition of velocities. Hint: See Example 3.5.1 and note that calculating the two elements in the first column of the block form of the product $\Lambda_1\Lambda_2$ is sufficient to yield the answer.

6.6. In Fizeau's experiment (see Example 3.5.2), light moves *perpendicular* to the direction of motion of the transparent medium.

[11] See Section 5.2.4 and page 218.

(a) Show that

$$v' = \sqrt{\left(\frac{c}{n}\right)^2 + v^2\left(1 - \frac{1}{n^2}\right)}.$$

(b) Show that for $v << c$, a complete drag—in which the classical LAV holds—yields

$$v' = \frac{c}{n} + \frac{nv^2}{2c}.$$

(c) Show that relativistically, for low velocity, the drag is not complete but is smaller by $v^2/(2nc)$.

6.7. Derive the velocity (6.23) directly from (6.18).

6.8. Use (6.25) and the trigonometric identity

$$\tan(\tfrac{1}{2}\varphi') = \frac{\sin\varphi'}{1 + \cos\varphi'}$$

to show that Equation (6.26) holds.

6.9. The headlights of a car send light only in the forward *hemisphere* as seen in the rest frame of the car. Show that the maximum angle as seen by a ground observer is $\tan^{-1}[1/(\gamma\beta)]$, or $\cos^{-1}\beta$.

6.10. Use Figure 6.1(d) and the law of sines to obtain

$$\frac{c}{\cos\theta'} = \frac{c\beta}{\sin(\theta' - \theta)}.$$

Show that this can also be written as

$$\tan\theta' = \frac{\beta + \sin\theta}{\cos\theta},$$

which is the non-relativistic version of (6.28).

6.11. Derive Equation (6.29). Hint: Substitute $\theta' = \theta + \epsilon$ on the left-hand side of (6.28), and expand both sides keeping only the first powers of ϵ and β; then equate the small terms on both sides.

6.12. From a classical perspective, since satellite S_2 of Figure 6.3 is moving away from R, the light it emits moves slower than c while the light from S_1 moves faster. To simplify the classical derivation of (6.39) and (6.40), let $y = 0$. Let d_1 and d_2 be the distances of S_1 and S_2 from R, respectively. Since time is universal classically, the emission of signals occurs at the same time according to O' as well as O. Use this to show that

$$\frac{d_1}{c + v} = \frac{d_2}{c - v}.$$

From this plus $d_1 + d_2 = L$, find d_1 and show that

$$x'_R = x'_1 + \frac{1}{2}L(1+\beta),$$

which is (6.40) for $\gamma \approx 1$ and $y = 0$. From this you can also find (6.39).

6.13. Show that

(a) Equation (6.41) reduces to the non-relativistic law when all velocities are small and to (3.26) when all velocities are in the same direction.
(b) Take the dot product of both sides of Equation (6.41) to find $|\vec{\alpha}'|^2 = \vec{\alpha}' \cdot \vec{\alpha}'$ in terms of unprimed quantities.
(c) From (b) calculate $\gamma_{\alpha'}$ and verify Equation (6.46).
(d) Show also that if $\vec{\alpha}$ is the velocity of light, so that $\vec{\alpha} \cdot \vec{\alpha} = 1$, then $\vec{\alpha}' \cdot \vec{\alpha}' = 1$ as well.

6.14. What are the coordinates of a particle's 4-velocity in its own rest frame? Use this and Equation (6.45) to obtain (6.42), the particle's 4-velocity in an arbitrary frame.

6.15. Provide the details of the proof of the statement: *A particle is massless if and only if it moves at light speed.*

6.16. Apply (6.53) to a photon moving in the $\vec{\beta}$-direction and use $|\vec{p}| = E/c$ to show that

$$E' = \sqrt{\frac{1-\beta}{1+\beta}}\,E.$$

Now use $E = hc/\lambda$ to find a formula for the relativistic Doppler shift.

6.17. Use (6.52)—which holds for any particle in any frame—and (6.54) to find the general relativistic law of addition of velocities (6.41).

6.18. Show that the 4-acceleration is orthogonal to the 4-velocity.

6.19. Differentiate the space component of the 4-velocity with respect to τ to get the second equation in (6.56). Now show directly that $u_0 a_0 - \vec{u} \cdot \vec{a} = 0$.

6.20. Derive Equation (6.58).

6.21. Derive Equation (6.60).

6.22. How long does it take a particle to attain a speed of $0.999c$, if its acceleration is 10 m/s^2? What is the answer based on Newtonian mechanics? How do the answers change if the ultimate speed of the particle is $0.99999c$?

6.23. Differentiate both sides of $E^2 = \vec{p} \cdot \vec{p} + m^2$ with respect to time and use $\vec{p} = E\vec{\alpha}$ to obtain (6.65).

6.24. Write $\vec{F}$ as $\vec{F}_\parallel + \vec{F}_\perp$ on both sides of (6.66) and derive (6.67).

6.25. Confine yourself to one dimension.

(a) Use $E^2 = p^2 + m^2$ for a moving object of mass m to show that $dE/dt = F\alpha$, where $F = dp/dt$ is the force and α is the speed of the object.

(b) Lorentz transform E, p, t, and x to another reference frame and use the result obtained in (a) to show that *in one dimension* force is invariant, i.e., it does not change when you go from one RF to another.

6.26. Define the power P consumed by an object as $P \equiv dE/dt = \vec{\alpha} \cdot \vec{F}$, where E is the relativistic energy of the object and $\vec{F}$ and $\vec{\alpha}$ are, respectively, the force acting on the object (the source of the power) and the velocity of the object. If P and P' are powers in the reference frames O and O', respectively, then

$$P' = \frac{P + \vec{\beta} \cdot \vec{F}}{1 + \vec{\beta} \cdot \vec{\alpha}},$$

where $\vec{\beta}$ is the velocity of O relative to O'. Prove this in three different ways:

(a) by writing the Lorentz transformation of dE' and dt' and dividing the first by the second;

(b) by noting that $P' = \vec{\alpha}' \cdot \vec{F}'$ and using the relativistic law of addition of velocities and the transformation rule for the three-force $\vec{F}$;

(c) by Lorentz transforming the zeroth component of the 4-force.

6.27. Provide the missing steps of (6.71) and (6.72) to arrive at (6.73).

CHAPTER 7

Relativistic Photography

Photography is the art of collecting photons emitted by an object on a light sensitive plate. Because in classical physics all objects move much slower than light, c can be assumed infinite and the flight time of photons from the object to the camera can be completely ignored. We can therefore assume that (barring the limitation on the shutter speed that causes the blurring of the image) the photograph of an object is not affected by its motion. In other words, we are tacitly assuming that the moving object is *instantaneously* frozen when its picture is taken.

7.1 HISTORY

The assumption of the instantaneous freezing of the object being photographed was carried over to relativity when the contraction of length was discovered. It was assumed that this effect was actually visualizable. Einstein in his 1905 paper writes, "A rigid body which, measured in a state of rest, has the form of a sphere, therefore has in a state of motion—viewed from a stationary system—the form of an ellipsoid of revolution."[Einstein 52, p. 48] This quote leaves the impression that a sphere would be *seen* as an ellipsoid of revolution with a shorter axis along the direction of motion. Lorentz even stated clearly in 1922 that the contraction could actually be *photographed*. [Terrell 59]

As early as 1924, however, Lampa, [Lampa 24] taking into account the motion of the photons as they left the moving object and were collected by a camera, suggested that such a contraction was not visualizable. But this idea remained largely unnoticed until 1959 when Penrose [Penrose 59] and Terrell [Terrell 59], using aberration formulas, showed independently that the photograph of a sphere remained a circle regardless of its speed. Actually, what Penrose and Terrell showed was that the *cone* converging on a *point observer* is a circular cone. As you'll see later, this does not necessarily mean that the *image* on a photographic plate is circular.

Sphere is a highly symmetric object. What if the moving object is not that symmetric? Terrell [Terrell 59] and Weisskopf [Weisskopf 60] analyzed the problem

Special Relativity. DOI:10.1016/B978-0-12-810411-8.00007-9

of the picture of a *distant* cube taken by a camera whose line of sight was perpendicular to one of the faces of the cube as well as to its direction of motion. They concluded that the photograph would be that of a rotated cube because of the exposure of the trailing side of the cube to the camera (see Problem 7.1).

A careful examination of the problem in the 1970s using the full machinery of Lorentz transformation demonstrated numerically that the image of a (not necessarily distant) relativistically moving sphere is an ellipse elongated along the direction of motion [Scott 70] and that the image of a cube is in general a distorted rectangle elongated in the direction of motion whose sides perpendicular to the direction of motion are curved. [Hickey 79]

7.2 PHYSICS OF PHOTOGRAPHY

Relativistic photography is a great medium in which to learn special relativity in general, and Lorentz transformation, in particular. Here you have light, an intrinsically relativistic object, emitted by a fast moving source, which forms an image in a camera in another frame, creating a situation that begs for Lorentz transformation!

In this chapter, I'll show how the finite velocity of photons and relativistic effects can combine to cause significant distortion of the image of a relativistically moving object on a photographic plate when compared to the same image taken by a camera that is stationary in the RF of the object. For example, ordinarily when the line joining a point of an object to the shutter of the camera is parallel to the photographic plate, that point does not form an image on the plate. However, if the object is moving relativistically, such an extreme case can occur even for a point whose (instantaneous) line to the aperture is *not parallel* to the photographic plate (see Remark C.1.1 on page 342 and Figure C.2).

To determine the shape of the image of an object, you need to know how to map the three-dimensional location of the sources of the photons captured by a camera onto the two-dimensional plane of its photographic plate. The details of the formation of a photograph can be complicated by the structure of the lens system of the camera. To simplify my discussion, I'll assume that

Note 7.2.1. *The lens is just a pinhole and the photograph of an object is determined by two things: (a) the shape of the cone of rays diverging from the pinhole onto the photographic plate, which has the same shape as the cone of rays converging onto the pinhole from the object, and (b) the angle at which the photographic plate cuts the axis of this cone.*

The outline of the image is the curve formed by the intersection of the cone and the photographic plate.

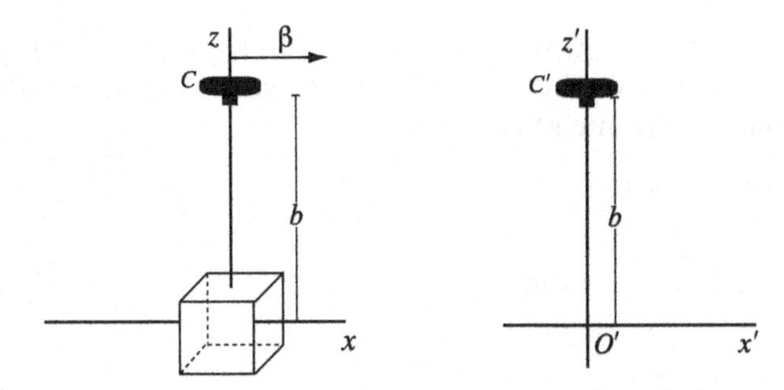

FIGURE 7.1 A snapshot of the reference frame O approaching O' some time in the past. The center of the object (a cube) shown is permanently fixed at the origin of O. At time $t = t' = 0$ both the origins and the cameras C and C' coincide. The pictures are also assumed to be taken at $t = t' = 0$.

I'll let O be the RF in which objects to be photographed are located and I'll place camera C in that RF. More specifically, I'll place the pinhole of C at P_0 with coordinates $(0, 0, b)$ in O; I'll let O move with speed β relative to another frame O' in the positive direction of the x'-axis (see Figure 7.1); and I'll assume that the corresponding axes of O and O' are parallel and that at $t = t' = 0$ the two origins coincide. I'll place a camera C' in O' whose pinhole is also at $(0, 0, b)$, so that at $t = t' = 0$, the cameras also coincide (because the coordinates perpendicular to the direction of motion don't change).[1] In all cases, the pictures are taken by C and C' at $t = t' = 0$. As usual, I'll set $c = 1$.

It is important to clarify a point that is oftentimes overlooked. If two events located *equidistant* from one camera occur simultaneously, the rays of light from those events reach *only that camera* simultaneously. The same two events cannot be simultaneous in the RF of the other camera. Since the second camera picks up the rays of light simultaneously by assumption, the earlier event must have occurred at a point farther away from the camera for an approaching object and closer to the camera for a receding object. Therefore, the two events *cannot be equidistant from the second camera.* I'll come back to this important point at the end of Section 7.5 where I talk about the "true" length of an object, for which the notion of simultaneity becomes crucial.

How to record length contraction.

Is there a way to *actually* observe (or record) the length contraction of a rod? Place the rod in the xy-plane parallel to the x-axis with $y = y_0$, and let it move along its length. Assume that the entire $x'y'$-plane is made of a light-sensitive material. To investigate conditions under which we see or record the length contraction of the rod, consider two events E_1 and E_2, the emission of light from the two ends of the rod, with coordinates $(x_1, y_0, 0, ct_1)$ and $(x_2, y_0, 0, ct_2)$

[1] If you are not comfortable with the idea of the cameras coinciding, I'll perform two experiments. In one, C takes the picture (at any time), and in the other, C is removed and C' takes the picture when the origins coincide.

in O and coordinates $(x'_1, y_0, 0, ct'_1)$ and $(x'_2, y_0, 0, ct'_2)$ in O'. The distances and the time intervals between the two events as measured by the two observers O and O' are related by (remember $c = 1$)

$$\begin{aligned} t'_2 - t'_1 &= \gamma[\beta(x_2 - x_1) + t_2 - t_1] \\ x'_2 - x'_1 &= \gamma[x_2 - x_1 + \beta(t_2 - t_1)]. \end{aligned} \tag{7.1}$$

The length is contracted if (and only if)

$$x'_2 - x'_1 = (x_2 - x_1)/\gamma.$$

Substituting this in the second equation of (7.1) yields

$$\left(1/\gamma^2 - 1\right)(x_2 - x_1) = \beta(t_2 - t_1),$$

or

$$t_2 - t_1 = -\beta(x_2 - x_1).$$

Inserting this in the first equation of (7.1) yields $t'_2 = t'_1$. Therefore,

Note 7.2.2. *The contraction of the distance between two events in O is* **recorded** *in O' if and only if the two events occur simultaneously according to O'.*

The Note implies that the length of the image of the rod in the light-sensitive $x'y'$-plane is contracted by the Lorentz factor γ only if the photons forming the image of the two ends are emitted simultaneously according to O'.[2] If the image is to represent the *actual* contracted length of the rod (as opposed to, say, its photograph in a camera), it is crucial that these photons are captured *immediately* by the $x'y'$-plane so I don't have to worry about their travel times, (thus the assumption that the rod is *in* the light-sensitive $x'y'$-plane).

The *photograph* of the rod in a camera (or its image in our eyes) records the photons emanating from the two ends of the rod (as well as the points in between) that reach the pinhole of the camera *simultaneously*. What exactly is the photograph the image of? To arrive at the aperture simultaneously, the photons from the trailing end of an *approaching* rod must have been emitted earlier than the leading end. Thus, the trailing photons make a mark in the $x'y'$-plane, the rod moves some distance, then the leading photons make their marks. The distance between these marks, which has nothing to do with the length of the rod, is the object being photographed and is larger than the rest length of the rod. For a receding rod the leading end emits the photons earlier. Therefore, the distance between the marks is shorter.

Keep in mind that these are all relativistic effects. If the object moves slowly, so that the laws of classical physics are relevant, all three images in the $x'y'$-plane

---

[2] This is the rigorous proof of the observation made on page 11.

(the contracted image, the longer image of an approaching rod, and the shorter image of a receding rod) are the same, because in classical physics the limit of $c \to \infty$ applies and whether the photons are captured simultaneously by the $x'y'$-plane or by the photographic plate, the marks made in the $x'y'$-plane are all the same.

7.3 PHOTOGRAPH OF TWO POINTS

The shape of the cone of Note 7.2.1 is determined by the direction of the velocity vectors of the image-forming photons originating from the surface of the object. Since this shape is specified by the angle between velocity vectors of the rays, I'll look at the "image" of two source points first.

Two point sources of light, P_1 and P_2, are at rest in reference frame O. At time $t = t' = 0$, the two cameras C and C' take pictures of P_1 and P_2, which are the spatial locations of the two *past events*, E_1 and E_2, of the emission of light. The angle subtended at the pinhole of each camera by the two points determines the photograph of the "object" consisting of P_1 and P_2. What is the relation between α_{12}, the angle subtended by P_1 and P_2 at C, and α'_{12}, the corresponding angle at C'? I'll find this relation in two ways: first by using the relativistic law of addition of velocities and then by invoking Lorentz transformations.

To avoid repeating the formulas for P_1 and P_2, I write either of them as P_i, where $i = 1, 2$, and let (c_{ix}, c_{iy}, c_{iz}), and $(c'_{ix}, c'_{iy}, c'_{iz})$ be the components of the photon velocities coming from P_i in the two RFs. Then, with $|\vec{c}| = 1$, I can express the angle formed by P_1 and P_2 at C as

$$\cos\alpha_{12} = c_{1x}c_{2x} + c_{1y}c_{2y} + c_{1z}c_{2z}. \tag{7.2}$$

The components of the velocity of photons from P_i in O' are

$$\begin{aligned} c'_{ix} &= \frac{c_{ix} + \beta}{1 + \beta c_{ix}} = \frac{c_{ix} + \beta}{1 - \vec{\beta}\cdot\hat{e}_i} \\ c'_{iy} &= \frac{c_{iy}}{\gamma(1 + \beta c_{ix})} = \frac{c_{iy}}{\gamma(1 - \vec{\beta}\cdot\hat{e}_i)}, \qquad i = 1, 2, \\ c'_{iz} &= \frac{c_{iz}}{\gamma(1 + \beta c_{ix})} = \frac{c_{iz}}{\gamma(1 - \vec{\beta}\cdot\hat{e}_i)}, \end{aligned} \tag{7.3}$$

with $\hat{e}_i = (-c_{ix}, -c_{iy}, -c_{iz})$ a unit vector directed from the pinhole to P_i. Now I substitute these in the formula equivalent to Equation (7.2):

$$\cos\alpha'_{12} = c'_{1x}c'_{2x} + c'_{1y}c'_{2y} + c'_{1z}c'_{2z} \tag{7.4}$$

and obtain

$$\begin{aligned} \cos\alpha'_{12} &= \frac{\gamma^2(c_{1x} + \beta)(c_{2x} + \beta) + c_{1y}c_{2y} + c_{1z}c_{2z}}{\gamma^2(1 - \vec{\beta}\cdot\hat{e}_1)(1 - \vec{\beta}\cdot\hat{e}_2)} \\ &= \frac{\gamma^2(c_{1x} + \beta)(c_{2x} + \beta) + \cos\alpha_{12} - c_{1x}c_{2x}}{\gamma^2(1 - \vec{\beta}\cdot\hat{e}_1)(1 - \vec{\beta}\cdot\hat{e}_2)}. \end{aligned}$$

I can write the numerator as

$$\begin{aligned}\text{Num} &= (\gamma^2-1)c_{1x}c_{2x} + \gamma^2\beta c_{1x} + \gamma^2\beta c_{2x} + \gamma^2\beta^2 + \cos\alpha_{12}\\ &= \gamma^2\beta^2 c_{1x}c_{2x} + \gamma^2\beta c_{1x} + \gamma^2\beta c_{2x} + (\gamma^2-1) + \cos\alpha_{12}\\ &= \gamma^2(1+\beta c_{1x})(1+\beta c_{2x}) - 1 + \cos\alpha_{12}\end{aligned}$$

if I use the useful identity $\gamma^2 - 1 = \gamma^2\beta^2$. Dividing this by the denominator gives me

$$\cos\alpha'_{12} = 1 - \frac{1-\cos\alpha_{12}}{\gamma^2(1-\vec{\beta}\cdot\hat{e}_1)(1-\vec{\beta}\cdot\hat{e}_2)}, \tag{7.5}$$

Relation between angles formed at the two pinholes.

because $\beta c_{1x} = -\vec{\beta}\cdot\hat{e}_1$ and $\beta c_{2x} = -\vec{\beta}\cdot\hat{e}_2$. This can be rewritten as

$$\sin(\alpha'_{12}/2) = \frac{\sin(\alpha_{12}/2)}{\gamma\sqrt{(1-\vec{\beta}\cdot\hat{e}_1)(1-\vec{\beta}\cdot\hat{e}_2)}}. \tag{7.6}$$

Now I'll show how to get this relation by employing LTs. The two events E_1 and E_2 of the emission of light signals from P_1 and P_2 have coordinates (t_i, x_i, y_i, z_i) and (t'_i, x'_i, y'_i, z'_i), $i = 1, 2$ in O and O', respectively. Denote the position vector of the pinhole of either camera by $\mathbf{r}_0 = \langle 0, 0, b\rangle$, with $b > 0$, and those of P_i, by $\mathbf{r}_i$ and $\mathbf{r}'_i$, $i = 1, 2$, and note that $t = 0 = t'$ at the reception of the signals at the cameras.[3] Then (again with $c = 1$) you should get $t_i = -|\mathbf{r}_i - \mathbf{r}_0|$ and $t'_i = -|\mathbf{r}'_i - \mathbf{r}_0|$. The LTs connecting the two events, with $i = 1, 2$, are therefore

$$\begin{aligned} t'_i &= -|\mathbf{r}'_i - \mathbf{r}_0| = \gamma(-|\mathbf{r}_i - \mathbf{r}_0| + \beta x_i)\\ x'_i &= \gamma(x_i - \beta|\mathbf{r}_i - \mathbf{r}_0|)\\ y'_i &= y_i\\ z'_i &= z_i.\end{aligned} \tag{7.7}$$

The angle between the lines joining camera C in O to the locations of E_1 and E_2 is given by

$$\cos\alpha_{12} = \frac{(\mathbf{r}_1-\mathbf{r}_0)\cdot(\mathbf{r}_2-\mathbf{r}_0)}{|\mathbf{r}_1-\mathbf{r}_0|\,|\mathbf{r}_2-\mathbf{r}_0|} = \frac{x_1x_2 + y_1y_2 + (z_1-b)(z_2-b)}{|\mathbf{r}_1-\mathbf{r}_0|\,|\mathbf{r}_2-\mathbf{r}_0|}, \tag{7.8}$$

and the corresponding angle for camera C' in O' is

$$\cos\alpha'_{12} = \frac{(\mathbf{r}'_1-\mathbf{r}_0)\cdot(\mathbf{r}'_2-\mathbf{r}_0)}{|\mathbf{r}'_1-\mathbf{r}_0|\,|\mathbf{r}'_2-\mathbf{r}_0|} = \frac{x'_1x'_2 + y'_1y'_2 + (z'_1-b)(z'_2-b)}{|\mathbf{r}'_1-\mathbf{r}_0|\,|\mathbf{r}'_2-\mathbf{r}_0|}. \tag{7.9}$$

Substitute Equations (7.7) and (7.8) in (7.9) and follow the steps similar to the ones that led to (7.5) to obtain (see Problem 7.3)

$$\sin(\alpha'_{12}/2) = \frac{\sin(\alpha_{12}/2)}{\gamma\sqrt{\left(1-\frac{\beta x_1}{|\mathbf{r}_1-\mathbf{r}_0|}\right)\left(1-\frac{\beta x_2}{|\mathbf{r}_2-\mathbf{r}_0|}\right)}}, \tag{7.10}$$

[3] In this chapter, I'll be using bold face for 3-vectors!

which is identical to (7.6) because

$$\hat{e}_i = \frac{\mathbf{r}_i - \mathbf{r}_0}{|\mathbf{r}_i - \mathbf{r}_0|} = \frac{\langle x_i, y_i, z_i - b\rangle}{|\mathbf{r}_i - \mathbf{r}_0|}, \quad i = 1, 2,$$

and $\vec{\beta} = \langle \beta, 0, 0\rangle$.

I have to emphasize that the LT approach contains more information than the velocity derivation. The former gives the actual location of the points from which the rays originate, while the latter approach specifies only the direction of the rays. I will therefore use LT to determine the shape of the objects being photographed.

Occasionally, we'll need the values of angles when an object is at a large distance from the camera, i.e., when the object's x-coordinate is large. If P_1 and P_2 are two points of the object determining its "size," then α_{12} and α'_{12} determine its angular size. Since only the x-coordinate is large, for $i = 1, 2$, I get

$$|\mathbf{r}_i - \mathbf{r}_0| \to |x_i|.$$

Substituting this in (7.10) and using the small angle approximation, you can obtain

$$\frac{\alpha'_{12}}{2} = \frac{\alpha_{12}/2}{\gamma\sqrt{(1 - \beta x_1/|x_1|)(1 - \beta x_2/|x_2|)}}$$

or

$$\frac{\alpha'_{12}}{2} = \sqrt{\frac{1 - \beta^2}{(1 - \beta x_c/|x_c|)^2}}\,\frac{\alpha_{12}}{2}$$

where x_c is any value between x_1 and x_2 and $x_1 \approx x_c \approx x_2$. Hence,

Angle between two pints when they are far away.

$$\alpha'_{12} = \sqrt{\frac{1 + \beta x_c/|x_c|}{1 - \beta x_c/|x_c|}}\,\alpha_{12}, \tag{7.11}$$

because $x_c/|x_c| = \pm 1$ and thus $\beta^2 = (\beta x_c/|x_c|)^2$.

Intuitive physical interpretation of Equation (7.11).

Since angles are related to sizes of the photographs of objects, Equation (7.11) has an intuitive physical interpretation. It implies that as a distant object approaches camera C' from $-\infty$ (with $\beta > 0$) or $+\infty$ (with $\beta < 0$), its photograph is smaller compared to its photograph in camera C stationary relative to the object, and larger when it recedes from C'. This is consistent with the fact that C' collects photons that are emitted *earlier*. Therefore, on approach, these photons were emitted farther away, while on recession, they were emitted at a closer distance.

The equivalence of (7.10) and (7.6) allows me to use LTs to determine the shape of the cone formed by the locus of the source points on the object that create the image on the photographic plate. The boundary of the image is then the curve obtained by the intersection of the photographic plate with this cone.

The coordinates (u, v) of a point of this curve are given by a formula which maps a three-dimensional source point P with coordinates (x, y, z) onto a two-dimensional photographic plate. I have derived this formula in Appendix C.1 for the case in which the camera is focused on a point Q with coordinates $(x_q, 0, 0)$. I reproduce the result here:

Obtaining the two coordinates of the image of a point from its three coordinates.

$$u' = \frac{bx' + x_q'(z-b)}{x_q'x' - b(z-b)} = \frac{b\gamma\left[x - \beta\sqrt{x^2+y^2+(z-b)^2}\right] + x_q(z-b)/\gamma}{x_q\left[x - \beta\sqrt{x^2+y^2+(z-b)^2}\right] - b(z-b)}$$

$$v' = \frac{y\sqrt{x_q'^2+b^2}}{x_q'x' - b(z-b)} = \frac{y\sqrt{x_q^2/\gamma^2+b^2}}{x_q\left[x - \beta\sqrt{x^2+y^2+(z-b)^2}\right] - b(z-b)}, \tag{7.12}$$

where $x_q' = x_q/\gamma$ and u' and v' are the coordinates in the plate of camera C'. You can obtain the coordinates (u, v) for camera C by setting $\beta = 0$. There is a caveat associated with these formulas:

Note 7.3.1. *Equation (7.12) is actually valid for a plane that is* outside *the camera. This avoids the unnecessary complication and confusion caused by the inversion of the image through the pinhole.*

7.4 PHOTOGRAPH OF A ROD

Rod oriented along the x-axis.

In this section, I'm going to look at the picture of a rod with an eye on the photograph of a cube, whose edges are parallel to the coordinate axes. Place a rigid straight rod of length $2a$ in O parallel to the x-axis and let (x, y_0, z_0) be the coordinates of an arbitrary point P of the rod with y_0 and z_0 constant. If P is to be photographed by the cameras C and C' at $t = t' = 0$, then the event E of the emission of light must have occurred in O at

$$t_P = -\sqrt{x^2 + y_0^2 + (b - z_0)^2}. \tag{7.13}$$

Therefore, P has space coordinates

$$x' = \gamma\left(x - \beta\sqrt{x^2 + y_0^2 + (b - z_0)^2}\right), \quad y' = y_0, \quad z' = z_0$$

in O'. Since the y'- and z'-coordinates are constant, the rod is parallel to the x'-axis in O', as it is in O. It doesn't have the same length, however.

What is the shape of the *photograph* of the rod? Substituting (x, y_0, z_0) in (7.12) and eliminating x in the two equations to solve v' in terms of u' gives

$$v' = \frac{x_q y_0}{\gamma\sqrt{x_q^2/\gamma^2 + b^2}\,(z_0 - b)}u' - \frac{b y_0}{\sqrt{x_q^2/\gamma^2 + b^2}\,(z_0 - b)}, \tag{7.14}$$

which is the equation of a straight line in the (u', v') plane of the photographic plate. The nonzero slope should be expected due to the mapping of a three-dimensional object onto a two dimensional plane. Even when the camera is stationary ($\beta = 0$, $\gamma = 1$) the line has a nonzero slope just as the image of the two tracks of a railroad appears nonparallel.

Now reorient the rod so that it is parallel to the y-axis. Let (x_0, y, z_0) be the coordinates of an arbitrary point of the rod with x_0 and z_0 constant. The point has a time coordinate given by an expression similar to (7.13), and space coordinates

Rod oriented along the y-axis.

$$x' = \gamma\left(x_0 - \beta\sqrt{x_0^2 + y^2 + (b - z_0)^2}\right), \quad y' = y, \quad z' = z_0$$

in O'. This is the equation of one half of a hyperbola in the $x'y'$-plane of O'. In fact, transferring γx_0 to the left-hand side and squaring, you should have no difficulty showing that

$$\frac{(x' - \gamma x_0)^2}{\gamma^2\beta^2\left[x_0^2 + (z_0 - b)^2\right]} - \frac{y'^2}{x_0^2 + (z_0 - b)^2} = 1. \tag{7.15}$$

To find the *photograph* of the rod, substitute (x_0, y, z_0) in (7.12) and obtain

$$u'(y) = \frac{b\gamma\left[x_0 - \beta\sqrt{x_0^2 + y^2 + (z_0 - b)^2}\right] + x_q(z_0 - b)/\gamma}{x_q\left[x_0 - \beta\sqrt{x_0^2 + y^2 + (z_0 - b)^2}\right] - b(z_0 - b)}$$

$$v'(y) = \frac{y\sqrt{x_q^2/\gamma^2 + b^2}}{x_q\left[x_0 - \beta\sqrt{x_0^2 + y^2 + (z_0 - b)^2}\right] - b(z_0 - b)}, \tag{7.16}$$

which is the parametric equation of a curve in the $u'v'$-plane with y as the parameter. In principle, I can eliminate y and write v' directly as a function of u', but the algebra will give me a headache! However, for $x_q = 0$, i.e., when the camera points to the origin, the calculation is much easier and the resulting curve is the hyperbola

A line perpendicular to the direction of motion is not photographed as a straight line segment!

$$\frac{\left(u' + \dfrac{\gamma x_0}{z_0 - b}\right)^2}{\gamma^2\beta^2\left[\dfrac{x_0^2}{(z_0 - b)^2} + 1\right]} - \frac{v'^2}{\dfrac{x_0^2}{(z_0 - b)^2} + 1} = 1, \tag{7.17}$$

in agreement with (7.15) and the fact that when C' points to the origin, except for a scale factor, the pair (u', v') is the same as (x', y'). I'll let you derive Equation (7.17).

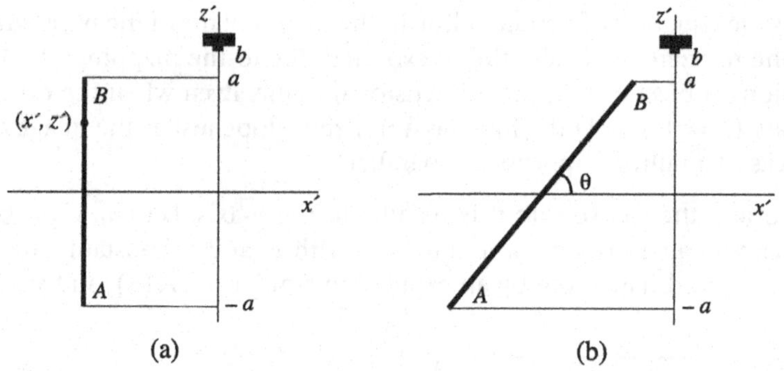

(a) (b)

FIGURE 7.2 (a) The location of the rod when the point (x', z') emits its ray. (b) How the rod appears to the camera.

Rod oriented along the z-axis.

Finally, I orient the rod along the z-axis and show an interesting effect. The z-axis is special because the camera is located on it. The arbitrary point now has coordinates (x_0, y_0, z) in O. Its space coordinates in O' are

$$x' = \gamma\left(x_0 - \beta\sqrt{x_0^2 + y_0^2 + (b-z)^2}\right), \quad y' = y_0, \quad z' = z. \tag{7.18}$$

I let you show that this is the equation of one half of the hyperbola

$$\frac{(x' - \gamma x_0)^2}{\gamma^2\beta^2\left(x_0^2 + y_0^2\right)} - \frac{(z' - b)^2}{x_0^2 + y_0^2} = 1. \tag{7.19}$$

The effect that I want to show is best demonstrated when the rod is on the z-axis. So, let's set $x_0 = 0 = y_0$ and note that Equation (7.18) becomes $x' = -\gamma\beta|b - z|$, which with $b > z$ (why?) yields

$$x' = -\gamma\beta(b - z) = \gamma\beta(z' - b),$$

or

$$z' = \frac{x' + b}{\gamma\beta}. \tag{7.20}$$

Equation (7.20) contains a surprising result that demonstrates the effect I wanted to show you. It describes a straight line with slope $\tan\theta = 1/(\gamma\beta)$, or $\cos\theta = \beta$, as shown in Figure 7.2(b). Furthermore, since the z-coordinates of the front and back of the rod are $z_f = +a$ and $z_b = -a$, and $\cos\theta = (x'_B - x'_A)/\overline{AB}$, we get

$$\beta = \frac{-\gamma\beta(b-a) + \gamma\beta(b+a)}{\overline{AB}} \quad \text{or} \quad \overline{AB} = 2\gamma a,$$

which says that the photographic image is that of a rod that is *stretched* by a factor of γ and *rotated* about the y'-axis (looking down from the y'-axis) by an angle $\theta = \cos^{-1}\beta$! This is entirely due to the relativistic motion of the rod and the fact that the camera captures light that was emitted earlier. In the non-relativistic limit $\beta \to 0$, and the rotation disappears.

Elongation and rotation of a rod lying along the z-axis.

You can find the shape of the *photograph* of the rod in the general case by substituting (x_0, y_0, z) in (7.12) and obtaining

$$u'(z) = \frac{b\gamma\left[x_0 - \beta\sqrt{x_0^2 + y_0^2 + (z-b)^2}\right] + x_q(z-b)/\gamma}{x_q\left[x_0 - \beta\sqrt{x_0^2 + y_0^2 + (z-b)^2}\right] - b(z-b)}$$

$$v'(z) = \frac{y_0\sqrt{x_q^2/\gamma^2 + b^2}}{x_q\left[x_0 - \beta\sqrt{x_0^2 + y_0^2 + (z-b)^2}\right] - b(z-b)}, \tag{7.21}$$

which is the parametric equation of a curve in the $u'v'$-plane with z as the parameter. The elimination of z, with the purpose of finding v' as a function of u', is much harder than in (7.16). Even when $x_q = 0$, you don't get a straightforward relation. In that case, the equations reduce to

$$u'(z) = \frac{\gamma\left[x_0 - \beta\sqrt{x_0^2 + y_0^2 + (z-b)^2}\right]}{b-z}$$

$$v'(z) = \frac{y_0}{b-z}, \tag{7.22}$$

which you can show to be equivalent to

$$y_0^2 u'^2 + \left(x_0^2 - \gamma^2\beta^2 y_0^2\right) v'^2 - 2\gamma x_0 y_0 u'v' - \gamma^2\beta^2 y_0^2 = 0. \tag{7.23}$$

This is the equation of a hyperbola with axes rotated relative to the u'- and v'-axes (see Problem 7.9).

For the special case of the rod lying along the z-axis, Equation (7.21) becomes

$$u'(z) = \frac{b\gamma\beta + x_q/\gamma}{\beta x_q - b}$$

$$v'(z) = 0, \tag{7.24}$$

which is independent of z, indicating that the image on the photographic plate is just a single point. Therefore, although the rod is rotated in the $x'z'$-plane, the rotation is not captured by the camera in O'.

7.5 PHOTOGRAPH OF A CUBE

Now that you have learned the photography of a rod, let's turn to a cube. Let's put the center of a stationary cube of side $2a$ in O at $(x_c, 0, 0)$ and have its sides

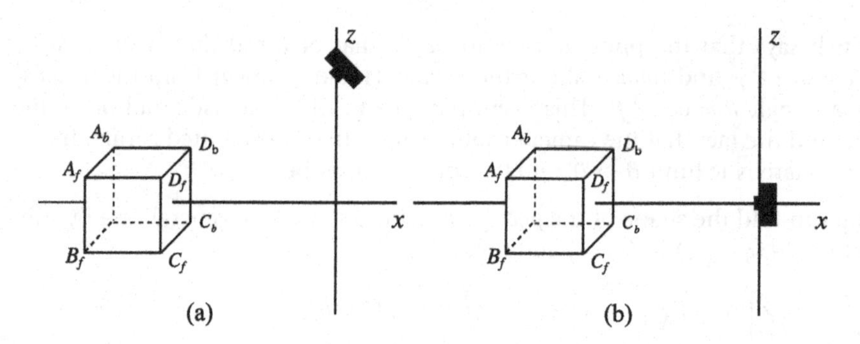

FIGURE 7.3 The center of the cube is at $(x_C, 0, 0)$, and both cameras aim at its center. (a) Cameras are at $(0, 0, b)$. (b) Cameras are at the origin.

parallel to the axes, as shown in Figure 7.3. Both cameras focus on the center of the cube. By (7.14), the four sides parallel to the x-axis of O are photographed as straight lines with nonzero slopes in C', while by (7.16) and (7.21), the other eight sides are curved.

For our subsequent discussion it is convenient to have the coordinates of the eight corners of the cube in O'. The photons from the front corners arriving at the pinhole of C at $t = 0$ left their source at t_f and those from the back corners at t_b, where

$$t_f = -\sqrt{2a^2 + (b-a)^2}, \qquad t_b = -\sqrt{2a^2 + (b+a)^2}. \tag{7.25}$$

Using these in (7.7) you get

$$\begin{aligned} x'_{A_f} &= x'_{B_f} = \gamma\left(-a - \beta\sqrt{2a^2 + (b-a)^2}\right) \\ x'_{C_f} &= x'_{D_f} = \gamma\left(a - \beta\sqrt{2a^2 + (b-a)^2}\right) \\ x'_{A_b} &= x'_{B_b} = \gamma\left(-a - \beta\sqrt{2a^2 + (b+a)^2}\right) \\ x'_{C_b} &= x'_{D_b} = \gamma\left(a - \beta\sqrt{2a^2 + (b+a)^2}\right), \end{aligned} \tag{7.26}$$

with y and z coordinates unchanged.

These expressions show that the back edge of the left face has an x'-coordinate that is *less than* the front edge:

$$x'_{A_b} - x'_{A_f} = \gamma\beta\left(\sqrt{2a^2 + (b-a)^2} - \sqrt{2a^2 + (b+a)^2}\right) < 0. \tag{7.27}$$

This is the rotation I talked about for the special case of Figure 7.2. The question now is, "Does the image of the left face appear on the photographic plate?" Equation (7.24) showed that, at least for the special case of a rod placed along the z-axis, it doesn't. What about the present more general case of a *side* (consisting of infinitely many "rods" parallel to the z-axis)?

To answer the question, I just need the u' coordinates of A_f and A_b. However, I might as well find the coordinates of all the eight corners of the cube on the plane of the photographic plate of C' so we can use them to construct photographs of a moving cube. For simplicity, I'll assume that the center of the cube is at the origin.[4] From (7.12) and $x_q = x_c = 0$, you should be able to find

$$\begin{aligned}
u'_{A_f} &= u'_{B_f} = -\frac{\gamma[a + \beta\sqrt{2a^2 + (b-a)^2}\,]}{b-a} \\
u'_{C_f} &= u'_{D_f} = \frac{\gamma[a - \beta\sqrt{2a^2 + (b-a)^2}\,]}{b-a} \\
u'_{A_b} &= u'_{B_b} = -\frac{\gamma[a + \beta\sqrt{2a^2 + (b+a)^2}\,]}{b+a} \\
u'_{C_b} &= u'_{D_b} = \frac{\gamma[a - \beta\sqrt{2a^2 + (b+a)^2}\,]}{b+a}
\end{aligned} \tag{7.28}$$

and

$$\begin{aligned}
v'_{A_f} &= -v'_{B_f} = \frac{a}{b-a} = v'_{D_f} = -v'_{C_f} \\
v'_{A_b} &= -v'_{B_b} = \frac{a}{b+a} = v'_{D_b} = -v'_{C_b}.
\end{aligned} \tag{7.29}$$

For $\beta = 0$, it is clear from these equations that

$$\begin{aligned}
u_{A_f} &= -\frac{a}{b-a}, \quad v_{A_f} = \frac{a}{b-a} \\
u_{A_b} &= -\frac{a}{b+a}, \quad v_{A_b} = \frac{a}{b+a},
\end{aligned}$$

so that $u_{A_b} > u_{A_f}$ and $v_{A_b} < v_{A_f}$, meaning that the upper left corner of the back face lies in the square formed by the front corners.[5] Similarly, you can show that the rest of the back corners lie in the square formed by the front corners, indicating that, for a stationary cube, the back face is hidden behind the front face. Hence, the picture of the cube in C looks like the top middle picture of Figure 7.4.

What about a moving cube? From (7.28), you should verify that $u'_{A_b} > u'_{A_f}$ and $v'_{A_b} < v'_{A_f}$ for all b and β (see Problem 7.12). Therefore, the photographic image of the left face hides behind the front face as shown in the bottom middle picture of Figure 7.4, and the rotation analogous to Figure 7.2 is not captured by C'.

7.5.1 Capturing Length Contraction With a Camera

Can we capture the length contraction of the cube in C' as it passes the origin of O'? To answer this question, I'll look at the length $\ell' \equiv u'_{D_f} - u'_{A_f} = u'_{C_f} - u'_{B_f}$

---

[4] It is not really just for simplicity. When the center of the cube is not at the origin, it is hard to tell whether the cube is rotated or not, because of the exposure of the side faces to the camera.

[5] Recall from Note C.1.1 that when the camera points at the origin, the photographic plate is parallel to the xy-plane. So, you can think of (u, v) as essentially (x, y).

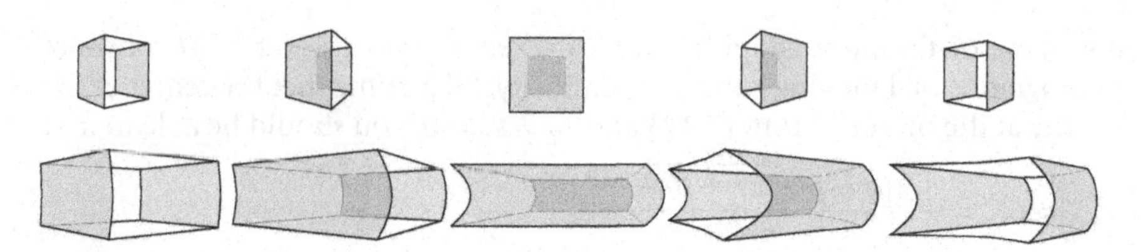

FIGURE 7.4 Snapshots of a stationary cube (top) and a moving cube (bottom) at distances (from left to right) $x_c = -4a, -2a, 0, 2a, 4a$ when the cameras are at $b = 3a$ and $\beta = 0.95$. The pictures are not drawn to scale.

of the photograph of the cube in C' compared to that in C. If you denote the latter by ℓ, then ℓ is obtained from ℓ' by setting $\beta = 0$. Therefore,

The image of a cube is *elongated* along the direction of motion!

$$\ell' = u'_{D_f} - u'_{A_f} = \frac{2\gamma a}{b-a} = \gamma \ell, \tag{7.30}$$

which is an *elongation* of the horizontal length of the photograph of the stationary cube by a factor of γ, as shown in the middle picture of Figure 7.4, which also shows the snapshots of the cube located at negative and positive values of x_c for a stationary cube as well as a cube moving at $\beta = 0.95$.

I should be careful when I talk about "lengths." In O, the two beams of light from A_f and D_f are emitted *simultaneously* and captured by camera C also *simultaneously*. Therefore, the image in C is truly the picture of the "length" of the front face. Camera C' is not capturing the light beams emitted simultaneously by A_f and D_f in O'. In fact,

$$t'_{A_f} = \gamma(-\beta a - \sqrt{2a^2 + (b-a)^2}), \quad t'_{D_f} = \gamma(\beta a - \sqrt{2a^2 + (b-a)^2}),$$

showing that the light from A_f was emitted earlier than that from D_f. Of course, I could arrange things so that A_f and D_f emit their light simultaneously according to O' as the center of the cube passes the origin. But then, C will not capture beams that were emitted simultaneously. My set-up from the beginning of this chapter has been to compare the photographs of the two cameras as both take pictures *at the same time* that the center of the cube passes the origin. In this situation, C captures the "true" length of $\overline{A_f D_f}$, but C' does not.

To compare the "true" lengths in the two cameras, I'll have to do *two* experiments. In one experiment, I let C catch two beams emitted simultaneously, and in the other I let C' do the same. In the first experiment, the "true" length is $u_{D_f} - u_{A_f}$, where these are the values given by (7.28) with $\beta = 0$:

How to capture length contraction with a camera.

$$\ell_{\text{true}} = u_{D_f} - u_{A_f} = \frac{a}{b-a} - \left(-\frac{a}{b-a}\right) = \frac{2a}{b-a}. \tag{7.31}$$

In the second experiment, I want $t'_{A_f} = t'_{D_f}$. I have the following set of LTs:

$$\begin{aligned} x'_{A_f} &= \gamma(-a + \beta t_{A_f}), \quad t'_{A_f} = \gamma(-\beta a + t_{A_f}) \\ x'_{D_f} &= \gamma(a + \beta t_{D_f}), \quad t'_{D_f} = \gamma(\beta a + t_{D_f}). \end{aligned} \tag{7.32}$$

Setting $t'_{A_f} = t'_{D_f}$ gives me $t_{D_f} - t_{A_f} = -2\beta a$. Now I calculate u'_{A_f} and u'_{D_f} using (7.12) with $x_q = 0$:

$$u'_{A_f} = \frac{bx'_{A_f}}{-b(z_{A_f} - b)} = \frac{\gamma(-a + \beta t_{A_f})}{b - a}, \qquad u'_{D_f} = \frac{bx'_{D_f}}{-b(z_{D_f} - b)} = \frac{\gamma(a + \beta t_{D_f})}{b - a}.$$

I can now evaluate the true length in C':

$$\begin{aligned} \ell'_{\text{true}} &= u'_{D_f} - u'_{A_f} = \frac{2\gamma a + \gamma\beta(t_{D_f} - t_{A_f})}{b - a} \\ &= \frac{2\gamma a + \gamma\beta(-2\beta a)}{b - a} = \frac{2\gamma a(1 - \beta^2)}{b - a} = \frac{2a}{\gamma(b - a)} = \frac{\ell_{\text{true}}}{\gamma}. \end{aligned}$$

The Lorentz length contraction is indeed captured by the camera!

A cube consists of a discrete set of (twelve) edges. Therefore, to find the image of a cube on a photographic plate I just needed to use (7.12) for this discrete set. To find the shape of the photograph of an object bounded by a *smooth* surface, such as a sphere, I first need to find the curve bounding the region of the surface, whose light forms the image. To find this curve, I have to draw lines from the pinhole of the camera *tangent* to the surface and calculate the coordinates of the point at which the line touches the surface. I have derived the general equation for this curve in Appendix C.2, and in Section 7.7 I'll use this method for the general case of a sphere centered at some arbitrary point on the x-axis.

7.6 PHOTOGRAPH OF A SPHERE AT THE ORIGIN

In this section, I consider the special case of a sphere centered at the origin of its rest frame. In addition to finding the curve outlining the region exposed to the camera, I'll examine the cone formed by the rays converging on the pinhole. This examination has a historical significance because Penrose—who for the first time proposed that the image of a moving sphere is a circle, not an ellipse flattened in the direction of its motion—looked *only* at the shape of the *cone* formed by the rays converging on a *point* observer. Since Penrose's argument is important for our subsequent discussion, I'll examine it in some detail here.

First note that the aberration formula (6.26) has an ingenious geometric interpretation first noticed by Penrose. Figure 7.5 shows observer O at the origin and at the center of a sphere of unit diameter. The three points Q_1, O, and Q_2 lie on the x-axis, along which O moves relative to the second observer O'. A light ray coming from a source cuts the sphere at a point P that is projected

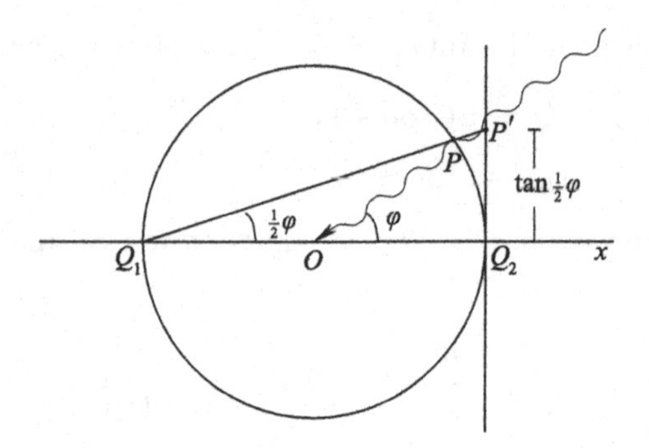

FIGURE 7.5 The sphere has unit diameter and is centered at the *point* observer O. A point P on the sphere (at which a light ray coming from a source cuts the sphere) is mapped onto a point P' in the plane tangent to the sphere at Q_2. The image P' is obtained by the intersection of the line Q_1P and the plane. This is called ***stereographic projection.***

onto the point P' in a plane tangent to the sphere at Q_2 by extending the line segment $\overline{Q_1P}$ until it intersects the tangent plane. This kind of projection is called a **stereographic projection.**

Stereographic projection.

Next note that the rays from the (illuminating) sphere located somewhere in the reference frame of O converge on him to form a *circular* cone. This cone intersects the sphere centered at O at a circle. Now, it is well known that the image of a circle under a stereographic projection is a circle.[6] Thus, the image of that circle on the tangent plane is a circle. Stated differently, each ray is described by a point P and a corresponding angle φ. As P varies over the circle, φ varies over a certain range of values. For each of these values, there is a point P' on the tangent plane. So as φ varies over its range, P' varies over a circle in the tangent plane.

Now consider O', who coincides with O at $t = t' = 0$. O' also draws her sphere of unit diameter and the tangent plane perpendicular to the common x-axis. The rays from the (illuminating) sphere converge on her to form a cone. She projects the intersection of this cone with her unit sphere onto her tangent plane. Since $\tan\frac{1}{2}\varphi$ on the right-hand side of (6.26) describes a circle as φ runs over its range of values, O' also gets a circle on her tangent plane, which has a different size than that of O due to the factor $\sqrt{(1-|\vec{\beta}|)/(1+|\vec{\beta}|)}$. The inverse image of this circle is also a circle. This means that the rays converging on O' cut her unit sphere at a circle, indicating that her cone is also circular.

Images of extended objects are necessarily extended. This means that in the discussion of image formation, we cannot assume that observers are points. We have to consider the *extended* eye or the *extended* camera of an observer

[6] Or a line under special conditions, which we ignore.

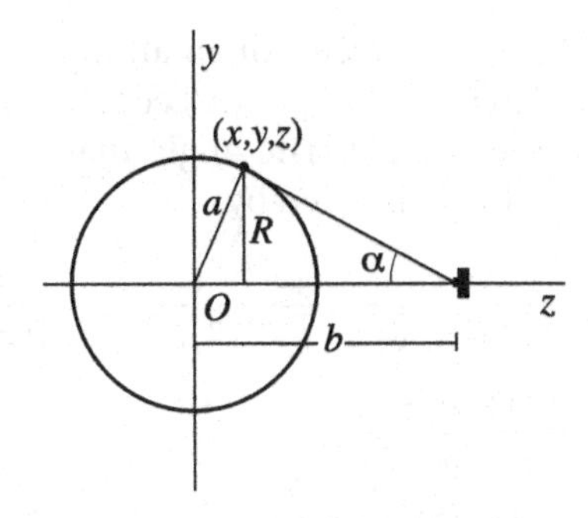

FIGURE 7.6 Cross section of the sphere with the yz-plane. The point $(x, y, z = a\sin\alpha)$ is the source of an arbitrary ray of light. The cone in O is tangent to the sphere at a circle of radius $R = a\cos\alpha$.

and see how the cones converging on these structures end up as images.[7] This requires the explicit construction of the cones, which I start doing now. I'll construct the cones in three different ways, each giving a different insight into the construction: using the aberration formula (6.25), using the relativistic law of addition of velocities (LAV), and using the equation of the curve bounding the region exposed to the camera. To obtain the faithful image of the sphere, I'll have to point the camera to its center.[8] Therefore, since the sphere is assumed to be at the origin of O, cameras C and C' will be pointing at their respective origins and take pictures when the origins coincide at $t = t' = 0$.

Both cameras *must* point to the center of the sphere.

7.6.1 Cone Construction via Aberration Formula

Consider a sphere of radius a centered at the origin of a reference frame O with a camera at $(0, 0, b)$ as shown in Figure 7.6. The light rays received at the pinhole form a cone making an angle $\alpha = \sin^{-1}(a/b)$ with the z-axis. The cone is tangent to the sphere and cuts it at a circle of radius $R = a\cos\alpha$ centered at $(0, 0, a\sin\alpha) = (0, 0, a^2/b)$.

Using the aberration formula, I want to find the cone of Figure 7.6 in the frame with respect to which the sphere is moving. The idea is to use Equation (6.25) for each ray of Figure 7.6. Note 6.3.3 tells me that the ray making an angle φ with the x-axis is "aberrated" to a ray making an angle φ' with the x-axis in the plane that the original ray makes with the x-axis. Therefore, for each ray, I find the plane it forms with the x-axis and calculate the components of a unit vector in that plane which makes an angle φ' with the x-axis. This unit vector defines one of the lines forming the cone in O'.

---

[7] Since image perception involves interpretations by the brain that may distort the actual image (as is the case in optical illusions, for example), I'll exclude the eye from my discussion and restrict myself to a simple pinhole camera.

[8] See Figure 7.8 and comments at the end of Section 7.7.1.

Let $\mathbf{r} = \langle x, y, a^2/b \rangle$ be the position vector of an arbitrary point of the intersection of the cone of Figure 7.6 with the sphere. Let $\mathbf{r}_0 = \langle 0, 0, b \rangle$ be the position vector of the pinhole of the camera. The angle that a ray originating from $(x, y, a^2/b)$ makes with the x-axis is given by

$$\cos\varphi = \left(\frac{\mathbf{r}_0 - \mathbf{r}}{|\mathbf{r}_0 - \mathbf{r}|}\right) \cdot \hat{\mathbf{e}}_x = -\frac{x}{\sqrt{x^2 + y^2 + (b - a^2/b)^2}} = -\frac{x}{b\cos\alpha}$$

$$\sin\varphi = \sqrt{1 - \cos^2\varphi} = \frac{\sqrt{b^2\cos^2\alpha - x^2}}{b\cos\alpha}. \tag{7.33}$$

In O', the angle is given by (6.25), or

$$\cos\varphi' = \frac{|\vec{\beta}|b\cos\alpha - x}{b\cos\alpha - |\vec{\beta}|x}, \quad \sin\varphi' = \frac{\sqrt{b^2\cos^2\alpha - x^2}}{\gamma(b\cos\alpha - |\vec{\beta}|x)}. \tag{7.34}$$

The plane containing the ray and the x-axis has a normal given by the cross product of $\mathbf{r}_0 - \mathbf{r} = \langle -x, -y, b\cos^2\alpha \rangle$ and $\langle 1, 0, 0 \rangle$. This normal has components $\langle 0, b\cos^2\alpha, y \rangle$. A vector $\langle \eta_x, \eta_y, \eta_z \rangle$ is parallel to (lies in) the plane if

$$\langle \eta_x, \eta_y, \eta_z \rangle \cdot \langle 0, b\cos^2\alpha, y \rangle = 0$$

or if $\eta_y = -\eta_z y/b\cos^2\alpha$. So the vector $\langle \eta_x b\cos^2\alpha, -\eta_z y, \eta_z b\cos^2\alpha \rangle$ is parallel to the plane, and

$$\hat{\mathbf{v}} = \frac{\langle \eta_x b\cos^2\alpha, -\eta_z y, \eta_z b\cos^2\alpha \rangle}{\sqrt{\eta_x^2(b\cos^2\alpha)^2 + \eta_z^2[y^2 + (b\cos^2\alpha)^2]}} \tag{7.35}$$

is a unit vector in that direction. For this to make an angle φ' with the x-axis, we must have

$$\cos\varphi' = \hat{\mathbf{v}} \cdot \hat{\mathbf{e}}_x = \frac{\eta_x b\cos^2\alpha}{\sqrt{\eta_x^2(b\cos^2\alpha)^2 + \eta_z^2[y^2 + (b\cos^2\alpha)^2]}}$$

or

$$\eta_z = \pm\frac{\eta_x b\cos^2\alpha}{\sqrt{y^2 + (b\cos^2\alpha)^2}} \tan\varphi'.$$

Substituting this in Equation (7.35) yields

$$\hat{\mathbf{v}} = \left\langle \cos\varphi', \mp\frac{y}{\sqrt{y^2 + (b\cos^2\alpha)^2}} \sin\varphi', \pm\frac{b\cos^2\alpha}{\sqrt{y^2 + (b\cos^2\alpha)^2}} \sin\varphi' \right\rangle,$$

and using

$$\begin{aligned} y^2 + (b\cos^2\alpha)^2 &= a^2\cos^2\alpha - x^2 + (b\cos^2\alpha)^2 \\ &= b^2\sin^2\alpha\cos^2\alpha + b^2\cos^4\alpha - x^2 = b^2\cos^2\alpha - x^2 \end{aligned}$$

and Equation (7.34), you get

$$\hat{\mathbf{v}} = \left\langle \frac{|\vec{\beta}|b\cos\alpha - x}{b\cos\alpha - |\vec{\beta}|x}, \mp\frac{y}{\gamma(b\cos\alpha - |\vec{\beta}|x)}, \pm\frac{b\cos^2\alpha}{\gamma(b\cos\alpha - |\vec{\beta}|x)} \right\rangle. \tag{7.36}$$

Let ϕ parametrize the circle of radius R, i.e., let $x \equiv R\cos\phi = a\cos\alpha\cos\phi$ and $y \equiv R\sin\phi = a\cos\alpha\sin\phi$, then (7.36) becomes

$$\hat{\mathbf{v}}(\phi,\beta) = \left\langle \frac{|\vec{\beta}| - \sin\alpha\cos\phi}{1 - |\vec{\beta}|\sin\alpha\cos\phi}, \mp\frac{\sin\alpha\sin\phi}{\gamma(1 - |\vec{\beta}|\sin\alpha\cos\phi)}, \pm\frac{\cos\alpha}{\gamma(1 - |\vec{\beta}|\sin\alpha\cos\phi)} \right\rangle. \tag{7.37}$$

For $\vec{\beta} = 0$, this reduces to

$$\hat{\mathbf{v}} = \langle -\sin\alpha\cos\phi, \mp\sin\alpha\sin\phi, \pm\cos\alpha \rangle. \tag{7.38}$$

This must have a positive z-component as you can see in Figure 7.6. So, I have to choose the upper sign in (7.37). Hence,

$$\hat{\mathbf{v}}(\phi,\beta) = \left\langle \frac{|\vec{\beta}| - \sin\alpha\cos\phi}{1 - |\vec{\beta}|\sin\alpha\cos\phi}, -\frac{\sin\alpha\sin\phi}{\gamma(1 - |\vec{\beta}|\sin\alpha\cos\phi)}, \frac{\cos\alpha}{\gamma(1 - |\vec{\beta}|\sin\alpha\cos\phi)} \right\rangle \tag{7.39}$$

is the direction of the light ray. To find the properties of the cone, it is more convenient to consider the negative of this vector. So, I define

$$\hat{\mathbf{e}}(\phi,\beta) = \left\langle -\frac{|\vec{\beta}| - \sin\alpha\cos\phi}{1 - |\vec{\beta}|\sin\alpha\cos\phi}, \frac{\sin\alpha\sin\phi}{\gamma(1 - |\vec{\beta}|\sin\alpha\cos\phi)}, -\frac{\cos\alpha}{\gamma(1 - |\vec{\beta}|\sin\alpha\cos\phi)} \right\rangle \tag{7.40}$$

as a typical unit vector on the surface of the cone emanating away for the pin-hole.

What is the shape of this cone? It turns out to be circular. How do I show this? First I let you show that

$$\cos\vartheta \equiv \hat{\mathbf{e}}(\phi,\beta)\cdot\hat{\mathbf{e}}(0,\beta) = 1 - \frac{\sin^2\alpha(1-\cos\phi)}{\gamma^2(1 - |\vec{\beta}|\sin\alpha\cos\phi)(1 - |\vec{\beta}|\sin\alpha)}$$

has a maximum at $\phi = 0$ and a minimum at $\phi = \pi$, meaning that $\hat{\mathbf{e}}_0 \equiv \hat{\mathbf{e}}(0,\beta)$ and $\hat{\mathbf{e}}_\pi \equiv \hat{\mathbf{e}}(\pi,\beta)$ are the two extremes of the cone, and if the cone is indeed circular, then the vector $\hat{\mathbf{e}}_0 + \hat{\mathbf{e}}_\pi$ should lie along its axis (why?), and the unit vector

$$\hat{\mathbf{e}}_{axis} = \frac{\hat{\mathbf{e}}_0 + \hat{\mathbf{e}}_\pi}{|\hat{\mathbf{e}}_0 + \hat{\mathbf{e}}_\pi|}$$

should define the direction of the axis. With

$$\hat{\mathbf{e}}_0 + \hat{\mathbf{e}}_\pi = \frac{2\cos\alpha}{1 - |\vec{\beta}|^2\sin^2\alpha}\left\langle -|\vec{\beta}|\cos\alpha, 0, -1/\gamma \right\rangle$$

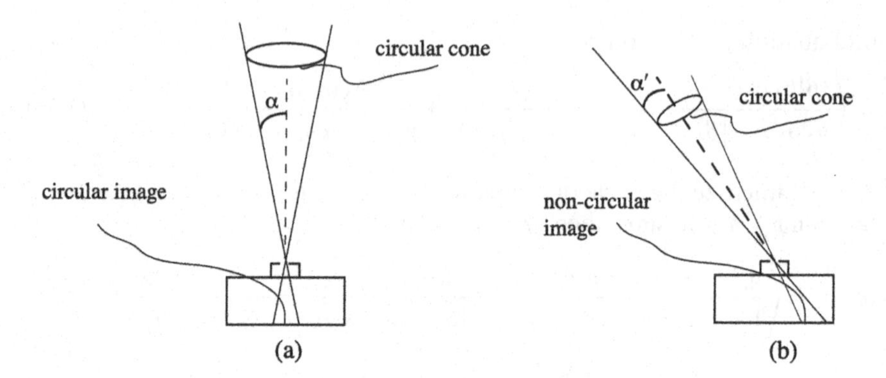

FIGURE 7.7 The cones are formed by the light rays coming from the illuminating sphere. (a) The camera is in the rest frame of the illuminating sphere. (b) The illuminating sphere moves from left to right relative to this camera. Since the camera captures light rays originated in the past, the source of the rays must be on the left of the camera.

it is easy to show that

$$\hat{\mathbf{e}}_{axis} = \frac{\left\langle -|\vec{\beta}|\cos\alpha, 0, -1/\gamma \right\rangle}{\left| \left\langle -|\vec{\beta}|\cos\alpha, 0, -1/\gamma \right\rangle \right|} = \frac{\left\langle -|\vec{\beta}|\cos\alpha, 0, -\sqrt{1-|\vec{\beta}|^2} \right\rangle}{\sqrt{1-|\vec{\beta}|^2 \sin^2\alpha}}. \tag{7.41}$$

Finally, the crucial step of proving that the cone is circular is to show that $\hat{\mathbf{e}}(\phi, \beta) \cdot \hat{\mathbf{e}}_{axis}$ is independent of ϕ. Indeed, a little calculation shows that

$$\hat{\mathbf{e}}(\phi, \beta) \cdot \hat{\mathbf{e}}_{axis} = \frac{\cos\alpha}{\sqrt{1-|\vec{\beta}|^2 \sin^2\alpha}} \equiv \cos\alpha', \tag{7.42}$$

where α' is the angle of the cone in O'. For $\vec{\beta} = 0$, we get $\hat{\mathbf{e}}_{axis} = -\langle 0, 0, 1 \rangle$ and $\cos\alpha' = \cos\alpha$, as expected.

For nonzero $\vec{\beta}$, Equation (7.41) shows that the axis of the cone points in the direction of negative values of x and z. This is as expected, because the rays are originating in the past when the sphere was on the negative x-axis. Figure 7.7 shows the cones in the frame of the sphere and in the frame in which the sphere is moving. Precisely because the cone in (b) is circular, it cannot form a circular image in the camera! In fact, from your experience with conic sections, you should deduce that the image is an ellipse.

> **Note 7.6.1.** *The image formed in a pinhole camera that points at the center of a relativistically moving sphere is an ellipse when the optical axis of the camera is perpendicular to the direction of motion of the sphere.*

7.6.2 Cone Construction via Relativistic LAV

The velocity of a ray in Figure 7.7(b) is just the velocity of the corresponding ray in Figure 7.7(a) obtained by the relativistic law of addition of velocities. The velocity of a light ray (being of unit length) of Figure 7.6 is given by

$$\vec{c} = \frac{\mathbf{r}_0 - \mathbf{r}}{|\mathbf{r}_0 - \mathbf{r}|} = \frac{\langle -x, -y, b - a^2/b \rangle}{\sqrt{x^2 + y^2 + (b - a^2/b)^2}}$$
$$= \frac{\langle -a\cos\alpha\cos\phi, -a\cos\alpha\sin\phi, b\cos^2\alpha \rangle}{b\cos\alpha}$$

or

$$\vec{c} = \langle -\sin\alpha\cos\phi, -\sin\alpha\sin\phi, \cos\alpha \rangle, \tag{7.43}$$

which is just Equation (7.38) with the right choice of sign. Using the relativistic law of addition of velocities (with $\beta > 0$ along the x-axis),

$$c'_x = \frac{c_x + \beta}{\beta c_x + 1}, \quad c'_y = \frac{c_y}{\gamma(\beta c_x + 1)}, \quad c'_z = \frac{c_z}{\gamma(\beta c_x + 1)},$$

you can obtain the velocities in O':

$$\vec{c}'(\phi, \beta) = \left\langle \frac{\beta - \sin\alpha\cos\phi}{1 - \beta\sin\alpha\cos\phi}, -\frac{\sin\alpha\sin\phi}{\gamma(1 - \beta\sin\alpha\cos\phi)}, \frac{\cos\alpha}{\gamma(1 - \beta\sin\alpha\cos\phi)} \right\rangle. \tag{7.44}$$

This is precisely the vector in (7.39).

7.6.3 Cone Construction via LT of Emission Events

Now I use the procedure of Appendix C.2 to find the equation of the curve that bounds the region of the sphere exposed to the camera. As a warm-up, I'll apply the procedure to the obvious case of the stationary sphere and re-derive what we already know. The stationary sphere has the parametric equations[9]

$$\begin{aligned} x &= a\sin\theta\cos\varphi \\ y &= a\sin\theta\sin\varphi \\ z &= a\cos\theta. \end{aligned} \tag{7.45}$$

Equations (C.14) and (C.15) of Appendix C.2 give two vectors tangent to the surface. For our sphere, these are

$$\vec{\theta} \equiv \left\langle \frac{\partial x}{\partial\theta}, \frac{\partial y}{\partial\theta}, \frac{\partial z}{\partial\theta} \right\rangle = \langle a\cos\theta\cos\varphi, a\cos\theta\sin\varphi, -a\sin\theta \rangle$$
$$\vec{\varphi} \equiv \left\langle \frac{\partial x}{\partial\varphi}, \frac{\partial y}{\partial\varphi}, \frac{\partial z}{\partial\varphi} \right\rangle = \langle -a\sin\theta\sin\varphi, a\sin\theta\cos\varphi, 0 \rangle,$$

---

[9] Do not confuse the azimuthal angle φ with that used previously in the aberration formulas. Unfortunately, there are only a finite number of Greek letters, and I have to stick to those that are commonly used!

giving the normal vector to the sphere

$$\vec{\theta} \times \vec{\varphi} = a^2 \sin\theta \langle \sin\theta\cos\varphi, \sin\theta\sin\varphi, \cos\theta\rangle, \quad \theta \neq 0, \pi. \tag{7.46}$$

Insert this and (7.45) in (C.17) to obtain

$$a^2 \sin\theta[a\sin^2\theta + \cos\theta(a\cos\theta - b)] = 0, \quad \theta \neq 0, \pi,$$

which has the solution $\cos\theta = a/b$, a result that is evident from Figure 7.6 (in which $\theta = \pi/2 - \alpha$). Substitute this in (7.45) to get

$$\begin{aligned} x &= a\sqrt{1-(a/b)^2}\cos\varphi \\ y &= a\sqrt{1-(a/b)^2}\sin\varphi \\ z &= a^2/b, \end{aligned} \tag{7.47}$$

which is the equation of a circle of radius $a\sqrt{1-(a/b)^2}$ centered at $(0, 0, a^2/b)$, consistent with the results of Section 7.6.1.

Image of sphere in camera C is a circle.

For future reference, I'll also find the equation of the curve on the photographic plate. With $\beta = 0$ and $x_c = 0$, (7.12) and (7.47) yield

$$\begin{aligned} u &= \frac{x}{b-z} = \frac{a\sqrt{1-(a/b)^2}\cos\varphi}{b-a^2/b} = \frac{a}{\sqrt{b^2-a^2}}\cos\varphi \\ v &= \frac{y}{b-z} = \frac{a\sqrt{1-(a/b)^2}\sin\varphi}{b-a^2/b} = \frac{a}{\sqrt{b^2-a^2}}\sin\varphi, \end{aligned} \tag{7.48}$$

which is a circle of radius $a/\sqrt{b^2-a^2}$.

What is the image of the moving sphere in camera C' when its center reaches the origin? The parametric equation of the sphere in O' is obtained by substituting (7.45) in the Lorentz transformation (7.7).[10] You should show that the result is

$$\begin{aligned} x' &= \gamma\left(a\sin\theta\cos\varphi - \beta\sqrt{a^2+b^2-2ab\cos\theta}\right) \\ y' &= a\sin\theta\sin\varphi \\ z' &= a\cos\theta \end{aligned} \tag{7.49}$$

and that Equations (C.14) and (C.15) of Appendix C.2 yield

$$\begin{aligned} \vec{\theta}' &= a\left\langle \gamma\cos\theta\cos\varphi - \frac{\beta\gamma b\sin\theta}{\sqrt{a^2+b^2-2ab\cos\theta}}, \cos\theta\sin\varphi, -\sin\theta\right\rangle \\ \vec{\varphi}' &= a\left\langle -\gamma\sin\theta\sin\varphi, \sin\theta\cos\varphi, 0\right\rangle. \end{aligned} \tag{7.50}$$

Take the cross product of these two vectors and assume that $\theta \neq 0, \pi$ to obtain

$$\frac{\vec{\theta}' \times \vec{\varphi}'}{a^2\sin\theta} = \left\langle \sin\theta\cos\varphi, \gamma\sin\theta\sin\varphi, \gamma\cos\theta - \frac{\beta\gamma b\sin\theta\cos\varphi}{\sqrt{a^2+b^2-2ab\cos\theta}}\right\rangle. \tag{7.51}$$

---

[10] Without the subscript i, of course!

Substitute (7.49) and (7.51) in (C.17) and simplify to get

$$(a - b\cos\theta)\left(1 - \frac{\beta a \sin\theta \cos\varphi}{\sqrt{a^2 + b^2 - 2ab\cos\theta}}\right) = 0, \tag{7.52}$$

which appears to have two solutions:

$$\cos\theta = \frac{a}{b} \quad \text{and} \quad \cos\varphi = \frac{\sqrt{a^2 + b^2 - 2ab\cos\theta}}{\beta a \sin\theta}.$$

However, for the fraction defining $\cos\varphi$, you can easily show that

$$\text{Numerator}^2 = (b - a\cos\theta)^2 + a^2\sin^2\theta$$

indicating that

$$\cos^2\varphi = \frac{(b - a\cos\theta)^2}{\beta^2 a^2 \sin^2\theta} + \frac{1}{\beta^2} > 1,$$

which has no solution. Thus, as expected, there is a unique solution for the contour of the image formed in C'. Substituting $\cos\theta = a/b$ in (7.49) yields

Contour of the region of moving sphere captured by the camera at the moment that it reaches the origin is an ellipse *elongated* along the direction of motion!

$$\begin{aligned} x' &= \gamma\left(a\sqrt{1 - (a/b)^2}\cos\varphi - \beta\sqrt{b^2 - a^2}\right) \\ y' &= a\sqrt{1 - (a/b)^2}\sin\varphi \\ z' &= a^2/b, \end{aligned} \tag{7.53}$$

i.e., the equation of an ellipse parallel to the $x'y'$-plane with its center at $(-\beta\gamma\sqrt{b^2 - a^2}, 0, a^2/b)$, a semi-major axis $\gamma a\sqrt{1 - (a/b)^2}$ along the direction of motion, and a semi-minor axis $a\sqrt{1 - (a/b)^2}$ perpendicular to the direction of motion. Thus the circular image of Equation (7.47) has expanded in the direction of motion by a factor of γ. This is consistent with the elongation of the image of the cube discussed earlier. As a check, note that (7.53) yields (7.47) when $\beta = 0$.

For future reference, I'll also find the equation of the curve on the photographic plate. With $x_q = x_c = 0$, (7.12) and (7.53) yield

Image of sphere in camera C' is an ellipse.

$$\begin{aligned} u' &= \frac{x'}{b - z} = \frac{\gamma\left(a\sqrt{1 - (a/b)^2}\cos\varphi - \beta\sqrt{b^2 - a^2}\right)}{b - a^2/b} \\ &= -\frac{b\beta\gamma}{\sqrt{b^2 - a^2}} + \frac{\gamma a}{\sqrt{b^2 - a^2}}\cos\varphi \\ v' &= \frac{y}{b - z} = \frac{a\sqrt{1 - (a/b)^2}\sin\varphi}{b - a^2/b} = \frac{a}{\sqrt{b^2 - a^2}}\sin\varphi, \end{aligned} \tag{7.54}$$

which is an ellipse of semi-major axis $\gamma a/\sqrt{b^2 - a^2}$ and semi-minor axis $a/\sqrt{b^2 - a^2}$, centered at $(-\gamma\beta b/\sqrt{b^2 - a^2}, 0)$ in the $u'v'$-plane.

It is worth noting that Equation (7.53) is the Lorentz transform of the corresponding equation for the stationary sphere (7.47). More accurately, as shown in Figure 7.6, all the light rays from the points of the curve of Equation (7.47) were emitted at the same time at

$$t = -\sqrt{x^2 + y^2 + (z - b)^2} = -\sqrt{a^2\left(1 - \frac{a^2}{b^2}\right) + \left(b - \frac{a^2}{b}\right)^2} = -\sqrt{b^2 - a^2}.$$

Therefore, (7.53) is indeed the space part of the Lorentz transform of the *emission events* located at the space points (7.47). This, in fact, is a general result, which I'll prove in Section 10.1.2 after you gain some familiarity with tensors:

Note 7.6.2. *The contour of the region of the Lorentz transformed surface forming the image in camera C' is the Lorentz transform of the contour of the region forming the image in camera C.*

This is not a trivial result, because there is no guarantee that when you Lorentz transform the surface and *then* apply the procedure of Appendix C.2, you get the Lorentz transform of the curve that you obtained when you applied the procedure of Appendix C.2 to the original surface.

When the light rays from the ellipse (7.53) converge on the pinhole of the camera in O', they form a cone with the pinhole as its vertex. A typical unit vector $\hat{\mathbf{e}}_{\text{ellipse}}$ along a line connecting the pinhole to this ellipse is given by

$$\hat{\mathbf{e}}_{\text{ellipse}} = \frac{\langle x', y', z' - b\rangle}{\sqrt{x'^2 + y'^2 + (z' - b)^2}}.$$

Equation (7.53) and

$$\sqrt{x'^2 + y'^2 + (z' - b)^2} = \frac{\sqrt{b^2 - a^2}}{b}\gamma(b - \beta a\cos\varphi)$$

yield

$$\hat{\mathbf{e}}_{\text{ellipse}} = \left\langle -\frac{b\beta - a\cos\varphi}{b - \beta a\cos\varphi}, \frac{a\sin\varphi}{\gamma(b - \beta a\cos\varphi)}, -\frac{\sqrt{b^2 - a^2}}{\gamma(b - \beta a\cos\varphi)}\right\rangle$$

or

$$\hat{\mathbf{e}}_{\text{ellipse}} = \left\langle -\frac{\beta - \sin\alpha\cos\varphi}{1 - \beta\sin\alpha\cos\varphi}, \frac{\sin\alpha\sin\varphi}{\gamma(1 - \beta\sin\alpha\cos\varphi)}, -\frac{\cos\alpha}{\gamma(1 - \beta\sin\alpha\cos\varphi)}\right\rangle, \tag{7.55}$$

which is identical to (7.40).

I have elaborated on the behavior of light rays converging onto the pinhole from the moving sphere and shown that although they are coming from an ellipse, they form a circular cone. Need I worry about what happens *inside* the camera after they diverge from the pinhole onto the photographic plate? No! The pinhole and the photographic plate are not moving relative to one another.

Note 7.6.3. *Once the camera collects the light rays at the pinhole, the shape of the image is determined entirely by the shape of the cone in the camera and the orientation of the photographic plate relative to that cone. It is irrelevant whether the source of the rays is moving or not.*

See Note 7.2.1 and Problem 7.26 for further insight.

It is interesting to note that the line from $(0, 0, b)$ in the direction of $\hat{\mathbf{e}}_{axis}$ of Equation (7.41) cuts the x-axis at $-|\vec{\beta}|\gamma b \cos\alpha$ or $-|\vec{\beta}|\gamma\sqrt{b^2 - a^2}$. This is the x-value of the Lorentz transformation of the event $(-\sqrt{b^2 - a^2}, 0, 0, 0)$, indicating that $\hat{\mathbf{e}}_{axis}$ points at the past location of the origin of the moving sphere.

Under what circumstances would the camera in Figure 7.7(b) produce a circular image? It is obvious that this happens only if the camera points in the direction of $\hat{\mathbf{e}}_{axis}$; if the picture is to be the *faithful* picture of the sphere, the camera must also point at its center. Both of these conditions hold only if the sphere moves directly toward (or away from) the camera.

Note 7.6.4. *The image of a relativistically moving sphere is a circle only if it moves directly toward or away from the camera.*

I'll discuss this more rigorously in Section 7.7.2.

7.7 PHOTOGRAPH OF A SPHERE OFF THE ORIGIN

In this section, I'll cover the general case of a sphere with its center fixed at an arbitrary point on the x-axis of O. Both cameras are focused on the center of the sphere. To find the image of this sphere, I must first find the equation of the contour of the region of the sphere whose photons form the image on the photographic plate.

The equation of a sphere centered at $(x_c, 0, 0)$ in O is

$$\begin{aligned} x &= x_c + a \sin\theta \cos\varphi \\ y &= a \sin\theta \sin\varphi \\ z &= a \cos\theta, \end{aligned} \tag{7.56}$$

giving the same $\vec{\theta}$, $\vec{\varphi}$, and $\vec{\theta} \times \vec{\varphi}$ as in Section 7.6.3. Inserting (7.46) and (7.56) in (C.17) yields

$$x_c \sin\theta \cos\varphi - b \cos\theta + a = 0. \tag{7.57}$$

Although derived for O, as I'll show below and as indicated in Note 10.1.12 (with ξ replaced by θ and η by φ), Equation (7.57) is also valid for O'. Therefore, it is worthwhile to derive the general u' and v' and then specialize to either case. I'll leave it for you to show that Equations (7.56) and (7.57) yield

$$x^2 + y^2 + (z-b)^2 = x_c^2 + b^2 - a^2$$

and that for $x_q = x_c$, the denominators of (7.12) become

$$\text{Den} = x_c^2 + b^2 - a^2 - \beta x_c \sqrt{x_c^2 + b^2 - a^2}.$$

Then when you substitute these and (7.56) in (7.12), you obtain

$$u' = \frac{b\gamma\left(x_c + a\sin\theta\cos\varphi - \beta\sqrt{x_c^2 + b^2 - a^2}\right) + x_c(a\cos\theta - b)/\gamma}{x_c^2 + b^2 - a^2 - \beta x_c\sqrt{x_c^2 + b^2 - a^2}}$$

$$v' = \frac{a\sqrt{x_c^2/\gamma^2 + b^2}\sin\theta\sin\varphi}{x_c^2 + b^2 - a^2 - \beta x_c\sqrt{x_c^2 + b^2 - a^2}}. \tag{7.58}$$

You can actually solve Equation (7.57) and find θ in terms of φ. I have outlined the procedure in Problem 7.29, where I ask you to derive the following formulas:

$$\sin\theta = \frac{b\sqrt{b^2 - a^2 + x_c^2\cos^2\varphi} - ax_c\cos\varphi}{b^2 + x_c^2\cos^2\varphi}$$

$$\cos\theta = \frac{x_c\cos\varphi\sqrt{b^2 - a^2 + x_c^2\cos^2\varphi} + ab}{b^2 + x_c^2\cos^2\varphi}. \tag{7.59}$$

Substituting these in (7.58) gives u' and v' in terms of φ, describing the parametric equation of the curve of the image.

Remark 7.7.1. When examining the image of a sphere in either of the cameras, you have to take into account the limitations of Equation (7.59). One limitation is that the camera cannot be inside the sphere, i.e., the distance between the pinhole of the camera and the center of the sphere must be larger than a, or $b^2 + x_c^2 > a^2$. Therefore, as long as $b > a$, this condition holds regardless of the value of x_c, and $\sin\theta$ and $\cos\theta$ are both well defined for all values of φ.

However, even if the condition $b^2 + x_c^2 > a^2$ holds, when either b or x_c is too small, $\sin\theta$ and $\cos\theta$ may not be defined because of the square root appearing in (7.59). It is clear that if φ is close to $\pi/2$ and $b < a$, the expression inside the square root can be negative even if $x_c > a$. This is already evident in (7.57), where at φ very close to $\pi/2$, the term $b\cos\theta$ may be too small to cancel a.

These remarks are important when we consider the head-on approach or recession of a sphere for which case I place the camera at the origin where $b = 0$. Then (7.59) becomes imaginary for $\cos\varphi < a/x_c$ and infinite at $\varphi = \pi/2$. In this case, the coordinates u' and v' of Equation (7.58) become unphysical and no image of the sphere is formed. ■

7.7.1 Stationary Sphere

With $\beta = 0$, Equations (7.58) and (7.59) yield

$$u = \frac{a(b^2 + x_c^2)\cos\varphi\sqrt{b^2 - a^2 + x_c^2\cos^2\varphi} + a^2 b x_c \sin^2\varphi}{(b^2 + x_c^2 - a^2)(b^2 + x_c^2\cos^2\varphi)}$$

$$v = \frac{a\sqrt{b^2 + x_c^2}\sin\varphi(b\sqrt{b^2 - a^2 + x_c^2\cos^2\varphi} - a x_c \cos\varphi)}{(b^2 + x_c^2 - a^2)(b^2 + x_c^2\cos^2\varphi)}. \quad (7.60)$$

If you plot this parametric equation, you get a circle. In fact, if you define a new parameter ϕ by (see Remark 7.7.2 coming up later)

$$\cos\phi = \frac{(b^2 + x_c^2)\cos\varphi\sqrt{b^2 - a^2 + x_c^2\cos^2\varphi} + a b x_c \sin^2\varphi}{\sqrt{b^2 + x_c^2 - a^2}(b^2 + x_c^2\cos^2\varphi)}, \quad (7.61)$$

you can show that

$$\sin\phi = \frac{\sqrt{b^2 + x_c^2}\sin\varphi(b\sqrt{b^2 - a^2 + x_c^2\cos^2\varphi} - a x_c \cos\varphi)}{\sqrt{b^2 + x_c^2 - a^2}(b^2 + x_c^2\cos^2\varphi)}. \quad (7.62)$$

and

Image of a stationary sphere centered at $(x_c, 0, 0)$ is a circle.

$$u = \frac{a}{\sqrt{b^2 + x_c^2 - a^2}}\cos\phi$$

$$v = \frac{a}{\sqrt{b^2 + x_c^2 - a^2}}\sin\phi. \quad (7.63)$$

This shows that the image of the bounding curve of the sphere (and therefore the image of the sphere itself) with center at $(x_c, 0, 0)$ is a circle of radius $a/\sqrt{b^2 + x_c^2 - a^2}$ when the camera is focused on the center of the sphere. Therefore, camera C records the picture of the sphere as a circle whose radius decreases with increasing x_c, as expected. Equation (7.63) also yields (7.48) when $x_c = 0$.

As I mentioned earlier, to capture the faithful image of a sphere, you need to point the camera to its center. If the camera is not pointing to the center, the image formed on the photographic plate will be an ellipse *even when the sphere is stationary*. For example, if the camera points to the origin while the center of the sphere is at $(x_c, 0, 0)$, Equations (7.12) and (7.56) give

Why it's important to point the cameras to the center of sphere.

$$u = \frac{x}{b - z} = \frac{x_c + a\sin\theta\cos\varphi}{b - a\cos\theta}$$

$$v = \frac{y}{b - z} = \frac{a\sin\theta\sin\varphi}{b - a\cos\theta} \quad (7.64)$$

when $\beta = 0$. With $\sin\theta$ and $\cos\theta$ given by Equation (7.59), Equation (7.64) describes the parametric equation of a curve that you can plot and show that

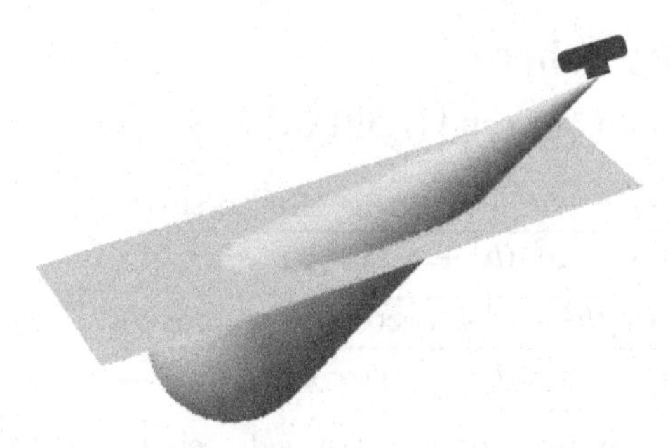

FIGURE 7.8 When the camera is not pointing to the center of the sphere, the image of the sphere is an ellipse because the plane of the photographic plate is not perpendicular to the axis of the *circular* cone formed by the sphere. The plane shown is parallel to the photographic plate.

it is indeed an ellipse. In fact, if you define

$$\cos\vartheta = \sqrt{\frac{b^2 + x_c^2 - a^2}{b^2 - a^2 + x_c^2\cos^2\varphi}}\cos\varphi$$
$$\sin\vartheta = \sqrt{\frac{b^2 - a^2}{b^2 - a^2 + x_c^2\cos^2\varphi}}\sin\varphi, \tag{7.65}$$

then you can show that (7.64) becomes

$$u = \frac{bx_c}{b^2 - a^2} + \frac{a\sqrt{b^2 + x_c^2 - a^2}}{b^2 - a^2}\cos\vartheta$$
$$v = \frac{a}{\sqrt{b^2 - a^2}}\sin\vartheta, \tag{7.66}$$

indicating that the image is an ellipse centered in the uv-plane at $(\frac{bx_c}{b^2-a^2}, 0)$ with semi-major and semi-minor axes, respectively, of

$$\frac{a\sqrt{b^2 + x_c^2 - a^2}}{b^2 - a^2} \quad \text{and} \quad \frac{a}{\sqrt{b^2 - a^2}}.$$

Figure 7.8 shows why the image is an ellipse when the camera is not pointing to the center of the sphere.

7.7.2 Moving Sphere

Now is the time to look at the full-fledged case of a moving sphere not necessarily placed at the origin. I'll presently examine the image of the sphere centered at $(x_c, 0, 0)$ in camera C'. Substituting (7.56) in (7.7) you should be able to show that

$$x' = \gamma\left(x_c + a\sin\theta\cos\varphi - \beta\sqrt{x_c^2 + a^2 + b^2 + 2a(x_c\sin\theta\cos\varphi - b\cos\theta)}\right)$$
$$y' = a\sin\theta\sin\varphi \tag{7.67}$$
$$z' = a\cos\theta.$$

Differentiating these with respect to θ and φ, you can find $\vec{\theta}'$ and $\vec{\varphi}'$ and show that (assuming $\theta \neq 0, \pi$) the component of $\vec{\theta}' \times \vec{\varphi}'$ is given by

$$\frac{(\vec{\theta}' \times \vec{\varphi}')_x}{a^2 \sin^2\theta} = \cos\varphi$$
$$\frac{(\vec{\theta}' \times \vec{\varphi}')_y}{a^2 \sin^2\theta} = \gamma \sin\varphi \left(1 - \frac{\beta x_c}{\sqrt{x_c^2 + a^2 + b^2 + 2a(x_c \sin\theta\cos\varphi - b\cos\theta)}}\right) \tag{7.68}$$
$$\frac{(\vec{\theta}' \times \vec{\varphi}')_z}{a^2 \sin\theta} = \gamma \left(\cos\theta - \frac{\beta(x_c\cos\theta + b\sin\theta\cos\varphi)}{\sqrt{x_c^2 + a^2 + b^2 + 2a(x_c \sin\theta\cos\varphi - b\cos\theta)}}\right).$$

If you substitute these components as well as the coordinates of (7.67) in Equation (C.17) and persevere long enough, you will be able to obtain

$$\gamma(x_c\sin\theta\cos\varphi - b\cos\theta + a) \times \left(1 - \frac{\beta(x_c + a\sin\theta\cos\varphi)}{\sqrt{x_c^2 + a^2 + b^2 + 2a(x_c \sin\theta\cos\varphi - b\cos\theta)}}\right) = 0. \tag{7.69}$$

In Problem 7.36 I have asked you to show that the second parentheses in (7.69) cannot be zero. Therefore, as I promised earlier, you obtain the same relation between θ and φ as in the stationary case (7.57), and Equations (7.58) and (7.59) describe the parametric equation of the image curve in C'.

If you substitute (7.59) in (7.58) and plot the resulting parametric equation, you get an ellipse regardless of the values of x_c, a, b, and β. Therefore, you should suspect that a reparametrization could simplify this parametric equation and make the properties of the ellipse more transparent. Up to here, I have pulled all the reparametrizations out of a hat! And you may have wondered how I got reparametrizations like (7.61) and (7.65). Sometimes, trial and error is the only way. Other times, experimenting with graphs and a little familiarity with geometry can be very helpful. Now I let you in on a neat secret!

All the ellipses you get have their major axes along the u'-axis. This should tell you that if you can find the two extreme values of u', you can determine both the coordinates of the center and the length of the major axis of the ellipse. In fact, if you denote the coordinates of the center by $(u'_c, 0)$ and the semi-major axis by A, then

$$u'_c = \frac{u'_{\text{max}} + u'_{\text{min}}}{2}, \quad A = \frac{|u'_{\text{max}} - u'_{\text{min}}|}{2},$$

where u'_{max} and u'_{min} are the points where the ellipse crosses the u'-axis on the right and left, respectively. The absolute value sign is necessary because A is a positive quantity. Now, the ellipse crosses the u'-axis when $v' = 0$, which, by the second equation in (7.58), corresponds to $\varphi = 0, \pi$. Hence,

$$u'_c = \frac{u'\,|_{\varphi=0} + u'\,|_{\varphi=\pi}}{2} = \frac{b\beta^2\gamma x_c\sqrt{x_c^2+b^2-a^2} - b\beta\gamma(x_c^2+b^2)}{(\sqrt{x_c^2+b^2-a^2}-\beta x_c)(x_c^2+b^2)}$$

$$A = \frac{\left|u'\,|_{\varphi=0} - u'\,|_{\varphi=\pi}\right|}{2} = \frac{a\gamma(b^2+x_c^2/\gamma^2)}{\left|\sqrt{x_c^2+b^2-a^2}-\beta x_c\right|(x_c^2+b^2)}. \tag{7.70}$$

If the image curve is indeed an ellipse, we should be able to right it as

$$\begin{aligned} u' - u'_c &= A\cos\phi \\ v' &= B\sin\phi, \end{aligned} \tag{7.71}$$

where $\cos\phi$ and $\sin\phi$ are functions of φ and B is independent of it. Substituting (7.70) and the first equations of (7.58) and (7.59) in the first equation of (7.71), you can calculate $\cos\phi$. As the calculation is cumbersome, a computer algebra program is very helpful. The result turns out to be the same as (7.61). Now evaluate $\sin\phi$ and substitute it in the second equation of (7.71) to find B. You should show that

Image of a moving sphere centered at $(x_c, 0, 0)$ is an ellipse *elongated* along the direction of motion!

$$B = \frac{a\sqrt{b^2+x_c^2/\gamma^2}}{\left|\sqrt{x_c^2+b^2-a^2}-\beta x_c\right|\sqrt{x_c^2+b^2}} = \sqrt{\frac{x_c^2+b^2}{x_c^2+\gamma^2 b^2}}\,A. \tag{7.72}$$

In the case of approach ($\beta x_c < 0$), the value of B is smaller than the radius in (7.63) of the image of the stationary sphere (show this!). This is *not* due to any motional shrinkage because B is perpendicular to the direction of motion. It is due to the fact that the light rays captured by the camera are coming from a location that is farther away than when the sphere is stationary.

Learning from Archimedes!

Remark 7.7.2. The preceding discussion illustrates a beautiful example of the connection between numerical and graphical analysis on the one hand and analytical formulation on the other. The latter is, of course, what we are after. The discovery of the fruitfulness of this connection is due to Archimedes, who experimented with vessels and liquids to prove some fundamental theorems of geometry. In the example discussed here, the graph helped me to discover that the curve was an ellipse, and having familiarity with the properties of an ellipse helped me find the center, semi-major, and semi-minor axes of the ellipse in terms of the relevant parameters of the problem. ■

Some interesting features of the image of a moving sphere emerge by examining Equations (7.70) and (7.72). Firstly, when the sphere is not moving, $\beta = 0$, $\gamma = 1$, and these equations yield

$$A_{\text{rest}} = B_{\text{rest}} = \frac{a}{\sqrt{x_c^2+b^2-a^2}}$$

as in Equation (7.63).

Secondly, at large distances from the camera, i.e., when $x_c >> b$, the image in C' approaches a circle of radius

$$r' = \lim_{|x_c|\to\infty} B = \lim_{|x_c|\to\infty} A = \frac{a}{|x_c|}\sqrt{\frac{1+(x_c/|x_c|)\beta}{1-(x_c/|x_c|)\beta}},$$

while the radius of the image in C approaches

$$r = \lim_{|x_c|\to\infty} B_{\text{rest}} = \lim_{|x_c|\to\infty} A_{\text{rest}} = \frac{a}{|x_c|} = r'|_{\beta=0}.$$

Therefore,

$$r' = \sqrt{\frac{1+(x_c/|x_c|)\beta}{1-(x_c/|x_c|)\beta}}\, r, \tag{7.73}$$

showing that the image in C' is smaller on approach (when x_c and β have opposite signs) and larger on recession (when x_c and β have the same sign). This is consistent with the small angle approximation formula (7.11).

There is an intuitive physical explanation for the fact that the sphere appears as a circle when far away. The elongation of the photograph of any object occurs when there is a sufficiently large separation between the emission of light from the two extremes (the trailing and leading points) of the object or, equivalently, when there is a sufficiently large difference between the distances from the pinhole to the two extremes of the object. When the object is far away, this difference is small, and the elongation is not pronounced.

Thirdly, if the cameras take the picture of the sphere when it is at the origin,

$$A = \gamma B = \frac{\gamma a}{\sqrt{b^2 - a^2}},$$

which agrees with Equation (7.54).

Finally, we can investigate conditions under which the image of the moving sphere is a circle. This happens if and only if $A = B$ in (7.71), which by (7.72) happens if and only if $x_c^2 + b^2 = x_c^2 + \gamma^2 b^2$, or $b^2(1-\gamma^2) = 0$. The $\gamma = 1$ solution corresponds to the stationary sphere, already discussed in Section 7.7.1. For the moving sphere, the only way to capture a circular image is to have $b = 0$, i.e., to move the camera to the origin, so that the sphere is moving directly toward or away from C'. Then the images in C and C' are circles of radii

Only a directly approaching or receding sphere has a circular image.

$$r' = \frac{a}{\gamma\left|\sqrt{x_c^2 - a^2} - \beta x_c\right|} \quad \text{and} \quad r = \frac{a}{\sqrt{x_c^2 - a^2}},$$

with

$$r' = \frac{\sqrt{x_c^2 - a^2}}{\gamma\left|\sqrt{x_c^2 - a^2} - \beta x_c\right|}\, r. \tag{7.74}$$

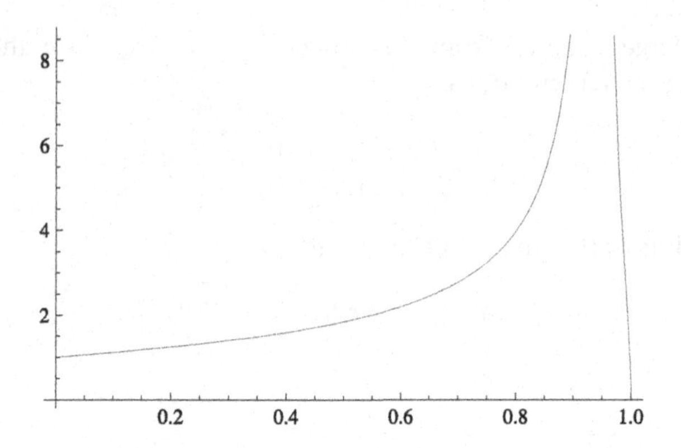

FIGURE 7.9 The behavior of the ratio r'/r as a function of β for $x_c = +3$.

Therefore, as already mentioned in Note 7.6.4, Penrose's argument that the image of a sphere is a circle applies only to the case where the sphere is either approaching directly toward or receding directly away from the camera.

From (7.74), it is quite obvious that $r' < r$ on approach ($\beta x_c < 0$). On recession ($\beta x_c > 0$), we expect r' to be larger than r because of the limiting case of Equation (7.11). However, this may not be the case for all values of β. In fact, the denominator of Equation (7.74) can be made arbitrarily small! This is the situation against which I warned in Remarks 7.7.1 and C.1.1, because $b = 0$ here. Thus, r'/r starts at 1, increases to infinity at $\beta = \sqrt{x_c^2 - a^2}/x_c$, decreases to 1 at $\gamma = 1 + 2(x_c/a)^2$, and continues to decrease for larger values of γ. Figure 7.9 shows this behavior for $x_c = +3$. Problem 7.14 looks at the behavior of a cube when directly approaching or receding from the camera. The behavior of a sphere is a little different from that of a cube because in the latter case, the camera collects photons coming from the single side facing it. Thus, the boundary from which image-forming photons emanate is a fixed square. In the case of the sphere, on the other hand, less and less of the surface gets photographed as the sphere approaches the camera.

7.8 PROBLEMS

7.1. Consider the cube of Figure 7.10(a), which moves with speed β along the positive x-axis relative to observer O'. Define the angle θ as $\sin\theta \equiv \beta$. It is argued[11] that—since $\overline{A_b A_f}$ is perpendicular to the direction of motion and thus does not change length—it takes light $2a/c$ to travel from A_b to A_f according to O'. During this time, the cube moves a distance of

$$d = (\beta c)(2a/c) = 2a\beta \equiv 2a \sin\theta,$$

[11] See, for example, [Weisskopf 60].

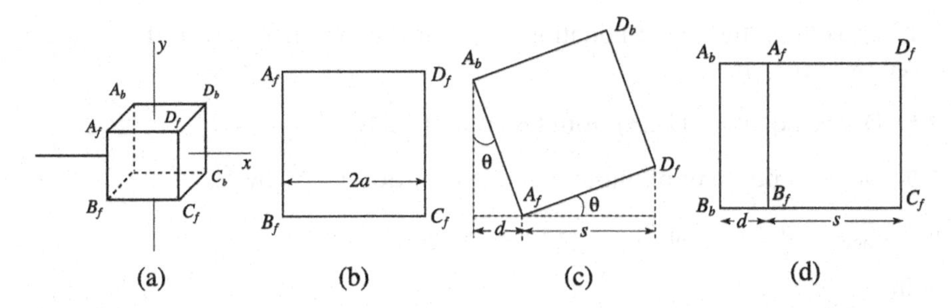

FIGURE 7.10 (a) The distant cube of side length $2a$ is centered at the origin of O, its rest frame. (b) The cube as it appears on the photographic plate of camera C in O. (c) The orientation of the top face of the cube when the cube is rotated counterclockwise by an angle $\theta = \sin^{-1}\beta$ about the y-axis. (d) The image of the rotated cube on the photographic plate of camera C in O.

exposing the trailing face of the cube to the camera. Furthermore, the top and bottom sides, $\overline{A_f D_f}$ and $\overline{B_f C_f}$, shrink from $2a$ to

$$s = 2a\sqrt{1-\beta^2} = 2a\sqrt{1-\sin^2\theta} = 2a\cos\theta.$$

As Figures 7.10(c) and 7.10(d) indicate, the combination of these two effects manifests itself as the appearance of a rotation on the photographic plate. To see what is wrong with this argument, do the following.

(a) Consider two events: emission of light from A_b and its arrival at A_f. Write the spacetime coordinates of these two events in O where the cube is at rest, assuming that the emission event takes place at time t_b.

(b) Lorentz transform these two events to O'.

(c) What is the time difference $t'_f - t'_b$ between the emission at A_b and arrival at A_f according to O'? How far does the cube move during this time according to O'? Do these results agree with the argument for rotation?

(d) You can derive the result of (c) without using Lorentz transformation. Simply note that according to O', the emission and arrival events occur at the two ends of the hypotenuse of a right triangle whose other two sides are of lengths $2a$ and $\beta(t'_f - t'_b)$.

7.2. Use the space coordinates of (7.7) to calculate $|\mathbf{r}'_i - \mathbf{r}_0|$ and show that the result agrees with the right-hand side of t'_i.

7.3. Substitute Equations (7.7) and (7.8) in (7.9) and follow the steps similar to the ones that led to (7.5) and (7.6) to obtain (7.10).

7.4. Derive Equations (7.14) and (7.15).

7.5. Let $x_q = 0$ and derive (7.17) from (7.16).

7.6. Derive Equation (7.19) from Equation (7.18).

7.7. Show that the line of Equation (7.20) makes an angle θ with the x'-axis given by $\cos\theta = \beta$.

7.8. Derive Equation (7.23) from Equation (7.22).

7.9. Define a new pair of photographic plate coordinates by

$$u' = u'_{\text{new}} \cos\theta + v'_{\text{new}} \sin\theta, \quad v' = -u'_{\text{new}} \sin\theta + v'_{\text{new}} \cos\theta$$

with

$$\sin\theta = \frac{x_0}{\sqrt{\gamma^2 y_0^2 + x_0^2}}, \quad \cos\theta = \frac{\gamma y_0}{\sqrt{\gamma^2 y_0^2 + x_0^2}}.$$

Note that the new axes are obtained from the old by a clockwise rotation of angle θ.

(a) Substitute these values in Equation (7.23) and simplify to show that that equation reduces to

$$(x_0^2 + y_0^2)u'^2_{\text{new}} - \beta^2\gamma^2 y_0^2 v'^2_{\text{new}} - \gamma^2\beta^2 y_0^2 = 0,$$

which is the equation of a hyperbola.

(b) What are the coordinates of the center of this hyperbola in the new coordinate system?

(c) What are the equations of the axes of the hyperbola in the new coordinate system? In the old coordinate system?

7.10. Derive Equations (7.25) and (7.26).

7.11. Derive Equations (7.28) and (7.29).

7.12. Using Equation (7.28), show that $u'_{A_b} > u'_{A_f}$ and $v'_{A_b} < v'_{A_f}$ for all b and β. Hint: Show that each of the two terms on the right-hand side of u'_{A_b} is larger than the corresponding term on the right-hand side of u'_{A_f}.

7.13. Recall that the aberration formula involves the angle with the direction of motion of a ray of light from a single point source. Therefore, if the aberration formula is to be interpreted as the angle between *two* light rays originating from an object (as is the case when image formation of the object is being considered), then one of those rays ought to be along the direction of motion. Show that when that is the case, then (7.5) reduces to the aberration formula (6.25).

7.14. Let's see what a cube looks like when it moves toward or away from the camera. Place the camera at the origin (i.e., let $b = 0$) and point it to the right face of the approaching cube as shown in Figure 7.3(b). Let the center of the cube have coordinates $(x_c, 0, 0)$ in O.

(a) Write down the coordinates of all the eight corners of the cube in O.

(b) From (a), (7.8), and (7.10) with $b = 0$, obtain

$$\sin(\alpha'_{D_f D_b}/2) = \frac{a}{\gamma\left[\sqrt{(x_c+a)^2+2a^2}-\beta(x_c+a)\right]},$$
$$\sin(\alpha'_{A_f A_b}/2) = \frac{a}{\gamma\left[\sqrt{(x_c-a)^2+2a^2}-\beta(x_c-a)\right]}, \qquad (7.75)$$

the angles formed by sides $\overline{D_f D_b}$ and $\overline{A_f A_b}$ (and the other three sides of the leading and trailing faces) at the pinhole.

(c) For $x_c < 0$ and $\beta > 0$, show that $\alpha'_{D_f D_b} > \alpha'_{A_f A_b}$, with the other three angles satisfying the same kind of inequality. Thus, the trailing face is hidden behind the leading face when the cube is approaching the camera.

(d) For $x_c < 0$ and $\beta < 0$, show that $\alpha'_{D_f D_b} > \alpha'_{A_f A_b}$ if and only if $|x_c|/a > |\beta|\sqrt{2\gamma^2+1}$.

(e) Set $\beta = 0$ in (7.75) to find $\sin(\alpha_{D_f D_b}/2)$. Then show that, for approach, $\alpha'_{D_f D_b} < \alpha_{D_f D_b}$, indicating that the picture in C' is smaller than in C.

(f) The case of recession is more complicated. The limiting case of (7.11) may lead you to believe that the C' image is larger. For the moment assume that and see what condition makes the assumption true. For definiteness, let $\beta > 0$ and $x_c > 0$. Then, for $\alpha'_{A_f D_f} > \alpha_{A_f D_f}$ to hold, you should have

$$\gamma\left[\sqrt{(x_c+a)^2+2a^2}-\beta(x_c+a)\right] < \sqrt{(x_c+a)^2+2a^2}.$$

Show that this is equivalent to

$$[(x_c+a)^2 - X_-][(x_c+a)^2 - X_+] > 0, \quad X_- < 0$$

for certain X_+ and X_- that you have to find. Thus, for the inequality to hold, $(x_c+a)^2$ must be larger than X_+. Now show that

$$\frac{(x_c+a)^2}{a^2} > \gamma - 1. \qquad (7.76)$$

When the cube is sufficiently far away, this condition holds, but not in general. Therefore, if the cube is close enough to the camera C' when the pictures are taken, the image in C can be larger.

(g) Show that (7.76) above holds also when $\beta < 0$ and $x_c < 0$.

(h) Plot the ratio

$$\frac{\sin(\alpha'_{D_f D_b}/2)}{\sin(\alpha_{D_f D_b}/2)}$$

for $x_c = +3a$ as a function of β for $-1 < \beta < 1$ to get a feel for the magnification of the image in C' relative to the image in C.

(i) Verify directly that, when $|x_c| \to \infty$, the ratio in (h) reduces to

$$\frac{\alpha'_{D_f D_b}}{\alpha_{D_f D_b}} \approx \sqrt{\frac{1+\beta|x_c|/x_c}{1-\beta|x_c|/x_c}},$$

as in (7.11).

7.15. Derive Equations (7.33) and (7.34).

7.16. Show that if $\hat{\mathbf{v}}$ of Equation (7.35) is to make an angle φ' with the x-axis, then

$$\eta_z = \pm \frac{\eta_x b \cos^2\alpha}{\sqrt{y^2 + (b\cos^2\alpha)^2}} \tan\varphi'.$$

7.17. With $\hat{\mathbf{e}}(\phi, \beta)$ defined as in Equation (7.40), show that

$$\hat{\mathbf{e}}(\phi,\beta)\cdot\hat{\mathbf{e}}(0,\beta) = 1 - \frac{\sin^2\alpha(1-\cos\phi)}{\gamma^2(1-|\vec{\beta}|\sin\alpha\cos\phi)(1-|\vec{\beta}|\sin\alpha)}.$$

Now prove that the dot product has a maximum at $\phi = 0$ and a minimum at $\phi = \pi$.

7.18. Provide all the missing steps leading to Equation (7.41).

7.19. Derive Equation (7.42).

7.20. Derive Equation (7.46).

7.21. Using (7.45) and (7.7), derive Equation (7.49).

7.22. Derive Equations (7.50) and (7.51).

7.23. Obtain Equation (7.52) by substituting (7.49) and (7.51) in (C.17).

7.24. Show that (x', y', z') of Equation (7.53) satisfy

$$\sqrt{x'^2 + y'^2 + (z'-b)^2} = \frac{\sqrt{b^2-a^2}}{b}\gamma(b - \beta a\cos\varphi).$$

7.25. Show that the line from $(0,0,b)$ in the direction of $\hat{\mathbf{e}}_{axis}$ of Equation (7.41) cuts the x-axis at $-|\vec{\beta}|\gamma\sqrt{b^2-a^2}$.

7.26. A point source produces a circular cone of light. It is placed at $(0,0,b)$ with its axis along the z-axis in reference frame O, moving with speed β in the positive direction of the x'-axis of O'. At $t = t' = 0$, when the two origins coincide, the source emits a pulse, producing a circular image of radius a in the xy-plane of O. What is the shape of the image in the $x'y'$-plane of O'? Hint: Look at the *events* of the arrival of light beams to the planes.

7.27. Derive Equation (7.57).

7.28. Starting with Equation (7.57), provide all the missing steps leading to (7.58).

7.29. Define $\tan\alpha \equiv b/(x_c \cos\varphi)$ and show that (7.57) can be written as

$$\sin\theta\cos\alpha - \cos\theta\sin\alpha + \frac{a\cos\alpha}{x_c\cos\varphi} = 0,$$

or

$$\sin(\theta - \alpha) = -\frac{a\cos\alpha}{x_c\cos\varphi}.$$

Now define

$$\sin\eta \equiv \frac{a\cos\alpha}{x_c\cos\varphi}$$

and show that $\theta = \alpha - \eta$. Using the trigonometric identity

$$\cos\alpha = \frac{1}{\sqrt{1+\tan^2\alpha}}$$

write η in terms of φ:

$$\sin\eta = \frac{a}{\sqrt{b^2 + x_c^2\cos^2\varphi}}.$$

Since $b > a$, this shows that our definition of $\sin\eta$ is consistent with the fact that $|\sin\eta| < 1$. With both α and η given in terms of φ, derive (7.59).

7.30. Show that with $\beta = 0$, Equations (7.58) and (7.59) give (7.60).

7.31. Show that with $\cos\phi$ defined as (7.61), Equation (7.62) follows and u and v of (7.60) could be expressed as in (7.63).

7.32. Derive Equation (7.66) from Equations (7.59), (7.64), and (7.65).

7.33. Substitute (7.56) in (7.7) and simplify to obtain (7.67).

7.34. Derive Equation (7.68).

7.35. Substitute Equations (7.67) and (7.68) in Equation (C.17) to get one expression. Multiply out the parentheses in Equation (7.69) to get another expression. Now show that the two expressions are equal.

7.36. For the second parentheses of (7.69) to be zero, you should have

$$\beta^2(x_c + a\sin\theta\cos\varphi)^2 = x_c^2 + a^2 + b^2 + 2a(x_c\sin\theta\cos\varphi - b\cos\theta).$$

(a) Show that this gives a quadratic equation in $\cos\varphi$ whose solution is

$$\cos\varphi = \frac{x_c/\gamma^2 \pm \sqrt{x_c^2/\gamma^2 + \beta^2(a^2+b^2-2ab\cos\theta)}}{\beta^2 a\sin\theta}.$$

(b) Verify that when x_c is positive (negative) and you choose the positive (negative) sign for the square root, $|\cos\varphi| > 1$ is trivially satisfied.

(c) For $x_c < 0$ and the positive sign for the square root, show that the inequality

$$\frac{-|x_c|/\gamma^2 + \sqrt{x_c^2/\gamma^2 + \beta^2[(b - a\cos\theta)^2 + a^2\sin^2\theta]}}{\beta^2 a\sin\theta} > 1$$

is equivalent to the inequality

$$(|x_c| - a\sin\theta)^2 + \gamma^2(b - a\cos\theta)^2 > 0,$$

which is obviously true. Therefore, $\cos\varphi > 1$.

(d) For $x_c > 0$ and the negative sign for the square root, $\cos\varphi$ is negative. Therefore, you have to prove the inequality

$$\frac{x_c/\gamma^2 - \sqrt{x_c^2/\gamma^2 + \beta^2[(b - a\cos\theta)^2 + a^2\sin^2\theta]}}{\beta^2 a\sin\theta} < -1.$$

Show that this leads to the same inequality as in (c), proving that $\cos\varphi < -1$.

7.37. Derive Equation (7.70) from Equation (7.58).

CHAPTER 8

Relativistic Interactions

There are many instances in nature and in laboratories in which particles moving at close to light speed interact with other particles. In a typical situation, two initial particles collide and give rise to two or more particles. This production of new particles is inherently relativistic as the initial energy of the colliding particles is converted into mass of the final product via $E = mc^2$.

8.1 CENTER OF MASS AND LAB FRAMES

The essential ingredient in any collision is conservation of total energy and total momentum, which can be stated succinctly in terms of the total 4-momenta before and after: $\mathbf{p}_{\text{tot}}^{\text{bef}} = \mathbf{p}_{\text{tot}}^{\text{aft}}$, where in each case, $\mathbf{p}_{\text{tot}}$ is the sum of the 4-momenta of all particles involved. The conservation of 4-momentum holds in all reference frames. In practice, two frames are especially important: the **center of mass** (CM) and **lab** frames.

The center of mass and lab frames.

The center of mass is defined as the reference frame in which the initial particles (and therefore, the final particles) have zero total 3-momentum. In the case of two initial particles, to which all the subsequent discussions are restricted, this means that they are moving in opposite directions with the same momentum. The lab frame is defined as the frame in which one of the particles, the **target**, is at rest, on which the other particle, the **projectile**, impinges. I'll use script letter $\mathcal{E}$ to denote energies in the lab frame.

Target and projectile in the lab frame.

Note that (see Note 6.1.1 for notation)

$$\begin{aligned}(\mathbf{p}_1 + \mathbf{p}_2)^2 &\equiv (\mathbf{p}_1 + \mathbf{p}_2) \bullet (\mathbf{p}_1 + \mathbf{p}_2) = \mathbf{p}_1 \bullet \mathbf{p}_1 + \mathbf{p}_2 \bullet \mathbf{p}_2 + 2\mathbf{p}_1 \bullet \mathbf{p}_2 \\ &\equiv \mathbf{p}_1^2 + \mathbf{p}_2^2 + 2\mathbf{p}_1 \bullet \mathbf{p}_2 = m_1^2 + m_2^2 + 2\mathbf{p}_1 \bullet \mathbf{p}_2 \qquad (8.1)\end{aligned}$$

is an invariant quantity that yields the same value in all reference frames. In the CM, where the energies of the initial particles are $E_{1\text{cm}}$ and $E_{2\text{cm}}$ and their momenta $\vec{p}_{1\text{cm}}$ and $\vec{p}_{2\text{cm}}$, I get

$$\mathbf{p}_1 + \mathbf{p}_2 = (E_{1\text{cm}} + E_{2\text{cm}}, \vec{p}_{1\text{cm}} + \vec{p}_{2\text{cm}}) \equiv (E_{\text{cm}}, \vec{0}),$$

Special Relativity. DOI:10.1016/B978-0-12-810411-8.00008-0

where E_{cm} is the total energy in the CM. Therefore, (8.1) gives

$$E_{\text{cm}}^2 = m_1^2 + m_2^2 + 2\mathbf{p}_1 \bullet \mathbf{p}_2. \tag{8.2}$$

And this equality holds in *any* inertial frame, by which I mean that I can go to any frame and calculate the dot product $\mathbf{p}_1 \bullet \mathbf{p}_2$. This may seem surprising, because $\mathbf{p}_1$ and $\mathbf{p}_2$ are different in different frames and I'll get different *expressions* involving energies and 3-momenta. However, once I put in *numbers* in those expressions, I'll get the same numerical value.

Let particle P_2 (don't confuse the "P" of the "particle" here with momentum) be in the lab frame. Then $\mathbf{p}_2 = (m_2, \vec{0})$ and $\mathbf{p}_1 = (\mathcal{E}_1, \vec{p}_1)$. Now evaluate the dot product of (8.2) in the lab frame to obtain

$$\mathbf{p}_1 \bullet \mathbf{p}_2 = (\mathcal{E}_1, \vec{p}_1) \bullet (m_2, \vec{0}) = \mathcal{E}_1 m_2.$$

This gives the formula

$$E_{\text{cm}}^2 = m_1^2 + m_2^2 + 2m_2\mathcal{E}_1, \tag{8.3}$$

which relates the lab energy of the incident particle to the total energy in the CM frame.

The significance of CM lies, among other things, in the determination of the minimum energy required to produce a particle. Consider two particles P_1 and P_2 that collide and form a single third particle P_3. Let the masses of the first two particles be m_1 and m_2. What is the mass M of the third particle? The conservation of 4-momentum in the present situation is

$$\mathbf{p}_1 + \mathbf{p}_2 = \mathbf{P}, \tag{8.4}$$

where $\mathbf{P}$ is the four-momentum of P_3. Since this is a 4-vector equation, all components must be equal. Separate the time and the space parts to get

$$\begin{aligned} p_{01} + p_{02} = P_0 \quad &\text{or} \quad E_1 + E_2 = E \\ \vec{p}_1 + \vec{p}_2 = \vec{P}&, \end{aligned} \tag{8.5}$$

which are simply the conservation of energy and 3-momentum. Square both sides of (8.4) to get

$$m_1^2 + m_2^2 + 2\mathbf{p}_1 \bullet \mathbf{p}_2 = M^2. \tag{8.6}$$

Comparing this with Equation (8.2) shows that $E_{\text{cm}} = M$ or $E_{1\text{cm}} + E_{2\text{cm}} = M$. Furthermore, the left-hand side of the second equation in (8.5) is zero in the CM. Therefore, in the CM, P_3 is produced at rest, meaning that its energy is just its mass M, the minimum energy it can have. Thus, to produce the particle, the sum of the energies of the initial particles *in the CM* must be at least equal to the mass of the particle produced. Equation (8.3), with $E_{\text{cm}} = M$, immediately yields the energy of the projectile particle P_1 in the lab frame:

Minimum energy for particle production.

$$\mathcal{E}_1 = \frac{M^2 - m_1^2 - m_2^2}{2m_2}. \tag{8.7}$$

The energy $\mathcal{E}_1$ is the minimum energy of the impinging particle (in the lab frame) needed to produce P_3.

You can also find the momentum and energy of P_3 in the lab frame. The second equation in (8.5) gives $\vec{P} = \vec{p}_1$, indicating that the final particle moves in the initial direction of P_1 in the lab. The magnitude of $\vec{P}$ can be calculated in terms of energies and masses:

$$|\vec{P}| = |\vec{p}_1| = \sqrt{\mathcal{E}_1^2 - m_1^2}. \tag{8.8}$$

The first equation in (8.5) gives the energy of P_3:

$$\mathcal{E} = \mathcal{E}_1 + m_2. \tag{8.9}$$

Combining Equations (8.8) and (8.9), you can obtain the speed of P_3:

$$V = \frac{|\vec{P}|}{\mathcal{E}} = \frac{\sqrt{\mathcal{E}_1^2 - m_1^2}}{\mathcal{E}_1 + m_2}. \tag{8.10}$$

Example 8.1.1. NUCLEAR EXCITATION
In some nuclear processes, a nucleus absorbs a photon and ends up in a higher energy state. This excitation is not unlike the atomic excitations, except that the energies involved are much larger. In (8.7), if you set $m_1 = 0$, and use the common notation of designating the initial mass of the nucleus by M and the final mass by M^*, then the energy of the photon becomes

$$\mathcal{E}_1 = \frac{M^{*2} - M^2}{2M} \quad \text{or} \quad M^{*2} = M^2 + 2M\mathcal{E}_1.$$

Therefore, M^* is smaller than the sum of the initial energies, $M + \mathcal{E}_1$. The reason is the conservation of 3-momentum: the final nucleus must be moving to conserve the 3-momentum of the initial photon, and this motion takes up some of the initial energy.

Because of the discreteness of the energy levels of a nucleus, the process of excitation takes place only if the initial photon has an energy that makes a transition to a higher level possible. Otherwise, the photon scatters off of the nucleus elastically. For example, the isotope lead-208 has an excitation energy of 2.51 MeV. If the incident photon has less than (or even equal to) this amount of energy, it will not be absorbed, but rather it will scatter off the lead nucleus elastically. If it has more than this energy, it will excite the nucleus and impart the remaining part of its energy to the nucleus as kinetic energy. ■

8.1.1 Lorentz Transformation of CM

Because of the importance of the CM frame, it is instructive to find the Lorentz transformation connecting the CM to an arbitrary frame. For this, you need the relative velocity of the CM with that frame. From the definition of the CM as the frame in which the sum of the 3-momenta of the initial particles is zero, you can find the speed of the CM relative to any other frame in which the two particles have energies E_1 and E_2 and momenta $\vec{p}_1$ and $\vec{p}_2$. Write the first equation of (6.54) for the two initial particles

$$E_1 = \gamma_{cm}(E_{1cm} + \vec{\beta}_{cm} \cdot \vec{p}_{1cm})$$
$$E_2 = \gamma_{cm}(E_{2cm} + \vec{\beta}_{cm} \cdot \vec{p}_{2cm}),$$

and add these equations and use $\vec{p}_{1cm} + \vec{p}_{2cm} = 0$ to get

$$E_1 + E_2 = \gamma_{cm}(E_{1cm} + E_{2cm})$$

or

$$\gamma_{cm} = \frac{E_1 + E_2}{E_{1cm} + E_{2cm}} = \frac{E_1 + E_2}{E_{cm}} = \frac{E_1 + E_2}{\sqrt{m_1^2 + m_2^2 + 2(E_1 E_2 - \vec{p}_1 \cdot \vec{p}_2)}}, \tag{8.11}$$

where in the last step, I used (8.2).

To find the velocity of the CM relative to an arbitrary frame, write the second equation of (6.54) for the two initial particles,

$$\vec{p}_1 = \vec{p}_{1cm} + \gamma_{cm}\vec{\beta}_{cm}\left(E_{1cm} + \frac{\gamma_{cm}}{\gamma_{cm}+1}\vec{\beta}_{cm} \cdot \vec{p}_{1cm}\right)$$
$$\vec{p}_2 = \vec{p}_{2cm} + \gamma_{cm}\vec{\beta}_{cm}\left(E_{2cm} + \frac{\gamma_{cm}}{\gamma_{cm}+1}\vec{\beta}_{cm} \cdot \vec{p}_{2cm}\right),$$

and add them as before to obtain

$$\vec{p}_1 + \vec{p}_2 = \gamma_{cm}\vec{\beta}_{cm}(E_{1cm} + E_{2cm}) = \gamma_{cm}\vec{\beta}_{cm}E_{cm}$$

Center of mass velocity in terms of initial momenta and energies.

or, after using (8.11),

$$\vec{\beta}_{cm} = \frac{\vec{p}_1 + \vec{p}_2}{E_1 + E_2}. \tag{8.12}$$

For the important case of the lab frame where $E_2 = m_2$ and $\vec{p}_2 = \vec{0}$, you get

$$\vec{\beta}_{cm} = \frac{\vec{p}_1}{\mathcal{E}_1 + m_2}, \quad \gamma_{cm} = \frac{\mathcal{E}_1 + m_2}{\sqrt{m_1^2 + m_2^2 + 2m_2\mathcal{E}_1}}, \tag{8.13}$$

where I used (8.3) to find γ_{cm}.

Now that you have the relative velocity of the CM and lab frames, you can find the energy of the projectile in the CM, E_{1cm}, in terms of its lab energy $\mathcal{E}_1$. The relevant Lorentz transformation is

$$E_{1cm} = \gamma_{cm}(\mathcal{E}_1 - \vec{\beta}_{cm} \cdot \vec{p}_1). \tag{8.14}$$

Substituting (8.13) in this equation, and doing a little algebra, you get

$$E_{1cm} = \frac{m_1^2 + m_2\mathcal{E}_1}{\sqrt{m_1^2 + m_2^2 + 2m_2\mathcal{E}_1}}. \tag{8.15}$$

This is an important formula when particle production is of interest. If $\mathcal{E}_1$ is the minimum lab energy for the production of a particle, then $E_{1\text{cm}}$ is the corresponding energy in the center of mass.

As a numerical example, suppose P_1 and P_2 are protons and you need $\mathcal{E}_1 = 1000$ GeV to produce a particle. Then

$$E_{1\text{cm}} = \frac{m_p^2 + m_p\mathcal{E}_1}{\sqrt{2m_p^2 + 2m_p\mathcal{E}_1}} = \sqrt{\frac{m_p^2 + m_p\mathcal{E}_1}{2}}$$

$$= \sqrt{\frac{(938.3)^2 + (938.3)(10^6)}{2}}\ \text{MeV} = 30.6\ \text{GeV}.$$

This shows a significant reduction in the required energy of the projectile. The target, being identical to the projectile will, of course, have the same amount of energy in the CM. However, the total energy of 61.2 GeV is only a fraction of the energy required in the lab. If the lab frame can somehow turn into the CM frame, particle production can be a lot cheaper. That is why from the early 70s on, with the construction of Fermilab in the US, *circular* accelerators have been colliding two beams of particles moving in opposite directions. For very high energies, in which masses are negligible, (8.15) gives

The advantage of colliding high energy beams of particles moving in opposite directions.

$$E_{\text{cm}} \sim 2E_{1\text{cm}} \sim \sqrt{2m_2\mathcal{E}_1}.$$

Thus, generally,

Note 8.1.2. *The energy needed in the CM frame goes as the square root of the energy needed in the lab frame.*

8.2 TWO-PARTICLE COLLISION

In this section, I consider a common collision with two particles in the initial state and two in the final. Let P_1 and P_2 be the initial particles, and P_3 and P_4, the final. The conservation of 4-momentum becomes

$$\mathbf{p}_1 + \mathbf{p}_2 = \mathbf{p}_3 + \mathbf{p}_4. \tag{8.16}$$

On the one hand, squaring both sides of (8.16) yields

$$m_1^2 + m_2^2 + 2\mathbf{p}_1 \bullet \mathbf{p}_2 = m_3^2 + m_4^2 + 2\mathbf{p}_3 \bullet \mathbf{p}_4, \tag{8.17}$$

which actually holds in any inertial frame. On the other hand, separating the time and the space parts gives the conservation of energy and momentum:

$$\begin{aligned} E_1 + E_2 &= E_3 + E_4 \\ \vec{p}_1 + \vec{p}_2 &= \vec{p}_3 + \vec{p}_4. \end{aligned} \tag{8.18}$$

I'll look at the collision in two cases: in the first case, one of the initial particles is at rest, so that the process takes place in the lab frame; the second case takes place in the center of mass frame and answers the question of what the minimum energy required of the first particle should be to produce the final two particles. For the first case, let P_2 be at rest. Then $\mathbf{p}_2 = (m_2, 0, 0, 0)$, and evaluating (8.17) under this assumption yields

$$m_1^2 + m_2^2 + 2m_2\mathcal{E}_1 = m_3^2 + m_4^2 + 2(\mathcal{E}_3\mathcal{E}_4 - \vec{p}_3 \cdot \vec{p}_4) \tag{8.19}$$

and Equation (8.18) becomes

$$\begin{aligned} \mathcal{E}_1 + m_2 &= \mathcal{E}_3 + \mathcal{E}_4 \\ \vec{p}_1 &= \vec{p}_3 + \vec{p}_4. \end{aligned}$$

Solving for $\mathcal{E}_4$ and $\vec{p}_4$ from these equations and substituting the results in (8.19) yields (after some algebra and using $\mathcal{E}_3^2 - |\vec{p}_3|^2 = m_3^2$)

$$m_1^2 + m_2^2 + 2m_2\mathcal{E}_1 = m_4^2 - m_3^2 + 2\mathcal{E}_3(\mathcal{E}_1 + m_2) - 2\vec{p}_1 \cdot \vec{p}_3. \tag{8.20}$$

Scattering angle.

Define the **scattering angle** of a final particle as the angle that that particle makes with the initial momentum of the projectile. Then, $\vec{p}_1 \cdot \vec{p}_3 = |\vec{p}_1||\vec{p}_3|\cos\theta_3$, where θ_3 is the scattering angle of P_3. Once the energy $\mathcal{E}_1$ of the projectile is known, Equation (8.20) gives the energy of P_3 as a function of the scattering angle ($|\vec{p}_1|$ and $|\vec{p}_3|$ are related to $\mathcal{E}_1$ and $\mathcal{E}_3$, respectively).

Example 8.2.1. COMPTON SCATTERING

Compton scattering.

The particle nature of light, which had been proposed by Einstein in his explanation of the photoelectric effect, was demonstrated by Compton in what is now called the **Compton scattering**. In this scattering, a photon of energy $\mathcal{E}$ is scattered off a stationary electron of mass m_e. The scattered photon is detected at an angle θ from the direction of the incident photon. What is the change in the wavelength of the photon as a function of θ?

In (8.20), let P_1 denote the incident photon, P_2 the stationary electron, P_3 the scattered photon, and P_4 the scattered electron. Let $\mathcal{E}'$ denote the energy of the scattered photon. Then, with $m_1 = m_3 = 0$, Equation (8.20) becomes

$$m_e^2 + 2m_e\mathcal{E} = m_e^2 + 2\mathcal{E}'(\mathcal{E} + m_e) - 2\mathcal{E}\mathcal{E}'\cos\theta,$$

or

$$m_e\mathcal{E} = \mathcal{E}'(\mathcal{E} + m_e) - \mathcal{E}\mathcal{E}'\cos\theta \Rightarrow m_e(\mathcal{E} - \mathcal{E}') = \mathcal{E}\mathcal{E}'(1 - \cos\theta).$$

Restore the factors of c and use the Planck-Einstein relation, $\mathcal{E} = hc/\lambda$, to obtain

$$m_e c^2\left(\frac{hc}{\lambda} - \frac{hc}{\lambda'}\right) = \left(\frac{hc}{\lambda}\right)\left(\frac{hc}{\lambda'}\right)(1 - \cos\theta),$$

which can be simplified to

$$\Delta\lambda \equiv \lambda' - \lambda = \frac{h}{m_e c}(1 - \cos\theta) \equiv \lambda_c(1 - \cos\theta), \tag{8.21}$$

where $\lambda_c = h/m_e c$ is called the **Compton wavelength** of the electron. By measuring the difference between the wavelengths of scattered and incident photons, Compton could verify Equation (8.21) and demonstrate that light had particle property. ■

Compton wavelength.

Example 8.2.2. ELASTIC SCATTERING OF PARTICLES OF EQUAL MASS
Equations (8.19) and (8.20) simplify considerably when all masses are equal. The first equation is used to calculate the "opening angle," i.e., the angle between the final scattered particles. In fact, (8.19) reduces to

$$m\mathcal{E}_1 = \mathcal{E}_3\mathcal{E}_4 - |\vec{p}_3||\vec{p}_4|\cos\theta_{34} \quad \text{or} \quad \cos\theta_{34} = \frac{\mathcal{E}_3\mathcal{E}_4 - m\mathcal{E}_1}{|\vec{p}_3||\vec{p}_4|}, \tag{8.22}$$

assuming that $|\vec{p}_3||\vec{p}_4| \neq 0$. Square the second equation of (8.22), use

$$|\vec{p}|^2 = \mathcal{E}^2 - m^2 = (\mathcal{E} - m)(\mathcal{E} + m),$$

and substitute for $\mathcal{E}_4$ from energy conservation to obtain

$$\begin{aligned} \cos^2\theta_{34} &= \frac{[\mathcal{E}_3(\mathcal{E}_1 - \mathcal{E}_3 + m) - m\mathcal{E}_1]^2}{(\mathcal{E}_3 - m)(\mathcal{E}_3 + m)(\mathcal{E}_1 - \mathcal{E}_3)(\mathcal{E}_1 - \mathcal{E}_3 + 2m)} \\ &= \frac{[(\mathcal{E}_1 - \mathcal{E}_3)(\mathcal{E}_3 - m)]^2}{(\mathcal{E}_3 - m)(\mathcal{E}_3 + m)(\mathcal{E}_1 - \mathcal{E}_3)(\mathcal{E}_1 - \mathcal{E}_3 + 2m)} \\ &= \frac{(\mathcal{E}_1 - \mathcal{E}_3)(\mathcal{E}_3 - m)}{(\mathcal{E}_3 + m)(\mathcal{E}_1 - \mathcal{E}_3 + 2m)}. \end{aligned} \tag{8.23}$$

In the non-relativistic case, where $\mathcal{E}_1 \approx m \approx \mathcal{E}_3$, this leads to a familiar result, namely that, when $\vec{p}_3 \neq \vec{0} \neq \vec{p}_4$, the opening angle is always 90° in the elastic scattering of two particles with the same mass, one of which is (initially) stationary. In the relativistic case, the angle is less than 90° because $\mathcal{E}_3 > m$ and $\mathcal{E}_1 > \mathcal{E}_3$, so that $\cos\theta_{34} > 0$.

Another interesting result, whose derivation is left as Problem 8.8, is

$$\tan\theta_3 \tan\theta_4 = \frac{2}{\gamma_1 + 1},$$

where θ_3 and θ_4 are the scattering angles of the scattered particles and γ_1 is the gamma factor of the projectile. This also reduces to the classical result, because in that case, $\gamma_1 \to 1$, and

$$\tan\theta_3 = \frac{1}{\tan\theta_4} = \cot\theta_4 = \tan\left(\frac{\pi}{2} - \theta_4\right) \Rightarrow \theta_3 + \theta_4 = \frac{\pi}{2}.$$

In the other extreme, the ultra-relativistic case, m could be neglected compared to the energies. Then Equation (8.23) gives $\cos^2\theta_{34} \approx 1$ indicating an opening angle of zero. In fact, the result applies even to non-equal masses, because the masses in (8.20) can be neglected compared to the energies even if they are not equal. This ultra-relativistic forward scattering is behind the **jet events** in high energy scattering of hadrons in which one of the constituent quarks would jettison in a forward direction carrying other quarks with it that eventually form hadrons. ■

Jet events in ultra-relativistic collisions.

8.2.1 Minimum Two-Particle Production Energy

For the second case of interest in collision theory, namely finding the minimum energy required for the production of two particles, you have to set

$E_3 = m_3$ and $E_4 = m_4$, because these are the minimum energies that the final particles could have. These two conditions automatically demand that $\vec{p}_3 = 0 = \vec{p}_4$. Therefore, $\vec{p}_1 = -\vec{p}_2$, i.e., you are in the CM frame. In this frame, in the most general case, the final two particles also have equal and opposite momenta because of the second equation in (8.18). However, for the minimum production energy each of these two final momenta is individually zero.

With the 4-momenta of P_3 and P_4 being, respectively, $(m_3, \vec{0})$ and $(m_4, \vec{0})$, the right-hand side of Equation (8.17) becomes $(m_3 + m_4)^2$. Thus, you get

$$m_1^2 + m_2^2 + 2\mathbf{p}_1 \bullet \mathbf{p}_2 = (m_3 + m_4)^2. \tag{8.24}$$

Although the right-hand side is evaluated in the CM, the result is the value of an *invariant* quantity. Had you evaluated it in any other frame, you would have gotten the same numerical value, even though the algebraic expression would not have been of the form $(m_3 + m_4)^2$. The point of this discussion is that (8.24) is still an invariant expression and you can evaluate the dot product on the left in any frame you wish.

Example 8.2.3. COSMIC RAY ENERGIES

An interesting application of (8.24) is when the second particle is a photon whose momentum is opposite (but not necessarily equal) to the first particle. This gives $\mathbf{p}_1 = (E_1, \vec{p}_1)$, $\mathbf{p}_2 = (E_2, \vec{p}_2)$, and

$$\mathbf{p}_1 \bullet \mathbf{p}_2 = E_1E_2 - \vec{p}_1 \cdot \vec{p}_2 = E_1E_2 + |\vec{p}_1|E_2 = E_2(E_1 + |\vec{p}_1|)$$

because $|\vec{p}_2| = E_2$. Substitute this in (8.24), and note that $m_2 = 0$, to obtain

$$m_1^2 + 2E_2(E_1 + |\vec{p}_1|) = (m_3 + m_4)^2,$$

or

$$E_1 + |\vec{p}_1| = \frac{(m_3 + m_4)^2 - m_1^2}{2E_2} = \frac{M^2}{2E_2}, \quad \text{where} \quad M^2 \equiv (m_3 + m_4)^2 - m_1^2. \tag{8.25}$$

Transfer E_1 to the right-hand side, square both sides, use $|\vec{p}_1|^2 = E_1^2 - m_1^2$, and simplify to obtain

$$E_1 = \frac{M^2}{4E_2} + \frac{m_1^2 E_2}{M^2}. \tag{8.26}$$

Cosmic rays are extremely energetic particles, mostly protons, bombarding the Earth's atmosphere. The energies of these protons have been measured to be up to 3×10^{20} eV, or more than 20 million times the highest energy available at the Large Hadron Collider! On their journey to Earth, these protons pass through the cosmic background radiation (see Section 12.5), an electromagnetic wave blanketing the entire universe, whose photons have energies of about 0.0002 eV in the rest frame of the Earth, or the solar system, or the Milky Way, or the local group of galaxies.[1] If the protons are energetic enough, and the photons are abundant enough, then the protons can be stopped

[1] All these frames are almost the same, because their velocities relative to one another are much smaller than the speed of light.

by photo-production of pions[2] in reactions such as

$$p+\gamma \to n+\pi^+ \quad \text{or} \quad p+\gamma \to p+\pi^0, \tag{8.27}$$

where p stands for proton, γ (not to be confused with the Lorentz factor!) for photon, and n for neutron.

What is the minimum energy required for the second process in (8.27) to go? Use (8.25) with $m_1 = m_3 = 938.3$ MeV, $m_4 = 135$ MeV, and $E_2 = 0.0002$ eV, to obtain

$$M^2 = (1.0733 \times 10^9 \text{ eV})^2 - (9.383 \times 10^8 \text{ eV})^2 = 2.71 \times 10^{17} \text{ (eV)}^2.$$

This gives a second term of (8.26), which is less than 0.0002 eV and therefore completely negligible compared to the first term. Thus, the minimum energy a proton needs to be stopped (remember that the proton on the right-hand side of (8.27) does not move when the projectile has minimum energy) when moving in the cosmic background radiation is

$$E_1 = \frac{2.71 \times 10^{17} \text{ (eV)}^2}{4 \times 0.0002 \text{ eV}} = 3.39 \times 10^{20} \text{ eV}. \tag{8.28}$$

Closeness of the source of cosmic rays.

If the cosmic ray protons have energies larger than this when they leave their source, and travel a long way, so that they have enough time to interact with the photons in the cosmic background radiation, they will lose almost all of their energy and will be almost stopped on their way and will not reach Earth. The mean free path[3] for the reactions in (8.27) is calculated to be about 10 million light years. It takes only several mean free paths to stop the protons. Since we *do* detect protons of energies calculated in (8.28), we have to conclude that they are coming from sources not more than a few 10 million light years away. This is only several times the scale of the local group of galaxies. The origin of the cosmic rays, therefore, is, on a cosmic scale, close to home. ■

8.2.2 Minimum Multi-Particle Production Energy

Equation (8.24) can be generalized to more than two final particles. What is required is that all the particles produced have zero momentum. So if you have $N-2$ particles at the end, the conservation of 4-momentum becomes

$$\mathbf{p}_1 + \mathbf{p}_2 = \mathbf{p}_3 + \mathbf{p}_4 + \cdots + \mathbf{p}_N. \tag{8.29}$$

Squaring both sides, you get the same left-hand side as before. For the right-hand side, note that since all momenta are zero,

$$\mathbf{p}_3 + \mathbf{p}_4 + \cdots + \mathbf{p}_N = (m_3 + m_4 + \cdots + m_N, \vec{0}).$$

---

[2] These are particles, denoted by π, that come in three varieties, π^+, π^-, and π^0, and were proposed by Hideki Yukawa to intermediate the nuclear force. The muon, discovered in 1936, was first mistakenly identified as the Yukawa particle. However, later it became clear that muons do not participate in strong nuclear force. The charged pions were discovered in 1947 and the neutral pion in 1950.

[3] The mean free path is the average distance a particle travels "freely" in a medium consisting of other particles before it interacts with one of the latter.

Thus,

$$(\mathbf{p}_3 + \mathbf{p}_4 + \cdots + \mathbf{p}_N) \bullet (\mathbf{p}_3 + \mathbf{p}_4 + \cdots + \mathbf{p}_N) = (m_3 + m_4 + \cdots + m_N)^2,$$

for minimum energy production. Therefore, you can write (8.24) more generally as

$$m_1^2 + m_2^2 + 2\mathbf{p}_1 \bullet \mathbf{p}_2 = (m_3 + m_4 + \cdots + m_N)^2. \tag{8.30}$$

The dot product on the left can be evaluated in any convenient frame. Rewrite the equation as

$$2\mathbf{p}_1 \bullet \mathbf{p}_2 = (m_3 + m_4 + \cdots + m_N)^2 - m_1^2 - m_2^2,$$

define the right-hand side to be M^2, and evaluate the left-hand side in the lab frame to obtain

$$\mathcal{E}_1 = \frac{M^2}{2m_2}, \quad M^2 \equiv (m_3 + m_4 + \cdots + m_N)^2 - m_1^2 - m_2^2. \tag{8.31}$$

This is the minimum energy that the incident particle must have to be able to produce the $N - 2$ final particles.

Example 8.2.4. DISSOCIATION OF A NUCLEUS
A nucleus has Z protons and N neutrons that are bound with a total binding energy BE. The mass $\mathcal{M}$ of the nucleus is therefore (remember that $c = 1$)

$$\mathcal{M} = Zm_p + Nm_n - BE.$$

What is the minimum energy of a particle of mass m_1 that can completely dissociate the nucleus at rest in the lab frame?

For this situation, (8.31) becomes

$$\mathcal{E}_1 = \frac{(Zm_p + Nm_n)^2 - m_1^2 - \mathcal{M}^2}{2\mathcal{M}},$$

which simplifies to

$$\mathcal{E}_1 = BE - \frac{m_1^2 - BE^2}{2\mathcal{M}}. \tag{8.32}$$

For $\mathcal{M} >> m_1$ and $\mathcal{M} >> BE$, which is usually the case, (8.32) reduces to

$$\mathcal{E}_1 \approx BE. \tag{8.33}$$

This is intuitively plausible because, in the CM, the energy needed to separate the nucleons has simply to overcome the BE. But because of its large mass, the rest frame of the nucleus (the lab frame) *is* the CM. ■

8.3 DECAYS

A decay process starts with one particle and ends up with two or more. In a two-particle decay, let the initial 4-momentum be $\mathbf{P}$ and the final four-momenta $\mathbf{p}_1$ and $\mathbf{p}_2$. Hence, the conservation of 4-momentum is $\mathbf{P} = \mathbf{p}_1 + \mathbf{p}_2$, which leads immediately to (8.6). For this situation, the lab frame is the decaying particle's rest frame. This means that $E = M$ and $\vec{p}_1 = -\vec{p}_2$. Thus the lab frame and the CM frame coincide. Using (8.5), you should be able to show that

$$E_1 = \frac{M^2 + m_1^2 - m_2^2}{2M}$$
$$E_2 = \frac{M^2 + m_2^2 - m_1^2}{2M}. \tag{8.34}$$

Thus, in a two-particle decay, the energy of the final particles are completely determined by the masses of the particles involved. In particular, if $m_1 = m_2$, then $E_1 = E_2 = \frac{1}{2}M$, as expected from symmetry considerations.

Example 8.3.1. THE MÖSSBAUER EFFECT
An important application of (8.34) is the nuclear gamma emission. If the first particle is a photon, and the notation is changed slightly, then (8.34) becomes

$$E_\gamma = \frac{M^{*2} - M^2}{2M^*}$$
$$E_M = \frac{M^{*2} + M^2}{2M^*}, \tag{8.35}$$

where M^* denotes the mass of the initial (excited) nucleus and M the mass of the final nucleus. The mass difference between the two nuclei is small and equal to the energy difference between the two states of the nucleus. Let $\Delta M = M^* - M$, and rewrite the first equation in (8.35) as

$$E_\gamma = \frac{(M^* - M)(M^* + M)}{2M^*} = \frac{\Delta M(2M^* - \Delta M)}{2M^*} = \Delta M\left(1 - \frac{\Delta M}{2M^*}\right).$$

Now suppose that you want to excite a second identical nucleus by impinging one of these photons on it. Of course you can't because these photons are deficient in the required energy by $(\Delta M)^2/(2M^*)$. For example, if the nucleus is the lead isotope 208 discussed in Example 8.1.1, then $\Delta M = 2.51$ MeV and E_γ, the energy of the photon, is deficient by

$$\frac{2.51^2}{2 \times 193750} = 1.63 \times 10^{-5} \text{ MeV} = 16.3 \text{ eV}.$$

This may be small compared to nuclear energies, but it is not small enough to be negligible in exciting a lead nucleus. In other words, if you send the photon emitted by an excited nucleus of lead to another lead nucleus, it would not be able to excite it because it would be 16.3 eV short of the required energy.

The shortage of energy is, of course, due to the recoil of the excited nucleus as it emits the photon. Is it possible to have recoilless emissions? In 1958, Mössbauer came up with the clever idea of embedding the excited nuclei in a crystal. Then the mass M^* is replaced by the mass of the crystal! This makes $\Delta M/2M^*$ truly negligible and the energy of the emitted photon equal to the energy difference between the two nuclear energy states. Mössbauer used this technique to map out the energy states of many nuclei, a task which was impossible before his invention. ■

Mössbauer effect.

Many nuclear decay processes have three final particles, with the 4-momenta satisfying $\mathbf{P} = \mathbf{p}_1 + \mathbf{p}_2 + \mathbf{p}_3$. Instead of squaring and writing the result in terms of energy and 3-momenta, I note that I can rewrite Equation (8.16) as $\mathbf{p}_2 = -\mathbf{p}_1 + \mathbf{p}_3 + \mathbf{p}_4$. Therefore, if I change $\mathbf{p}_1 = (E_1, \vec{p}_1)$ to its negative, m_2 to M, and alter the remaining subscripts accordingly, then (8.20) turns into the equation relevant for a decay:

$$m_1^2 + M^2 - 2M\mathcal{E}_1 = m_3^2 - m_2^2 + 2\mathcal{E}_2(-\mathcal{E}_1 + M) + 2\vec{p}_1 \cdot \vec{p}_2, \tag{8.36}$$

where the decaying particle is assumed to be in the lab frame.

In a nuclear beta decay, a neutron in the nucleus of an element decays into a proton, an electron, and an antineutrino. So, in (8.36), $M = m_n$, $m_1 = m_p$, $m_2 = m_e$, and $m_3 = m_\nu$. Because of the slight difference in the mass of a neutron and a proton, the fact that neutrinos are essentially massless, and the fact that the mass of the electron is much smaller than the mass of a proton or a neutron, Equation (8.36) simplifies to

$$m_n^2 - m_n\mathcal{E}_p \approx \mathcal{E}_e(-\mathcal{E}_p + m_n) + |\vec{p}_p||\vec{p}_e|\cos\theta, \tag{8.37}$$

where θ is the angle between the electron and proton 3-momenta.

History of the discovery of neutrino.

Historically, because of the extreme difficulty in detecting neutrinos, it was believed that beta decay was a two-particle decay and that the electron had to have a definite single energy given by Equation (8.34). The variation of the energy of the electron with angle prompted some physicists to propose that in beta decay energy was not conserved. It was Wolfgang Pauli who proposed a 3-body decay and postulated the existence of a hitherto unknown and undetected particle, which Enrico Fermi christened *neutrino*, the "little neutron."

Another reason to abandon the concept of velocity-dependent mass!

In a decay process, it is natural to let the decaying particle sit in the lab frame, in which the initial energy is just the mass of the particle. Obviously, a particle at rest cannot decay into particles one of which is more massive than the original particle. What if the decaying particle is accelerated, so that it now has more energy? Can it now decay into more massive particles? If you note that a decay is an *event*, whose universality is guaranteed by Note 1.2.1, then Note 3.3.2 tells you that since the decay cannot happen in the rest frame of the decaying particle, it cannot happen at all. This simple argument becomes extremely complicated if mass is allowed to have velocity dependence!

In a thermonuclear fusion, two light nuclei with kinetic energies much smaller than their masses combine to form a heavier nucleus, whose KE is also small compared to its mass. Because of their essentially zero kinetic energies, the two initial light nuclei could be regarded as a single decaying particle of mass equal to the combined masses of the particles. In the special, but important, case of two nuclei fusing to form a heavier nucleus plus two other particles, the process is similar to the decay of one particle into three. More specifically, going to the CM of the initial nuclei, the conservation of 4-momentum

$$\mathcal{E} + \mathcal{E}' = \mathcal{E}'' + \mathcal{E}_1 + \mathcal{E}_2$$

$$\vec{P}'' = -(\vec{p}_1 + \vec{p}_2)$$

becomes

$$M + M' = M'' + \mathcal{E}_1 + \mathcal{E}_2$$
$$\vec{P}'' = -(\vec{p}_1 + \vec{p}_2). \tag{8.38}$$

In particular, the largest energy that either of the fast moving particles could have is when it carries the entire momentum. For example, the largest value for $\mathcal{E}_2$ is obtained when $\vec{p}_1 = 0$. Then $\mathcal{E}_1 = m_1$ and

$$\mathcal{E}_{2\max} = M + M' - M'' - m_1 \equiv \Delta M \equiv M_{\text{initial}} - M_{\text{final}}, \tag{8.39}$$

where ΔM is the mass defect, which is the difference between the initial mass and the total mass of the stationary final particles.

Proton-proton cycle.

In the **proton-proton cycle**, occurring in the interior of stars similar to the Sun, four protons turn into helium and other particles. This, however, does not happen in one step. First two protons of mass m_p fuse to form a deuteron of mass m_D,[4] a positron of mass m_e, and an almost massless neutrino. With these symbols, the first equation in (8.38) becomes

$$2m_p = m_D + \mathcal{E}_e + \mathcal{E}_\nu. \tag{8.40}$$

The largest value of neutrino energy has been measured and it agrees very well with the value given in (8.39):

$$\mathcal{E}_{\nu\max} = 2m_p - m_D - m_e = 2 \times 938.272 - 1875.613 - 0.511 = 0.42 \text{ MeV}.$$

The measurement also shows that the initial protons have very small kinetic energies, much smaller than 0.42 MeV (otherwise this extra KE would show up in $\mathcal{E}_{\nu\max}$). This creates a dilemma!

Protons are positively charged spherical particles with a repulsive potential energy given by $U(r) = ke^2/r$, where r is the distance between their centers. As they approach each other in their CM and get so close that r becomes twice the proton radius r_p (so that the two protons touch each other), the potential energy becomes $U(2r_p) = ke^2/(2r_p)$. For the two protons to fuse, their initial combined KE should be larger than $U(2r_p)$. This will be enough to overcome the potential barrier and push them into the "well" of attractive strong nuclear force. Figure 8.1 shows that classically, for the two protons to end up in the strong nuclear well, they must have a minimum total energy $E_C = U(2r_p)$. This means that far away, when $r >> r_p = 0.85 \times 10^{-15}$ m and $U \approx 0$, their combined KE must be

$$KE = E_C = \frac{ke^2}{2r_p} = \frac{9 \times 10^9 \times (1.6 \times 10^{-19})^2}{2 \times 0.85 \times 10^{-15}} = 1.36 \times 10^{-13} \text{ J} = 0.85 \text{ MeV},$$

shooting $\mathcal{E}_{\nu\max}$ up to $0.42 + 0.85 = 1.27$ MeV, in complete disagreement with observation!

[4] Before the deuteron is formed, the two protons form a **diproton**, which immediately decays into the final products.

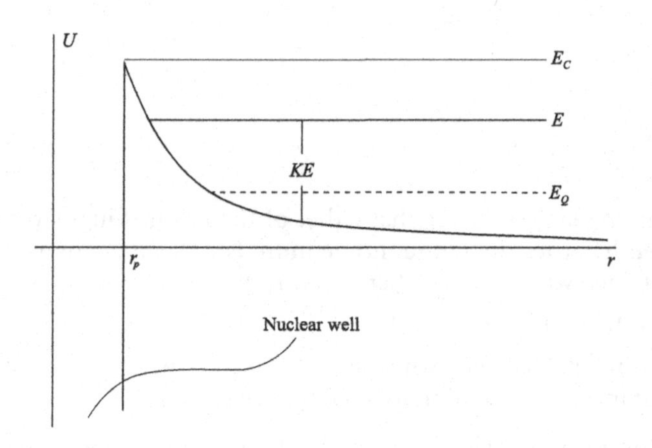

FIGURE 8.1 A particle of energy E has a KE equal to $E - U$. Classically, the energy needed for protons to end up in the nuclear well is E_C. However, quantum tunneling allows the protons to "tunnel" through the potential barrier at energy E_Q, even though classically this cannot happen.

The British astronomer, Arthur Eddington, actually proposed the proton-proton cycle as the nuclear process fueling the Sun at its core in the 1920s. However, because of the above calculation and the fact that each proton should carry a KE of 0.425 MeV (half of the KE above), the process requires a temperature given by

$$KE_{\text{avg}} = \tfrac{3}{2}k_B T, \quad \text{or} \quad T = \frac{0.425 \text{ MeV}}{\frac{3}{2} \times 8.62 \times 10^{-11} \text{ MeV/K}} = 3.3 \times 10^9 \text{ K}.$$

Quantum tunneling resolves the enigma of proton-proton cycle.

This is over two orders of magnitude higher than the actual core temperature of the Sun! You can argue that even at $KE_{\text{avg}} < 0.425$ MeV, there are protons with sufficient KE to make the process go. However, the argument does not lower the temperature sufficiently. It took the weird, ghostly property of "tunneling" in the quantum theory of 1925–1930 to resolve the issue. According to this theory, there is a finite probability for a particle to penetrate to the other side of a potential barrier—the region disallowed by classical physics—even though it has an energy E_Q less than the maximum value of a potential barrier (see Figure 8.1).

8.4 SPECIAL RELATIVITY, SPIN, AND ANTIMATTER

One of the greatest achievements of modern physics is the unification of special relativity with quantum theory. This was done in 1928 by one of the greatest mathematical physicists of the last century, Paul Adrian Maurice Dirac. Erwin Schrödinger's original plan in 1926 was to apply de Broglie's idea of the wave nature of the electron to special relativity. But after he realized that the idea did not work for the hydrogen atom, he downgraded his plan to the non-relativistic regime.

8.4.1 Dirac Equation

The problem with a "straightforward" unification of the quantum theory with special relativity was the relation between energy and momentum. In classical physics, this relation reads $E = p^2/2m$; in special relativity it reads $E^2 = p^2 + m^2$. The transition from classical to quantum physics is obtained by replacing E with essentially the time derivative $\partial/\partial t$ and $\vec{p}$ with the gradient ∇. Then the energy-momentum equation becomes a partial differential equation involving a (wave) function universally denoted by Ψ. These partial differential equations look like

$$\begin{aligned} \text{non-relativistic equation:} \quad & \left(E - \frac{\vec{p}\cdot\vec{p}}{2m}\right)\Psi = 0 \\ \text{relativistic equation:} \quad & \left(E^2 - \vec{p}\cdot\vec{p} - m^2\right)\Psi = 0 \end{aligned} \tag{8.41}$$

with $E \to \partial/\partial t$ and $\vec{p} \to \vec{\nabla}$.

The second equation in (8.41) is of second order in time. This gave rise to solutions of the partial differential equation that appeared to have negative probabilities for the physical properties of the electron. Therefore, the task was to come up with a differential equation that had only first-order time derivatives. That is why Schrödinger concentrated on the non-relativistic limit of the equation that is first-ordered in E.

Dirac's idea was to factor out $E^2 - p^2$ into $(E - p)(E + p)$ and to somehow incorporate the mass term in the factorization. Since it is not possible to include m in a straightforward factorization, Dirac came up with the most ingenious idea of multiplying E and $\vec{p}$ by quantities to be determined by certain consistency conditions. More precisely, he considered an equation of the form[5]

$$(\beta E + \vec{\alpha}\cdot\vec{p} + m)\Psi = 0, \tag{8.42}$$

and demanded that β and $\vec{\alpha}$ be chosen in such a way that the equation

$$(\beta E + \vec{\alpha}\cdot\vec{p} - m)(\beta E + \vec{\alpha}\cdot\vec{p} + m)\Psi = 0 \tag{8.43}$$

reduces to the second equation in (8.41). If $\vec{\alpha}$ were an ordinary vector and β an ordinary number, then the multiplication of the two parentheses on the left would yield

$$\beta^2 E^2 + 2\beta\vec{\alpha}\cdot\vec{p}E + (\vec{\alpha}\cdot\vec{p})^2 - m^2. \tag{8.44}$$

This would require that $\beta^2 = 1$ and $\beta\vec{\alpha} = 0$, because the second term of (8.44) does not exist in the second equation of (8.41). Therefore, $\vec{\alpha} = 0$, and the third term in (8.44) would be zero instead of $\vec{p}\cdot\vec{p}$.

Dirac had the genius to realize that β and $\vec{\alpha}$ had to be *non-commuting* matrices. Multiplying the two parentheses in (8.43) and being careful not to commute

---

[5] Do not confuse $\vec{\alpha}$ and β with any kind of speed. They are completely unrelated!

β and $\vec{\alpha}$, I obtain

$$\beta^2 E^2 + (\beta\vec{\alpha} + \vec{\alpha}\beta) \cdot \vec{p}E + (\vec{\alpha} \cdot \vec{p})^2 - m^2.$$

For this equation to be equal to the second equation of (8.41), I need to have

$$\beta^2 = 1, \quad \beta\vec{\alpha} + \vec{\alpha}\beta = 0, \quad (\vec{\alpha} \cdot \vec{p})^2 = -\vec{p} \cdot \vec{p}. \tag{8.45}$$

To see what conditions are required for the third equation in (8.45) to hold, write its left-hand side as

$$(\alpha_x p_x + \alpha_y p_y + \alpha_z p_z)(\alpha_x p_x + \alpha_y p_y + \alpha_z p_z),$$

and expand the product without commuting the α's (you can commute p's among themselves, of course, and with α's). Then set it equal to $-\vec{p} \cdot \vec{p}$ to obtain the following relations

$$\begin{aligned} \alpha_x^2 p_x^2 + \alpha_y^2 p_y^2 + \alpha_z^2 p_z^2 &= -p_x^2 - p_y^2 - p_z^2, \\ (\alpha_x\alpha_y + \alpha_y\alpha_x) p_x p_y &= 0 \\ (\alpha_x\alpha_z + \alpha_z\alpha_x) p_x p_z &= 0 \\ (\alpha_y\alpha_z + \alpha_z\alpha_y) p_y p_z &= 0. \end{aligned} \tag{8.46}$$

These equations will hold independently of the components of the 3-momentum only if the following conditions are satisfied:

$$\begin{gathered} \alpha_x^2 = \alpha_y^2 = \alpha_z^2 = -1, \\ \alpha_x\alpha_y = -\alpha_y\alpha_x, \quad \alpha_x\alpha_z = -\alpha_z\alpha_x, \quad \alpha_y\alpha_z = -\alpha_z\alpha_y. \end{gathered} \tag{8.47}$$

Succinctly stated, β squares to 1 and anticommutes[6] with all the components of $\vec{\alpha}$. All components of $\vec{\alpha}$ square to -1, and different components anticommute.

Dirac matrices and Dirac equation.

Dirac showed that if β and the components of α are to satisfy these conditions, then they have to be 4×4 matrices, and he constructed these matrices, which now bear his name. He could then substitute these, as well as the time derivative for E and the gradient operator for $\vec{p}$ in (8.42), and come up with one of the most important differential equations discovered in the last century, the **Dirac equation**.

If β, α_x, α_y, and α_z are 4×4 matrices, then Ψ had better be a column vector with four elements. Dirac could identify the first two elements as the electron wave function. The fact that the electron needs two functions rather than one (as given in the Schrödinger equation) was in itself a remarkable achievement of the Dirac equation: the electron has a spin of $\frac{1}{2}$ and, therefore, two spin components.[7] Although the spin of the electron was discovered by Otto Stern

[6] Meaning that switching the order of multiplication introduces a minus sign.

[7] The spin of a particle is a multiple of $\hbar \equiv h/2\pi$, with h the Planck constant. In general if a particle has spin s (meaning that its spin is $\hbar s$), then it has $2s + 1$ spin components.

and Walther Gerlach in 1922, it was a mysterious quantity with no classical counterpart and no clear understanding of it was available until Dirac came up with his equation. Spin is a physical property of particles that can be understood only when relativity and quantum mechanics are unified.

Spin can be understood only when relativity and quantum mechanics are unified.

8.4.2 Discovery of Antimatter

If the first two components of Ψ describe an electron, which particle is represented by the other two components? Some very general mathematical arguments showed that the particle must have the same mass as the electron, but its charge must be exactly opposite to that of an electron. No such particle existed in 1928. However, in 1932, unaware of Dirac's prediction, Carl Anderson found a particle that acted exactly like an electron except for its electric charge, which was positive. He called it the **positron**, which was the antimatter counterpart of the electron. The most distinguishing feature of the positron is that if it comes close to an electron, the pair completely annihilate themselves into electromagnetic radiation.

Prediction and discovery of antielectron.

Is it possible that other particles have *their* antimatter counterpart? Is there an antiproton to be discovered? The Second World War interrupted the pursuit of this question. However, once the war was over and the second generation of accelerators was constructed, the answer could be found.

What sort of energy is needed to produce an antiproton? More specifically, if you collide a projectile proton with a target proton to create an antiproton, how much energy should the projectile proton have? It turns out that protons are **baryons**, and the baryonic "charge" is a conserved quantity. This means that if you assign a +1 to baryons and −1 to antibaryons, then the total baryon number before the collision must equal the total after the collision. Therefore, if you have two protons at the beginning, the total baryon number is +2. If you create an antiproton (with a baryon number −1) at the end, then you have to create 3 protons with it. Thus, for the production of an antiproton by colliding two protons, you have to create—at a minimum—3 protons and one antiproton at the end. For the present consideration, Equation (8.31) becomes

Baryon number and baryonic charge.

$$\mathcal{E}_1 = \frac{(m_3 + m_4 + m_5 + m_6)^2 - m_1^2 - m_2^2}{2m_2} = \frac{(4m_p)^2 - 2m_p^2}{2m_p} = 7m_p, \tag{8.48}$$

or about 6.5 GeV.

Equation (8.48) assumes that no other particles are produced beside the three protons and an antiproton. This is unrealistic because in any high energy collision, pions are produced abundantly. These are particles (see Example 8.2.3) with a mass of about 140 MeV and a large probability of production in a high energy collision. Pion production increases the minimum energy. Luckily there is an easy mechanism to lower the energy in (8.48). If instead of a proton the target is a nucleus, the minimum energy is reduced by about a factor of 2 (see

Problem 8.23). Thus, even in the presence of pions, it is possible to produce antiprotons with a reasonable amount of energy.

As early as 1946, cyclotron engineers at the University of California Radiation Laboratory (UCRL) were designing an accelerator that could reach energies up to 10 GeV. However, convincing the politicians at the Atomic Energy Commission (AEC) turned out to be tough, especially since Brookhaven National Laboratory (BNL) was also seeking funding for a similar accelerator. After two years of negotiation by all three parties involved, in April 1948 the AEC approved both a 3 GeV Cosmotron, to be built at BNL, and a 6 GeV Bevatron, to be built at UCRL.[8] The name Bevatron was used to mark a milestone in accelerator energies at which the 10^9 record was broken.

Discovery of antiproton.

The design of the Bevatron started in 1949, but after the Soviet Union's testing of its atomic bomb in the same year, the Bevatron project was interrupted until 1952 to allow its scientific crew to work on another project prioritized by the military. Two years after the work was resumed, the construction of the Bevatron was completed, and by early 1955 it was operational. Accelerating protons to energies in the range of 4.2–6.2 GeV in several runs of the Bevatron, Owen Chamberlain, Emilio Segré, and their group discovered a negatively charged particle with exactly the same mass as the proton. A few months later, using the same facility at UCRL, the antineutron was discovered. The universality of antimatter has been now confirmed as the antiparticle of every known particle has been produced.

Note 8.4.1. ***Notation:*** *If a particle is represented by the letter p, then it is customary to denote its antiparticle by $\bar{p}$. If a particle is represented by the letter q^- (or q^+), then it is customary to denote its antiparticle by q^+ (or q^-).*

8.5 PROBLEMS

8.1. Two identical particles of mass m approach each other along a straight line with speed $v = \beta c$ as measured in the lab frame. Show that the energy of one particle as measured in the rest frame of the other is

$$\frac{1+\beta^2}{1-\beta^2}mc^2.$$

8.2. Find an expression for the magnitude of the momentum (8.8), energy (8.9), and velocity (8.10) of the produced particle in the lab frame all in terms of only masses m_1, m_2, and M.

---

[8] The name Bevatron comes from the acronym BeV, standing for billion electron volt. The word "billion" has a double meaning. Using the short-scale usage used in the US, it means 10^9. In the long-scale usage used in Britain and other countries, it means 10^{12}. While BeV was common (but confusing) in the early 1950s, it was eventually replaced by the SI prefixed GeV.

8.3. Insert (8.13) in (8.14) to derive (8.15). Another way of obtaining this result is to use Equation (8.3) with $E_{cm} = E_{1cm} + E_{2cm}$, and the fact that

$$E^2_{2cm} = E^2_{1cm} - m_1^2 + m_2^2,$$

which you should derive.

8.4. Suppose that in the elastic scattering of two identical relativistic particles in the lab frame, one of the final particles is produced at rest. Use (8.22) and the conservation of 4-momentum to show that the other final particle moves with the same velocity as the initial incident particle.

8.5. A photon of momentum p_γ hits a macroscopic object of mass M initially at rest, gets absorbed, and sets the object in motion. All motions are in one dimension. Write (8.18) for this process assuming that the mass of the macroscopic object does not change. What is the final momentum of the macroscopic object? What is wrong? How can you resolve the issue?

8.6. A photon of momentum p_γ hits a perfectly reflecting (i.e., the photon does not lose any of its initial energy upon reflection) macroscopic object of mass M initially at rest. All motions are in one dimension. Write (8.18) for this process. Can you assume that the mass of the macroscopic object does not change? If not, write an expression connecting the final mass with M and p_γ.

8.7. Derive (8.20) by transferring $\mathbf{p}_3$ to the left-hand side of (8.16) and squaring both sides.

8.8. A particle of mass m and energy E_1 collides with an identical stationary particle. The two particles scatter with momenta $\vec{p}_3$ and $\vec{p}_4$ making angles θ_3 and θ_4 with the direction of motion of the incident particle.

(a) Use (8.20) to show that

$$\cos\theta_3 = \frac{(E_1 + m)(E_3 - m)}{|\vec{p}_1||\vec{p}_3|}.$$

(b) Using the square of (a) and a similar result for θ_4 show that

$$\tan^2\theta_3 = \frac{2m(E_1 - E_3)}{(E_1 + m)(E_3 - m)}$$

and

$$\tan^2\theta_4 = \frac{2m(E_1 - E_4)}{(E_1 + m)(E_4 - m)} = \frac{2m(E_3 - m)}{(E_1 + m)(E_1 - E_3)}.$$

(c) Take the product of the last two results to show that

$$\tan\theta_3 \tan\theta_4 = \frac{2m}{E_1 + m} = \frac{2}{\gamma_1 + 1}$$

where γ_1 is the gamma factor for the incident particle.

8.9. Show that in the ultra-relativistic case, Equation (8.20) leads to the angle between the final two particles being zero.

8.10. A particle of mass m and relativistic energy $4mc^2$ collides with another stationary particle of mass $2m$ and sticks to it. What is the mass of the resulting composite particle?

8.11. An electron of kinetic energy 1 GeV strikes a positron (antielectron) at rest and the two particles annihilate each other and produce two photons, one moving in the forward direction (the direction that electron had before collision) and the other in the backward direction. What are the energies of the two photons? The mass (times c^2) of the electron and positron are the same and equal to 0.511 MeV.

8.12. An electron of energy E and momentum $\vec{p}$ strikes a positron (antielectron) at rest. The two particles annihilate each other and produce two photons of energies $E_{\gamma 1}$ and $E_{\gamma 2}$.

(a) Use (8.20) to express the energy of each photon in terms of E, the mass of the electron (or positron) m_e, and the photon's scattering angle.

(b) Show that the photons have the same scattering angle if and only if they have the same energy.

(c) Prove that when the photons have the same scattering angles θ, then

$$\cos\theta = \sqrt{\frac{E - m_e}{E + m_e}} = \frac{|\vec{p}|}{2E_\gamma},$$

where E_γ is the energy of either photon.

(d) Show that for an ultra-relativistic incident electron, both photons move in the forward direction.

8.13. An electron of energy E, momentum $\vec{p}$, and mass m_e strikes a positron (antielectron) at rest. The two particles annihilate each other and produce two photons. Transfer the collision to the center of mass and do the following:

(a) Find the energy and momentum of each photon in the CM in terms of E, $\vec{p}$, and m_e.

(b) Transfer back to the rest frame of the positron and show the results in Problem 8.12.

8.14. A particle of mass m and energy E collides with an identical particle at rest. The collision results in the formation of a single particle. Show that the mass and the speed of the formed particle are, respectively, $\sqrt{2m(E+m)}$ and $\sqrt{(E-m)/(E+m)}$.

8.15. A photon of energy E is absorbed by a stationary nucleus of mass M. The collision results in an excitation of the nucleus. Show that the mass and the speed of the excited nucleus are, respectively, $\sqrt{M(2E+M)}$ and $E/(E+M)$.

8.16. Derive Equation (8.26).

8.17. Show that (8.32) follows from the equation before it.

8.18. Derive Equation (8.34) from (8.5).

8.19. Obtain Equation (8.34) by transferring $\mathbf{p}_1$ or $\mathbf{p}_2$ to the left-hand side of $\mathbf{P} = \mathbf{p}_1 + \mathbf{p}_2$ and squaring both sides.

8.20. The interior of a star like the Sun has a temperature of about 15–20 million Kelvin. Using the familiar statistical physics formula $\langle KE \rangle = \frac{3}{2} k_B T$, with k_B the Boltzmann constant, estimate the kinetic energy of a proton in the interior of the Sun. What is the speed β of this proton?

8.21. Derive Equation (8.45).

8.22. Derive Equations (8.46) and (8.47).

8.23. A proton of mass $m_p = 938.3$ MeV collides with a stationary nucleus of mass M to produce antiprotons. The minimum number of particles at the end is two protons, one antiproton, and the nucleus.

(a) Show that the minimum amount of energy for this production is

$$\mathcal{E}_{\min} = 3m_p + \frac{4m_p^2}{M}.$$

(b) If pions of mass $m_\pi \approx 140$ MeV are also produced, this minimum energy increases. Show that if N pions are added to the outcome, then

$$\mathcal{E}_{\min} = 3m_p + Nm_\pi + \frac{(3m_p + Nm_\pi)^2 - m_p^2}{2M}.$$

As a check for your answer, make sure you get (a) as a special case.

(c) Find $\mathcal{E}_{\min}$ for $N = 12$ when the initial proton impinges on a copper nucleus of mass $M = 59{,}150$ MeV.

8.12 Show that (8.72) follows from the equation before it.

8.13 Derive Equation [illegible] from [illegible].

8.1[illegible] Obtain Equation [illegible] by transferring [illegible] to the left-hand side of [illegible] and squaring both sides.

8.[illegible] The [illegible]

[illegible]

[illegible]

(a) Show that [illegible] for this process is

[illegible]

(b) [illegible] Show that if [illegible] photons are [illegible] the [illegible], then

[illegible]

[illegible]

[illegible] mass $M = 52$ [illegible] eV.

CHAPTER 9

Interstellar Travel

The length contraction and time dilation effects work in favor of traveling to exoplanets of neighboring stars. Once the rocket carrying the crew attains a speed of $\beta = 0.99$, it would take only 14 years (plus whatever it takes to decelerate the rocket for landing) to go to a star that is 100 light years away. So, it is possible to send a crew to the star and bring them back in just one generation. Or so it seems! The sticky point is attaining the desired speed. That requires energy. How much? What you have learned in the previous chapter can provide some answers.

9.1 MASSIVE PARTICLE PROPULSION

A rocket is propelled by ejecting matter, usually in the form of gas. Let $\vec{\beta}_g$ and $\vec{\beta}_g'$ be the velocities with which the gas is ejected from the rocket as measured by the rocket and Earth observers, respectively. To avoid complications, I'll assume that the gas is ejected from the rocket with constant speed, so that $|\vec{\beta}_g|$ does not change. Let $\vec{\beta}$ be the instantaneous velocity of the rocket relative to the Earth's RF. At some instant, the rocket ejects an infinitesimal amount dm_g of gas and is propelled slightly in the opposite direction. If dm_g is ejected from the back of the rocket opposite to the direction of motion, the rocket accelerates, and if it is ejected from the front of the rocket in the direction of motion, the rocket decelerates.

With M the mass of the rocket and the payload, the conservation of 4-momentum becomes

$$d(M\gamma) + dm_g\gamma_g' = 0$$
$$d(M\gamma\vec{\beta}) + dm_g\gamma_g'\vec{\beta}_g' = 0.$$

Solving the first equation for $dm_g\gamma_g'$ and substituting in the second yields

$$d(M\gamma\vec{\beta}) - d(M\gamma)\vec{\beta}_g' = 0. \tag{9.1}$$

Since $\vec{\beta}_g$ is assumed to be constant, you need to express $\vec{\beta}_g'$ in terms of $\vec{\beta}_g$. In the following discussion, I'll make the assumption that all velocities lie along

Special Relativity. DOI:10.1016/B978-0-12-810411-8.00009-2

the same line, so that the motion of the rocket is one-dimensional. Then the relativistic law of addition of velocities (6.18) gives

$$\vec{\beta}_g' = \frac{\vec{\beta} + \vec{\beta}_g}{1 + \vec{\beta} \cdot \vec{\beta}_g}.$$

Substituting this in (9.1) yields

$$(1 + \vec{\beta} \cdot \vec{\beta}_g) d(M\gamma\vec{\beta}) - (\vec{\beta} + \vec{\beta}_g) d(M\gamma) = 0.$$

With all velocities being along the path of motion, $\vec{\beta}_g$ can be opposite to the direction of $\vec{\beta}$ (acceleration) or in the same direction as $\vec{\beta}$ (deceleration). So, I'll get rid of the vector signs and write the previous equation as

$$(1 \mp \beta\beta_g) d(M\gamma\beta) - (\beta \mp \beta_g) d(M\gamma) = 0, \tag{9.2}$$

where the upper (lower) sign is for acceleration (deceleration). You can show that

$$d(M\gamma) = \gamma dM + M\beta\gamma^3 d\beta$$
$$d(M\gamma\beta) = \gamma\beta dM + M\gamma^3 d\beta.$$

Substitute these in (9.2) and simplify to obtain

$$\pm\beta_g dM + M\gamma^2 d\beta = 0, \quad \text{or} \quad \frac{dM}{M} = \mp\frac{d\beta}{\beta_g(1-\beta^2)}. \tag{9.3}$$

Integrating both sides of (9.3) with the assumption that $M = M_0$ when $\beta = 0$ yields

$$\int_{M_0}^{M} \frac{du}{u} = \mp\frac{1}{\beta_g}\int_0^{\beta} \frac{du}{1-u^2},$$

leading to

$$\ln\left(\frac{M}{M_0}\right) = \pm\frac{1}{2\beta_g}\ln\left(\frac{1-\beta}{1+\beta}\right)$$

or

$$\frac{M}{M_0} = \left(\frac{1-\beta}{1+\beta}\right)^{\pm 1/2\beta_g}, \tag{9.4}$$

with the positive sign corresponding to acceleration and the negative sign to deceleration.

What does the preceding analysis tell us about the feasibility of space travel? It should be obvious that to expedite the attainability of speeds close to the speed of light, the exhaust velocity β_g should be as large as possible. Chemical processes are too slow to achieve such desired speeds in a reasonable amount of time. Therefore, only nuclear and high energy processes can be viable candidates.

Consider nuclear fusion in which hydrogen turns into helium, a process that takes place inside all stars and is responsible for the energy emitted by them. In the actual fusion, four protons combine and produce one helium nucleus plus a few other particles that carry some of the energy of the original protons. The initial protons move so slowly that you can neglect their kinetic energy. Thus, the mass of four protons provides the energy carried away by the final particles.

Nuclear fusion, although done routinely in the lab, is not, as of now, commercially available as a source of energy. Nevertheless, let's assume that we can fuse hydrogen atoms easily and cheaply to extract energy from their mass.[1] Furthermore, let's assume that *all* the energy available in the hydrogen mass is carried away only by the helium nuclei. Let's go even further and assume that you can guide all these energetic helium nuclei toward the exhaust pipe without the expenditure of any extra energy. With the mass of the hydrogen atom being 938 MeV, this means that each helium atom carries $4 \times 938 = 3752$ MeV of energy. Since the mass of helium is 3728 MeV, this corresponds to $\gamma_g = E/m = 3752/3728 = 1.00644$ and $\beta_g = 0.113$. Therefore, for the rocket to reach a speed of $\beta = 0.99$, according to Equation (9.4), you must have

$$\frac{M}{M_0} = \left(\frac{1-0.99}{1+0.99}\right)^{1/0.226} = 1.478 \times 10^{-9}$$

or $M_0/M = 6.76 \times 10^8$. This means that the rocket's payload ought to be more than 670 million times the weight of the rocket itself!

Any round-trip interstellar journey has four accelerating parts: accelerating from Earth to the desired speed, decelerating to land on the target planet, accelerating on the way back, and decelerating to land on Earth. The process of deceleration is identical to that of acceleration except that the exhaust will take place in front of the rocket instead of its back. Let M_0 be the payload mass as the rocket takes off from Earth and M_f the mass as it lands back on Earth. Then

$$\frac{M_0}{M_f} = \left(\frac{M_0}{M_1}\right)\left(\frac{M_1}{M_2}\right)\left(\frac{M_2}{M_3}\right)\left(\frac{M_3}{M_f}\right) = r^4 \tag{9.5}$$

where M_1 is the payload at the end of the first acceleration, M_2 the payload at landing, M_3 the payload at the end of the second acceleration, and r is the common ratio of the initial and final masses at each stage. For the data furnished above, you get

$$\frac{M_0}{M_f} = (6.76 \times 10^8)^4 = 2 \times 10^{35}.$$

An interstellar round-trip requires an initial payload of over 100 million times the mass of the Sun!

[1] See also E. Purcell in [Cameron 63].

If the final mass of the payload is to be one ton, then the initial mass ought to be 2×10^{38} kg, which is over 100 million times the mass of the Sun!

For a one-way trip, say sending a robot to be stationed permanently on a distant planet, the ratio reduces to

$$\frac{M_0}{M_f} = (6.76 \times 10^8)^2 = 4.58 \times 10^{17},$$

so that for a one-ton payload, the initial mass would be 4.58×10^{20} kg, which is almost as massive as Tethys, one of Saturn's moons!

What about the other nuclear reaction, fission? Can it improve the situation? The most efficient fission process is that of uranium-235. A slow neutron is absorbed by uranium, turning it into an excited state of uranium-236, which decays into two daughter nuclei, two or three neutrons, and some photons. The energy released is about 200 MeV, most of which is carried away by the daughter nuclei because of their Coulomb repulsion. However, this large chunk of energy does not boost the nuclei to appreciable speeds because of their large masses. In fact, in this case $\beta_g \approx 0.03$. So exhausting these heavy nuclei from the back of the rocket cannot compete with fusion. What about the neutrons? The neutrons' share of energy is about 5 MeV. Assuming that each neutron gets 2 MeV, the neutron gas will have a γ of $(939+2)/939 = 1.002$, which is also smaller than that of the fusion.

9.2 PHOTON PROPULSION

Both fission and fusion reactions produce photons, which move at the ultimate desirable speed. Disregarding the practical difficulty of guiding the photons to the end of the rocket for ejection, would a photon gas improve the propulsion of the rocket? The percentage of the energy that photons carry in fusion and fission is very small. This means that only a small fraction of the fuel turns into photons. However, there are other sources of photons, such as lasers and matter-antimatter annihilation, that may be viable for space travel. So let's investigate the propulsion of a rocket by photons in some detail.

Starting with the conservation of 4-momentum in which the subscript γ refers to a photon,

$$d(M\gamma) + E_\gamma = 0$$
$$d(M\gamma\vec{\beta}) + \vec{p}_\gamma = 0,$$

and noting that $|\vec{p}_\gamma| = E_\gamma$, I get

$$d(M\gamma) \pm d(M\gamma\beta) = 0, \quad \text{or} \quad d(M\gamma \pm M\gamma\beta) = 0$$

where I have restricted motion to a straight line (no vector sign). The plus sign comes from the case where $\vec{p}_\gamma$ is opposite to the direction of motion, therefore when the rocket is accelerating, and the minus sign corresponds to the case when the rocket is decelerating. The last equation can be trivially integrated:

$$M\gamma \pm M\gamma\beta = C.$$

The constant of integration is determined by the condition that M and β, which are functions of time, satisfy $M(0) = M_0$ when $\beta(0) = \beta_0$. Then $C = M_0\gamma_0(1 \pm \beta_0)$ and

$$\frac{M}{M_0} = \frac{\gamma_0(1 \pm \beta_0)}{\gamma(1 \pm \beta)} = \left[\frac{\gamma_0(1+\beta_0)}{\gamma(1+\beta)}\right]^{\pm 1}, \tag{9.6}$$

where the last equality comes from the identity $\gamma(1-\beta) = [\gamma(1+\beta)]^{-1}$. It is interesting to note (see Problem 9.4) that (9.3) reduces to (9.6) when $\beta_g = 1$, i.e., when the "gas" (now an aggregate of photons) moves at the speed of light! Making the dependence on time explicit and separating the acceleration and deceleration parts, I get

$$\begin{aligned} &\text{for acceleration} \quad m(t) \equiv \frac{M(t)}{M_0} = \frac{\gamma_0(\beta_0+1)}{\gamma(t)[\beta(t)+1]} \\ &\text{for deceleration} \quad m(t) \equiv \frac{M(t)}{M_0} = \frac{\gamma(t)[\beta(t)+1]}{\gamma_0(\beta_0+1)} \end{aligned} \tag{9.7}$$

Example 9.2.1. How practical is a round-trip journey using photons as fuel? Consider such a journey to Alpha Centauri, the nearest star at a distance of 4.3 light yeas, at a speed of $0.9c$. A very light rocket of bare mass 100 kg carries a 100-kilogram astronaut plus the payload. The round trip takes about 3.8 years (of proper time). If the astronaut consumes only one kilogram (half the average) of food per day, 1400 kg of food is needed for the journey. To accurately calculate the fuel burned for the propulsion of the part of the mass consisting of only food, I'll have to include 1400 kg for the first day, 1399 kg for the second day, etc. A less accurate, but reasonable, way is to take the average of the food mass at the beginning and the end and consider that as the contribution of food to the unconsumed mass remaining at the end of the trip. Since at the end of the trip there is no food left, the average is 700 kg, and this is the contribution of food to the final left-over mass M_f. So, $M_f = 900$ kg.

For a one-way journey, Equation (9.7) gives, very generally, a ratio of the final to initial mass of (see Problem 9.5)

$$\frac{M_f}{M_{\text{in}}} = \frac{1}{[\gamma_0(\beta_0+1)]^2},$$

so that for a round trip,

$$\frac{M_f}{M_{\text{in}}} = \frac{1}{[\gamma_0(\beta_0+1)]^4}.$$

For $\beta_0 = 0.9$ and $M_f = 900$ kg, the initial mass has to be

$$M_{\text{in}} = M_f[\gamma_0(\beta_0+1)]^4 = 900 \times 361 = 324900 \text{ kg},$$

yielding an initial fuel mass of $M_0 = 324000$ kg. The most efficient fuel is the one that has no residual mass at the end of the trip. This means that by the time the rocket lands on Earth, all the fuel has completely turned into photons. There is only one physical process that can accomplish this: matter-antimatter annihilation.

The "inconvenience" of antimatter!

How feasible is it to provide 324 tonnes of matter and antimatter, or 162 tonnes of antimatter? Because of its tendency to annihilate ordinary matter, antimatter needs to be confined. The only available means of confinement is electromagnetic bottling. This requires the antimatter to be charged. But the Coulomb repulsion of charged antimatter puts a stringent limit on how much of it can be confined. The largest storage of antimatter in the world at CERN has contained only about 10 nanograms in its entire history! Furthermore, as a source of energy, matter-antimatter annihilation is very inefficient: the efficiency of an antimatter "engine" is 10^{-9}. Therefore, if you *could* confine large amounts of antimatter, for each Joule of energy that it provides, you'll have to put in one billion Joules. So for the round trip journey above, you need

$$324000 \times (3 \times 10^8)^2 \times 10^9 = 2.9 \times 10^{31}$$

Joules of energy. At its current rate of 4×10^{20} J/yr, this is equivalent to over 70 billion years of world consumption of energy! ■

The lesson of Example 9.2.1 is that interstellar travel, if possible at all at this stage of our civilization, seems to be limited to sending probes without worrying about returning them. To investigate the details of this one-way trip, I need both velocity and distance as a function of (Earth) time. I let you solve (9.7) for $\beta(t)$ in terms of $m(t)$ to obtain

$$\text{for acceleration} \quad \beta(t) = \frac{dx}{dt} = \frac{\gamma_0^2(\beta_0+1)^2 - m^2(t)}{\gamma_0^2(\beta_0+1)^2 + m^2(t)}$$
$$\text{for deceleration} \quad \beta(t) = \frac{dx}{dt} = \frac{\gamma_0^2(\beta_0+1)^2 m^2(t) - 1}{\gamma_0^2(\beta_0+1)^2 m^2(t) + 1}, \tag{9.8}$$

and then integrate the above to get

$$\text{for acceleration} \quad x(t) = \int_0^t \frac{\gamma_0^2(\beta_0+1)^2 - m^2(u)}{\gamma_0^2(\beta_0+1)^2 + m^2(u)}\, du$$
$$\text{for deceleration} \quad x(t) = \int_0^t \frac{\gamma_0^2(\beta_0+1)^2 m^2(u) - 1}{\gamma_0^2(\beta_0+1)^2 m^2(u) + 1}\, du \tag{9.9}$$

where $x(t)$ is given in units of light seconds, each light second being 3×10^8 m, of course, and each light year being 31.5 million light seconds.

9.2.1 Uniform Depletion

M_f is the mass left at the end of the trip.

M_0 is the **fuel mass** at take-off.

To be able to integrate (9.9), I need to know m as a function of time. I'll consider a uniform depletion of the fuel. Let M_f denote the mass that is not consumed. It includes the masses of the rocket as well as everything else that is left over at the end of the trip. In other words, it includes everything except fuel. Let the rocket be loaded with fuel of mass M_{fuel} that depletes uniformly at a constant rate k. Denoting by M_0 the fuel mass at take-off and by T_0 the time in which the fuel is completely depleted, the total mass can be written as

$$M_{\text{tot}}(t) = M_f + M_{\text{fuel}}(t) = M_f + M_{\text{fuel}}^{\text{beg}} - kt = M_f + M_{\text{fuel}}^{\text{beg}} - (M_0/T_0)t,$$

where $M_{\text{fuel}}^{\text{beg}}$ is the amount of fuel left at the beginning of the motion at the stage of the trip under consideration. For the accelerating stage $M_{\text{fuel}}^{\text{beg}} = M_0$. For the decelerating stage,

$$M_{\text{fuel}}^{\text{beg}} = M_0(1 - t_{\text{acc}}/T_0), \quad t_{\text{acc}} \le T_0$$

is the amount of fuel left over from the accelerating stage that lasted t_{acc} seconds (assuming no fuel is lost in the constant-velocity cruising stage). With $m(t) \equiv M_{\text{tot}}(t)/(M_f + M_{\text{fuel}}^{\text{beg}})$, I can now write

$$m(t) = 1 - m_0^{\text{beg}} t, \quad m_0^{\text{beg}} = \frac{M_0}{M_f + M_{\text{fuel}}^{\text{beg}}}, \quad t \le 1 \tag{9.10}$$

where t is now measured in units of T_0.

Let's start our journey by first accelerating from rest to some final cruising speed β_0! Substituting (9.10) in the first equation of (9.8) with $\beta_0 = 0$, you get

$$\beta(t) = \frac{dx}{dt} = \frac{1 - (1 - m_0 t)^2}{1 + (1 - m_0 t)^2}, \quad m_0 = \frac{M_0}{M_f + M_0}. \tag{9.11}$$

Plugging in the first equation of (9.9) and doing the integral gives

$$x(t) = \frac{\pi}{2m_0} - t - \frac{2}{m_0}\tan^{-1}(1 - m_0 t). \tag{9.12}$$

The time needed to accelerate to β_0 is

$$t_{\text{acc}} = \frac{\gamma_0(1 + \beta_0) - 1}{m_0\gamma_0(1 + \beta_0)}, \tag{9.13}$$

during which time the rocket travels a distance of

$$x(t_{\text{acc}}) = \frac{\pi}{2m_0} - \frac{\gamma_0(1 + \beta_0) - 1}{m_0\gamma_0(1 + \beta_0)} - \frac{2}{m_0}\tan^{-1}\left[\frac{1}{\gamma_0(1 + \beta_0)}\right]. \tag{9.14}$$

We have cruised virtually all the interstellar distance with a speed of β_0. We are now approaching the destination exoplanet. Let's decelerate to a safe landing speed (of zero)! First we have to know how much fuel is left. Inserting (9.13) in (9.10) you get

$$m_0^{\text{beg}} = \frac{M_0}{M_f + M_0(1 - t_{\text{acc}})} = \frac{m_0}{1 - m_0 t_{\text{acc}}} = m_0\gamma_0(\beta_0 + 1)$$

and

$$m(t) = 1 - m_0\gamma_0(\beta_0 + 1)t.$$

Now plug this in the second equation of (9.8) to get

$$\beta(t) = \frac{dx}{dt} = \frac{\gamma_0^2(1 + \beta_0)^2[1 - m_0\gamma_0(1 + \beta_0)t]^2 - 1}{\gamma_0^2(1 + \beta_0)^2[1 - m_0\gamma_0(1 + \beta_0)t]^2 + 1}. \tag{9.15}$$

Substituting in the second integral of (9.9) and integrating yields

$$x(t) = t + 2\frac{\tan^{-1}\left\{\gamma_0(1+\beta_0)\left[1 - m_0\gamma_0(1+\beta_0)t\right]\right\} - \tan^{-1}[\gamma_0(1+\beta_0)]}{m_0\gamma_0^2(1+\beta_0)^2}. \tag{9.16}$$

If the rocket is to come to rest upon landing, it has to travel for a time t_{dec} such that $\beta(t_{\text{dec}}) = 0$. Setting (9.15) equal to zero, and solving for t, you get

$$t_{\text{dec}} = \frac{\gamma_0(1+\beta_0) - 1}{m_0\gamma_0^2(1+\beta_0)^2} = \frac{t_{\text{acc}}}{\gamma_0(1+\beta_0)}, \tag{9.17}$$

during which time the rocket travels a distance of

$$x(t_{\text{dec}}) = \frac{\gamma_0(1+\beta_0) - 1 + \pi/2 - 2\tan^{-1}\left[\gamma_0(1+\beta_0)\right]}{m_0\gamma_0^2(1+\beta_0)^2}. \tag{9.18}$$

To achieve the highest possible speed β_0 in a one-way probe mission, β_0 must be chosen in such a way that in the process of deceleration the landing speed reaches zero as the last drop of fuel is used. This requires $t_{\text{acc}} + t_{\text{dec}}$ to be equal to one.[2] Setting the sum of (9.13) and (9.17) equal to one and solving for β_0 yields

$$\beta_0 = \frac{1 - m_f}{1 + m_f}, \qquad m_f \equiv \frac{M_f}{M_0 + M_f}, \tag{9.19}$$

and therefore,

$$t_{\text{acc}} = \frac{1}{1 + \sqrt{m_f}}, \qquad t_{\text{dec}} = \frac{\sqrt{m_f}}{1 + \sqrt{m_f}}. \tag{9.20}$$

Note that m_f is the fractional mass of everything except the fuel at launch.

These are interesting results! Equation (9.19) reveals that you can accelerate the rocket arbitrarily close to the speed of light as long as you can make a rocket that is arbitrarily light. For such a limiting case, (9.20) shows that $t_{\text{acc}} \approx 1$ and $t_{\text{dec}} \approx 0$, meaning that almost all the fuel is burnt during acceleration while deceleration require very little fuel. (Convince yourself that this makes sense!)

The distance covered during acceleration is obtained by plugging t_{acc} of (9.20) in (9.12). It's a good exercise for you to show that what you get is

$$x_{\text{acc}} = \frac{\sqrt{m_f} - 2\tan^{-1}(\sqrt{m_f}) + \pi/2 - 1}{1 - m_f}. \tag{9.21}$$

Similarly by (9.16) and (9.20), the distance during deceleration is

$$x_{\text{dec}} = \frac{\sqrt{m_f} + m_f[\pi/2 - 1 - 2\tan^{-1}(1/\sqrt{m_f})]}{1 - m_f}. \tag{9.22}$$

---

[2] Remember that t is measured in units of T_0, so that at $t = 1$ no fuel is left.

For a very light rocket, $x_{dec} \to 0$ and $x_{acc} \to \pi/2 - 1 = 0.57$ light second. In this limiting case, both x_{acc} and x_{dec} are negligible compared to interstellar distances of at least several light years. In fact, this is true for all m_f: Plot $x_{acc} + x_{dec}$ as a function of m_f and note that it is a decreasing function with its maximum of 0.57 light second occurring at $m_f = 0$. Therefore, regardless of the mass of the rocket, during the entire trip the rocket is practically moving at constant velocity.

During the entire trip the rocket is practically moving at constant velocity.

Example 9.2.2. Suppose we want to send a probe to an exoplanet with a cruising speed of $0.9c$. Equation (9.19) gives a final fractional mass m_f of 0.0526. If the residual mass M_f is to be one kilogram, then M_0 has to be 18 kilograms. You have to remember that M_f includes the "ashes" of the fuel. If we use a fuel whose ashes are more than one kilogram, the project fails. If it is exactly one kilogram, the spacecraft will consist of only fuel: no craft, no survey equipment, no transmitters to send information! The mission will be a futile exercise.

So assume that the mass of the craft plus all the communication gadgets is one kilogram. Therefore, there is no room for any "ash." This means that the fuel must be matter and antimatter. With the current efficiency of a matter-antimatter "engine," to produce 9 kilograms of antimatter, we need

$$9 \times (3 \times 8)^2 \times 10^9 = 8.1 \times 10^{26} \text{ Joules}$$

of energy, or more than 2 million years of world consumption of energy at its current rate! ■

9.2.2 Laser Propulsion

Matter-antimatter annihilation, because it is "ashless," would be ideal for space travel. However, the inaccessibility of antimatter makes this ideal a thing of the distant future. What about other sources of EM waves? Lasers produce light, and they seem to be getting more and more powerful. So let's investigate this possibility.

Lasing is an *atomic* process, in which photons with an energy of several electron volts are emitted from certain atoms. Let E_γ be the energy of the photons, and M_{atom} the mass of the lasing atom. Then $\kappa \equiv (E_\gamma/c^2)/M_{atom}$ is the ratio of the equivalent mass of the photons (the fuel) to the mass of the atom, the "ash." So, the total mass of the spacecraft is $M_f + M_{las} + \kappa M_{las}$, in which I have separated the mass of the lasing material M_{las} from the rest of the residual mass M_f. Then

$$m_0 = \frac{\kappa M_{las}}{M_f + M_{las} + \kappa M_{las}} = \frac{\kappa M_{las}}{M_f + M_{las}(1+\kappa)}.$$

The largest theoretical value of m_0, and therefore its most efficient value, is obtained by letting $M_f \to 0$. Thus,

$$m_{0max} = \frac{\kappa}{1+\kappa}. \tag{9.23}$$

Example 9.2.3. The largest κ occurs for the lightest lasing atom. The lightest atom is, of course, hydrogen. But it doesn't lase. Nevertheless, let's assume that it does, and in the process of lasing it emits the highest energy photon of 13.6 eV. The mass of hydrogen is essentially that of a proton: 938 MeV. Therefore, the largest possible κ in existence is

$$\kappa_{\max} = \frac{13.6}{938 \times 10^6} = 1.45 \times 10^{-8},$$

with a corresponding m_0 of

$$m_{0\max} = \frac{\kappa_{\max}}{1 + \kappa_{\max}} \approx \kappa_{\max} = 1.45 \times 10^{-8}.$$

Problem 9.16, which is based on the assumption of uniform depletion, now shows that the largest possible speed attainable is an abysmal 2.175 m/s! Problem 9.17 gives the same result without any assumption about how the fuel is depleted.

By contrast, taking one kilogram of matter and an equal amount of antimatter and mounting them on a small 0.1-kg probe gives

$$m_f = \frac{0.1}{0.1 + 2} = 0.0476,$$

because, unlike lasers, the entire matter-antimatter combination *is* the fuel. Substituting this in (9.19) yields $\beta_0 = 0.963$, making matter-antimatter annihilation the fuel of choice, once antimatter becomes available commercially ... in the very very distant future, if at all! ■

Interstellar travel on fuel-carrying rockets is next to impossible.

The moral of all the preceding discussion is that interstellar travel on fuel-carrying rockets appears to be next to impossible. Our only hope of learning about other worlds seems to be by electromagnetic signal communication. That too has its limitations, because any directed signal, as concentrated as it may be, spreads out. A circular laser beam is really a cone with an angle θ given by $\sin\theta = 1.22\lambda/D$, where λ is the wavelength of the laser light and D the diameter of the aperture of the laser gun. For a blue laser with an aperture diameter of 1 cm, $\theta = 49$ microradians. Such a laser spreads out to a diameter of about 19 km by the time it reaches the Moon!

To make θ as small as possible, you need a laser with a large aperture producing EM waves of very short wavelength. Let's assume that laser guns could be made that emit x-rays from a 10-cm aperture. For this laser, θ would be about 1 nanoradian. Aim this laser to (a planet of) Alpha Centauri, the nearest star, 4 light years away. When the beam reaches the star system, it has a diameter of 46000 km. A laser gun of this nature that is a thousand times stronger than a regular laser gun puts out about 5×10^{15} photons per second. When they reach the star system these photons spread out over an area of 1.7×10^{15} m^2, donating only three photons per second to each square meter of that star system. The actual number is much lower because the photons, on their way to the star system, interact with the interstellar gas and many of them are absorbed by it.

As feeble as this may be, once run over a long period of time, it might give the "inhabitants" of the system's exoplanets (if they have reached a level of intelligence as advanced as ours) a sufficient number of photons—encoded with an easily identifiable greetings—to find them and send us a "Hello!" in return. ... But do we really have to wait for this remotely possible "Hello"?

9.3 GROUND-BASED PROPULSION

On April 12, 2016, flanked by Stephen Hawking and other high-profile supporters, Yuri Milner, the Russian internet entrepreneur announced his most ambitious investment in science: $100 million toward a research program to send robotic probes to Alpha Centauri within a generation. Milner has called the project "Breakthrough Starshot." It intends to squeeze all the key components of a robotic probe—cameras, sensors, maneuvering thrusters and communications equipment—into tiny gram-scale "nanocraft." These would be small enough to boost to enormous speeds using other technologies the project plans to help develop, including a ground-based kilometer-scale laser array capable of beaming 100-gigawatt laser pulses through the atmosphere for a few minutes at a time, and atoms-thin, meter-wide "light sails" to reflect those beams and boost the nanocraft to higher speeds.[3]

A research program to send robotic probes to Alpha Centauri within a generation.

9.3.1 The Physics

What is the fundamental physics behind this project? A photon of 4-momentum $(E, \vec{p})$ hits a light sail attached to a spacecraft of total mass M (which includes the mass of the light sail) of 4-momentum $(E_{\text{sail}}, \vec{P}_{\text{sail}})$ moving at an instantaneous velocity $\vec{\beta}$ and imparts some momentum to it. As a result, a photon of 4-momentum $(E_{\text{ref}}, \vec{p}_{\text{ref}})$ gets reflected from the light sail and the 4-momentum of the craft changes to $(E'_{\text{sail}}, \vec{P}'_{\text{sail}})$. Since everything takes place in one dimension, I'll remove the vector signs. Equation (8.18) can be written as

$$\begin{aligned} E - E_{\text{ref}} &= E'_{\text{sail}} - E_{\text{sail}} = M\gamma' - M\gamma \equiv M\delta\gamma \\ p + p_{\text{ref}} &= P'_{\text{sail}} - P_{\text{sail}} = M\gamma'\beta' - M\gamma\beta \equiv M\delta(\beta\gamma), \end{aligned} \tag{9.24}$$

because $\vec{p}_{\text{ref}}$ is opposite to the direction of $\vec{p}$.[4] Add the two equations and note that $E = p$ and $E_{\text{ref}} = p_{\text{ref}}$. Then

$$2E = M\delta[(1+\beta)\gamma]. \tag{9.25}$$

Now assume that each ground-based laser emits Δn photons in Δt seconds. The laser emits the ith photon at t_i, which by (9.25) changes the quantity

[3] I thank Dane Hassani for bringing the *Scientific American* article (http://bit.ly/1SLGRKx) to my attention.

[4] At this point, I am assuming that the mass of the craft does not change. I'll come back to the question later and show that the assumption is legitimate (see Remark 9.3.2 coming up later).

$M[(1+\beta)\gamma]$ by $2E_i$ *at the reflection site*. Rewrite (9.25) for the ith photon as

$$E_i = \tfrac{1}{2} M\delta_i^{\text{ref}}[(1+\beta)\gamma], \tag{9.26}$$

where the superscript is a reminder that the change on the right-hand side takes place at the light sail. The $(i+1)$th photon, emitted at $t_i + \delta t_i$, does *not* arrive at the light sail δt_i seconds after the ith one! How much later does it arrive? Suppose that the light sail is at x_i when the laser emits the ith photon, and it moves Δt_i by the time the ith photon arrives. Then the ith photon covers a distance of $x_i^{\text{ref}} \equiv c\Delta t_i$ while the sail covers a distance of $v\Delta t_i$, and[5]

$$x_i + v\Delta t_i = c\Delta t_i = x_i^{\text{ref}} \quad \text{or} \quad \Delta t_i = \frac{x_i}{1-\beta} = x_i^{\text{ref}}.$$

Therefore, the time of reflection of the ith photon is

$$t_i^{\text{ref}} = t_i + \Delta t_i = t_i + \frac{x_i}{1-\beta}.$$

With a similar expression for the $(i+1)$th photon and $\delta x_i \equiv x_{i+1} - x_i = \beta\delta t_i$, you get

$$\delta t_i^{\text{ref}} = \delta t_i + \frac{\delta x_i}{1-\beta} = \delta t_i + \frac{\beta\delta t_i}{1-\beta} = \frac{\delta t_i}{1-\beta}. \tag{9.27}$$

Remark 9.3.1. I have to say a few words about Equation (9.25). It is coming directly from Equation (8.16), which is in terms of 4-vectors. In fact, I can write it as

$$\mathbf{p} - \mathbf{p}_{\text{ref}} = \mathbf{P}'_{\text{sail}} - \mathbf{P}_{\text{sail}},$$

if I define the 4-momenta of the light sail before and after the reflection as $\mathbf{P}_{\text{sail}}$ and $\mathbf{P}'_{\text{sail}}$, respectively. Dealing with 4-vectors has the advantage of being valid in all reference frames. Problem 9.18 shows the **Lorentz covariance** of (9.25): If the equation holds in one RF, then it holds in all RFs. Some authors prefer to work only with the second equation of (9.24) because it is more directly related to force. While this is preferable classically, it is not Lorentz covariant. If you define the reflection coefficient R as $p_{\text{ref}} = Rp$, as is done in the literature, then the second equation of (9.24) becomes

Meaning of Lorentz covariance.

$$(1+R)p = (1+R)E = P'_{\text{sail}} - P_{\text{sail}} = M\delta(\beta\gamma). \tag{9.28}$$

While the left-hand side of this equation transforms the same way as the left-hand side of (9.25), the right-hand side transforms differently! The equation, therefore, is not Lorentz covariant. In fact, Problem 9.20 shows that the reflection coefficient cannot be defined, because its definition would lead to

$$\frac{1-R}{1+R} = \beta, \quad \text{or} \quad R = \frac{1-\beta}{1+\beta},$$

which makes no sense at all, making R velocity dependent in a specific way! ■

[5] I'm switching back and forth between $c = 1$ and $c = c$ for clarity (I hope!).

Lasers are characterized by the power they emit. So, I'm going to calculate the energy delivered in $\Delta t = \sum_{i=1}^{\Delta n} \delta t_i$ to the light sail by adding (9.26) over i.

$$\Delta E = \sum_{i=1}^{\Delta n} E_i = \tfrac{1}{2} M \sum_{i=1}^{\Delta n} \delta_i^{\mathrm{ref}}[(1+\beta)\gamma] \equiv \tfrac{1}{2} M \Delta^{\mathrm{ref}}[(1+\beta)\gamma]. \tag{9.29}$$

Dividing both sides by Δt and letting $\Delta t \to dt$ gives me P, the laser power, on the left, and

$$\tfrac{1}{2} M \frac{\Delta^{\mathrm{ref}}[(1+\beta)\gamma]}{\Delta t} = \tfrac{1}{2} M \frac{\Delta^{\mathrm{ref}}[(1+\beta)\gamma]}{(1-\beta)\Delta t^{\mathrm{ref}}} = \tfrac{1}{2} M \frac{1}{1-\beta} \frac{d}{dt}[(1+\beta)\gamma]$$

on the right. Thus the equation of motion of the spacecraft becomes

$$\frac{1}{1-\beta} \frac{d}{dt}[(1+\beta)\gamma] = \frac{2P}{M}, \tag{9.30}$$

where P is the fraction of the total laser power P_{laser} reflected from the light sail. When the light sail is close, all of the laser power hits it and $P = P_{\mathrm{laser}}$. However, as the sail moves away, due to the spread of the laser beam, the sail gets immersed in the beam and P becomes smaller than P_{laser}.

9.3.2 The Equation of Motion

Let $\Omega(x)$ be the solid angle subtended by the light sail at the distance x. Let Ω_{laser} be the solid angle of the cone defined by the laser beam. If this beam spreads out to an angle θ_{laser}, then

$$\Omega_{\mathrm{laser}} = \int_0^{2\pi} d\varphi \int_0^{\theta_{\mathrm{laser}}} \sin\theta \, d\theta \, d\varphi = 2\pi(1 - \cos\theta_{\mathrm{laser}}) \approx \pi\theta_{\mathrm{laser}}^2$$

for small θ_{laser}. Diffraction theory in optics gives

$$\sin\theta_{\mathrm{laser}} = 1.22 \frac{\lambda}{D},$$

where λ is the wavelength of the laser and D is the diameter of the opening through which the beam emerges. For $\lambda << D$, $\sin\theta_{\mathrm{laser}} \approx \theta_{\mathrm{laser}}$ and

$$\Omega_{\mathrm{laser}} \approx \pi \left(1.22 \frac{\lambda}{D}\right)^2 = \left(\frac{2.16\lambda}{D}\right)^2. \tag{9.31}$$

Let the light sail be a square of side L. It can be shown that its solid angle at a distance x from its center on a line perpendicular to it is (see Problem 9.21)

$$\Omega(x) = 4 \tan^{-1} \left(\frac{L^2}{2x\sqrt{2L^2 + 4x^2}} \right),$$

which, for $x >> L$, reduces to

$$\Omega(x) \approx \frac{L^2}{x^2}. \tag{9.32}$$

At a certain distance x_0 the two solid angles become equal. This distance is given by

$$\Omega_{\text{laser}} = \left(\frac{2.16\lambda}{D}\right)^2 = \frac{L^2}{x_0^2} \quad \text{or} \quad x_0 = \frac{LD}{2.16\lambda}. \tag{9.33}$$

I can now write the power delivered to the light sail with N lasers in the array:

$$P = \begin{cases} N P_{\text{laser}} & \text{if } x < x_0, \\ N P_{\text{laser}}[\Omega(x)/\Omega_{\text{laser}}] & \text{if } x > x_0. \end{cases} \tag{9.34}$$

Using (9.32), (9.33), and (9.30), the complete equation of motion of the spacecraft becomes

$$\frac{1}{1-\beta}\frac{d}{dt}[(1+\beta)\gamma] = \frac{2N P_{\text{laser}}}{M} \times \begin{cases} 1 & \text{if } x < x_0, \\ (x_0/x)^2 & \text{if } x > x_0. \end{cases} \tag{9.35}$$

On the left, I have the derivative with respect to time. On the right (at least in the second case), I have x. I go around the problem by employing the usual trick of changing the differentiation variable from t to x,

$$\frac{d}{dt}[(1+\beta)\gamma] = \frac{d}{dx}[(1+\beta)\gamma]\frac{dx}{dt} = \beta\frac{d}{dx}[(1+\beta)\gamma],$$

and write (9.35) as

$$\frac{\beta}{1-\beta}d[(1+\beta)\gamma] = \frac{2N P_{\text{laser}}}{M} \times \begin{cases} dx & \text{if } x < x_0, \\ x_0^2\dfrac{dx}{x^2} & \text{if } x > x_0. \end{cases} \tag{9.36}$$

But $d\gamma = \gamma^3\beta d\beta$ and

$$d(\gamma\beta) = \gamma d\beta + \beta d\gamma = \gamma d\beta + \beta^2\gamma^3 d\beta = \gamma d\beta\,\underbrace{(1+\beta^2\gamma^2)}_{=\gamma^2} = \gamma^3 d\beta.$$

Therefore, the left-hand side of (9.36) becomes

$$LHS = \frac{\beta(1+\beta)\gamma^3}{1-\beta}d\beta = \frac{\beta\gamma d\beta}{(1-\beta)^2} = d\left[\frac{2\beta-1}{3(1-\beta)^2\gamma}\right].$$

(Verify the last equality!) Now I can write (9.36) in a way that can be trivially integrated:

$$d\left[\frac{2\beta-1}{3(1-\beta)^2\gamma}\right] = \frac{2N P_{\text{laser}}}{M} \times \begin{cases} dx & \text{if } x < x_0, \\ d\left(-\dfrac{x_0^2}{x}\right) & \text{if } x > x_0. \end{cases} \tag{9.37}$$

Remark 9.3.2. QUESTION OF CHANGE IN SPACECRAFT'S MASS
If you assume that the mass of the spacecraft changes, then (9.24) becomes

$$E - E_{\text{ref}} = E'_{\text{sail}} - E_{\text{sail}} \equiv \delta(M\gamma) = M\delta\gamma + \gamma\delta M$$
$$p + p_{\text{ref}} = P'_{\text{sail}} - P_{\text{sail}} \equiv \delta(M\beta\gamma) = M\delta(\beta\gamma) + \beta\gamma\delta M,$$

turning (9.25) into

$$2E = M\delta[(1+\beta)\gamma] + (1+\beta)\gamma\delta M. \tag{9.38}$$

The actual creation of a particle of mass m requires the photon to have an energy of mc^2 (see Example 12.1.1). For the lightest particle, the electron, this energy is more than 0.5 MeV, placing the photon in the shortest wavelength region of gamma rays, and beyond the capability of any laser.[6]

So, let's consider the case where δM is the equivalent mass of the internal (or heat) energy produced in the light sail by the photons.[7] To get an estimate of the magnitude of this energy, let the sail be made up of aluminum, and take the extreme condition of so huge a heat energy as to cause the melting of aluminum at 660° C. How does the first term of (9.38) compare with the second term?

The heat absorbed by a substance of mass M and specific heat c_{sh} when its temperature changes by ΔT is $Mc_{sh}\Delta T$, with an equivalent mass of $\delta M = Mc_{sh}\Delta T/c^2$. So, the right-hand side of (9.38) becomes

$$\left\{\delta[(1+\beta)\gamma] + (1+\beta)\gamma\frac{c_{sh}\Delta T}{c^2}\right\}Mc^2 = \left(\gamma^2\delta\beta + \frac{c_{sh}\Delta T}{c^2}\right)(1+\beta)\gamma Mc^2, \tag{9.39}$$

where I used the results obtained immediately after Equation (9.36) and restored the factor of c. The two expressions in the parentheses on the right-hand side are dimensionless quantities. For aluminum, $c_{sh} = 900$ J/(kg K) and $c^2 = 9 \times 10^{16}$ J/kg. So, for the heat required to melt aluminum (from, say, 0° C), the second term gives

$$\frac{c_{sh}\Delta T}{c^2} = \frac{900 \times 660.3}{9 \times 10^{16}} = 6.6 \times 10^{-12}. \tag{9.40}$$

To estimate the first term, refer to the first case of (9.35) and write it as

$$\frac{1}{1-\beta}\delta[(1+\beta)\gamma] = \frac{2NP_{\text{laser}}\delta t}{M},$$

from which—after restoring the factors of c and restricting to one laser—you get

$$\delta\beta = \frac{(1-\beta)^2}{\gamma}\frac{2P_{\text{laser}}\delta t}{Mc^2}.$$

The first term of (9.39), therefore, becomes

$$\gamma^2\delta\beta = \gamma(1-\beta)^2\frac{2P_{\text{laser}}\delta t}{Mc^2}.$$

---

[6] Although neutrinos are lighter than electrons, because they are neutral, they cannot participate in electromagnetic interactions involving photons.

[7] The first term in (9.38) takes care of the "external," or bulk energy of the sail.

Assuming that it takes one second ($\delta t = 1$ s) to melt one gram ($M = 0.001$ kg) of aluminum with a one-GW laser ($P_{\text{laser}} = 10^9$ W), you get

$$\gamma^2 \delta\beta = \gamma(1-\beta)^2 \frac{2 \times 10^9}{0.001 \times (3 \times 10^8)^2} = 2.22 \times 10^{-5} \gamma(1-\beta)^2.$$

For any reasonable value of β, this is at least five to six orders of magnitude larger than the second term (9.40), and we can safely assume that the mass of the spacecraft remains constant. ■

9.3.3 The Formal Solution

For the first case of (9.37), assuming that the spacecraft starts from rest at $x = 0$, you easily obtain x as a function of β:

$$x = \frac{Mc^3}{2NP_{\text{laser}}}\left[\frac{2\beta - 1}{3(1-\beta)^2\gamma} + \frac{1}{3}\right], \quad x < x_0, \tag{9.41}$$

where I have restored the factors of c for convenience. When the spacecraft ends the first stage of its journey, it has a speed β_0 at x_0, with

$$x_0 = \frac{Mc^3}{2NP_{\text{laser}}}\left[\frac{2\beta_0 - 1}{3(1-\beta_0)^2\gamma_0} + \frac{1}{3}\right]. \tag{9.42}$$

Now it begins its second stage. Integration of the second case of (9.37) yields

$$\frac{x_0^2}{x} = 2x_0 - \frac{Mc^3}{2NP_{\text{laser}}}\left[\frac{2\beta - 1}{3(1-\beta)^2\gamma} + \frac{1}{3}\right], \quad x > x_0, \tag{9.43}$$

where I chose the constant of integration so that (9.42) holds. Equation (9.43) can also be written as

$$\frac{x_0}{x} = 2 - \frac{(2\beta - 1)/[3(1-\beta)^2\gamma] + \frac{1}{3}}{(2\beta_0 - 1)/[3(1-\beta_0)^2\gamma_0] + \frac{1}{3}}, \quad x > x_0, \tag{9.44}$$

which has an interesting result. Rewrite it as

$$\frac{(2\beta - 1)/[3(1-\beta)^2\gamma] + \frac{1}{3}}{(2\beta_0 - 1)/[3(1-\beta_0)^2\gamma_0] + \frac{1}{3}} = 2 - \frac{x_0}{x},$$

and note that $x \geq x_0 > 0$. Moreover, a plot of the left-hand side reveals that it is monotone increasing for $0 \leq \beta < 1$. Therefore, the ratio on the left is always less than 2, getting closer and closer to 2 as x increases. This means that there is a terminal speed β_∞ reached only when $x \to \infty$. This speed is given by

$$\frac{2\beta_\infty - 1}{3(1-\beta_\infty)^2\gamma_\infty} + \frac{1}{3} = 2\left[\frac{2\beta_0 - 1}{3(1-\beta_0)^2\gamma_0} + \frac{1}{3}\right]. \tag{9.45}$$

The reason for this behavior is that as the light sail gets farther and farther away, by (9.34), the power delivered to it gets smaller and smaller, and fewer and fewer photons bounce from it. Therefore, the push on it becomes smaller and smaller. At infinity, the power delivered,

$$P = N P_{\text{laser}} \frac{\Omega(x)}{\Omega_{\text{laser}}} = N P_{\text{laser}} \frac{x_0^2}{x^2},$$

goes to zero; there is no push, and the speed remains unchanged.

I'm also interested in β and x as functions of time. I can find β as a function of time for $x < x_0$ directly from (9.35). You can easily verify that the left-hand side can be written as

$$\frac{d}{dt}\left[\frac{(1+\beta)(2-\beta)\gamma}{3(1-\beta)}\right],$$

and thus, with factors of c restored,

$$\frac{2N P_{\text{laser}}}{Mc^2} t = \frac{(1+\beta)(2-\beta)\gamma}{3(1-\beta)} - \frac{2}{3}, \tag{9.46}$$

with the constant of integration chosen so that $\beta = 0$ at $t = 0$. I can actually solve β in terms of t. It involves solving a cubic equation (which is doable); but the solution is no more transparent than the implicit relation (9.46). So, I'm going to leave it as is. Now that I have (in principle) β as a function of time, I can substitute it in (9.41) and find x as a function of time.

The second case of (9.35) is more complicated because of the presence of x on the right-hand side. All I can do is write the solution as an integral. The left-hand side can be simplified to

$$\frac{\gamma}{(1-\beta)^2} \frac{d\beta}{dt},$$

and (9.44) can be inserted on the right-hand side. The result is the differential equation

$$\frac{1}{\sqrt{1-\beta^2}\,(1-\beta)^2} \frac{d\beta}{dt} = \frac{2N P_{\text{laser}}}{M} \left\{2 - \frac{(2\beta-1)/[3(1-\beta)^2\gamma] + \frac{1}{3}}{(2\beta_0-1)/[3(1-\beta_0)^2\gamma_0] + \frac{1}{3}}\right\}^2,$$

whose solution, with the assumption that $\beta = \beta_0$ at $t = 0$, can be written as an integral:

$$\int_{\beta_0}^{\beta} \frac{du}{\sqrt{1-u^2}\,(1-u)^2} \left\{2 - \frac{(2u-1)\sqrt{1-u^2}/[3(1-u)^2] + \frac{1}{3}}{(2\beta_0-1)/[3(1-\beta_0)^2\gamma_0] + \frac{1}{3}}\right\}^{-2} = \frac{2N P_{\text{laser}}}{Mc^2} t, \tag{9.47}$$

where again I have restored the factors of c. This (highly implicit) solution of β as a function of time can now be inserted on the right-hand side of (9.44) to give x as a (even more highly implicit) function of time for $x > x_0$.

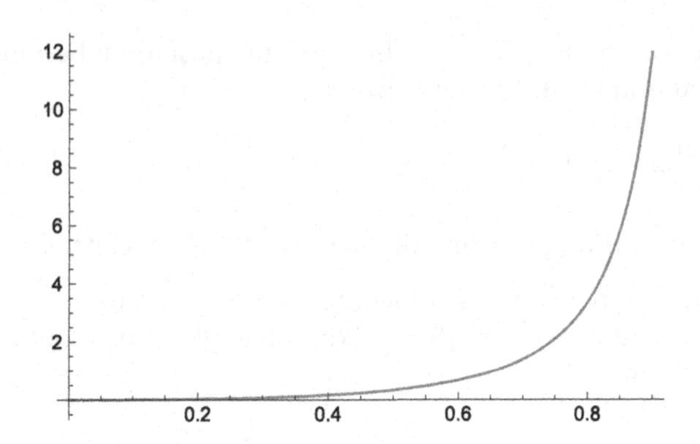

FIGURE 9.1 Plot of the function $(2\beta_\infty - 1)/[3(1-\beta_\infty)^2\gamma_\infty] + \frac{1}{3}$.

9.3.4 Numerical Examples

Let's put in some numbers and get a feel for the abstract analysis developed so far. First note that x_0, as given in (9.33), is completely determined by the lasers and the light sail, the two essential components of the project. It is desirable to have as large an x_0 as possible, because after x_0, the light sail receives fewer and fewer photons and cannot be accelerated as much. Furthermore, x_0 determines β_0 and, by (9.45), β_0 determines β_∞, the terminal speed of the spacecraft.

It is convenient to have a relation that directly relates the parameters of the project to our target terminal velocity. Using (9.33), (9.42), and (9.45), you can show that

$$LD = \frac{2.16\lambda M c^3}{4N P_{\text{laser}}}\left[\frac{2\beta_\infty - 1}{3(1-\beta_\infty)^2\gamma_\infty} + \frac{1}{3}\right]. \tag{9.48}$$

Let me remind you that L is the length of the square light sail that reflects the laser beam and gives momentum to the craft, and D is the diameter of the opening of the laser. The smaller the LD, the more practical the project. This means that the spacecraft must be as light as practical, and each laser must have as short a wavelength and as large a power output as possible, and the array should contain as many of these lasers as feasible. Figure 9.1 shows the behavior of the function in the square bracket in (9.48) as a function of β_∞. You note that for low velocities, the function affects the value of L very little. However, it becomes important once higher velocities become the target.

As an example, say we want to accelerate a spacecraft of 100 kg to a terminal speed of $0.9c$. Then even with (yet to be invented) gamma-ray lasers with $\lambda = 10^{-12}$ m and P_{laser} of 100 GW, we would have

$$LD = \frac{2.16 \times 10^{-12} \times 100 \times (3\times 10^8)^3}{4N \times 100 \times 10^9}\left(\frac{0.8}{3 \times 0.01 \times 2.29} + \frac{1}{3}\right) = \frac{174{,}600}{N}\ \text{m}^2.$$

So, for a light sail square of length one meter and $D = 1$ cm, we need over 17 million lasers, highly impractical! So, let's reduce the mass of the spacecraft to 1 kg. Then

$$LD = \frac{2.16 \times 10^{-12} \times 1 \times (3 \times 10^8)^3}{4N \times 100 \times 10^9} \left(\frac{0.8}{3 \times 0.01 \times 2.29} + \frac{1}{3} \right) = \frac{1746}{N} \text{ m}^2,$$

reducing the number of lasers to over 170 thousand. Although still impractical, I'll look at the details of the motion anyway, assuming the array contains 100 lasers. I can find β_0 by numerically solving (9.45) with $\beta_\infty = 0.9$. The solution is $\beta_0 = 0.852$. Substituting this value in (9.46), I get

$$\begin{aligned} t &= \frac{Mc^2}{2NP_{\text{laser}}} \left[\frac{(1+\beta_0)(2-\beta_0)\gamma_0}{3(1-\beta_0)} - \frac{2}{3} \right] = 4.24 \frac{Mc^2}{NP_{\text{laser}}} \\ &= 3.8 \times 10^4 \text{ s} = 10.56 \text{ hours} \end{aligned}$$

as the time required for the craft to reach 85.2% the speed of light. How long does it take the craft to reach to within 99% of its terminal speed?[8] The answer is in the following numerical equation coming from (9.47):

$$\underbrace{\int_{0.852}^{0.891} \frac{du}{\sqrt{1-u^2}(1-u)^2} \left\{ 2 - \frac{(2u-1)\sqrt{1-u^2}/[3(1-u)^2] + \frac{1}{3}}{5.94} \right\}^{-2}}_{=18.4439} = \frac{2NP_{\text{laser}}}{Mc^2} t,$$

or,

$$t = 9.222 \frac{Mc^2}{NP_{\text{laser}}} = 8.3 \times 10^4 \text{ s} = 23 \text{ hours}.$$

So, just over a day after its launch, the spacecraft will be cruising the rest of its journey at 90% the speed of light. Such a spacecraft will reach our nearest neighbor, Alpha Centauri, in a little over five years.

This is all science fiction! There are no gamma-ray lasers. Even if they were invented in the distant future, coming up with 170 thousand of them is impossible. And *any* laser continuously delivering 100 GW of power is hard to come by.

Now you can appreciate why the Breakthrough Starshot project seeks to develop gram-sized nanocraft and powerful lasers to achieve a modest terminal speed of $0.2c$. With a mass of 10 grams and the possibility of 100-GW x-ray lasers with $\lambda = 1$ nm, you get

$$LD = \frac{2.16 \times 10^{-9} \times 0.01 \times (3 \times 10^8)^3}{4N \times 100 \times 10^9} \left(\frac{-0.6}{3 \times 0.64 \times 1.02} + \frac{1}{3} \right) = \frac{39.31}{N} \text{ m}^2, \tag{9.49}$$

[8] I can't ask for a speed of $0.9c$ because it happens only at $x = \infty$ and therefore $t = \infty$.

which is not completely out of reach. So, let's analyze the motion of such nanocraft with an array of 100 lasers. For $\beta_\infty = 0.2$, Equation (9.45) yields 0.147782 for β_0. Then, (9.46) gives a value of 7.8 seconds for the nanocraft to finish the first leg of its trip. For the rest of the trip, you'll have to solve the following numerical equation:

$$\int_{0.147782}^{0.198} \frac{du}{\sqrt{1-u^2}(1-u)^2} \left\{ 2 - \frac{(2u-1)\sqrt{1-u^2}/[3(1-u)^2] + \frac{1}{3}}{0.0135566} \right\}^{-2} = \frac{2NP_{\text{laser}}}{Mc^2} t$$

to find the time the nanocraft needs to reach 99% of its terminal speed. The value of the integral is 1.43892, which gives a value of about 76 seconds for t. Therefore, in less than a minute and a half, the nanocraft will be cruising its journey at (very close to) 20% the speed of light. A trip to Alpha Centauri, 4.3 light years away, will take 21.5 years.

9.3.5 Nonrelativistic Limits

Terminal speeds close to the speed of light are unreachable with today's technology. The reasonable values of terminal speed such as the one sought after in the Breakthrough Starshot project are low enough that nonrelativistic limits of the equations can be applied with fairly accurate results. So, let's take a look at this limit.[9]

Expand the function of β on the right-hand side of (9.41) to get

$$\frac{2\beta - 1}{3(1-\beta)^2\gamma} + \frac{1}{3} \approx \frac{\beta^2}{2}.$$

Then, the equation becomes

$$x \approx \frac{Mc^3}{4NP_{\text{laser}}}\beta^2 = \frac{Mc^3}{4NP_{\text{laser}}}\frac{v^2}{c^2} = \frac{Mc}{4NP_{\text{laser}}}v^2. \tag{9.50}$$

This is the equation connecting distance to the velocity of a *uniformly accelerated* motion with acceleration

$$a = \frac{2NP_{\text{laser}}}{Mc}.$$

I can verify this conclusion directly by looking at how the velocity changes with time. Expanding the right-hand side of (9.46),

$$\frac{(1+\beta)(2-\beta)\gamma}{3(1-\beta)} - \frac{2}{3} \approx \beta,$$

and plugging it in the equation, I get

$$\frac{2NP_{\text{laser}}}{Mc^2}t \approx \beta = \frac{v}{c} \quad \text{or} \quad v \approx \frac{2NP_{\text{laser}}}{Mc}t,$$

---

[9] You can also obtain the nonrelativistic equation of motion from the beginning by ignoring relativity (except that photons have momenta given by $p = E/c$). Problems 9.22 and 9.23 ask you to derive the nonrelativistic equation of motion from scratch.

verifying that the motion is indeed uniformly accelerated with the acceleration given above.

The nonrelativistic limit of the second stage of the journey is obtained from (9.44). The first nonzero term in the expansion of both the numerator and denominator is of second order. Therefore,

$$\frac{x_0}{x} = 2 - \frac{\frac{1}{2}\beta^2}{\frac{1}{2}\beta_0^2} \quad \text{or} \quad \beta = \frac{dx}{dt} = \beta_0\left(2 - \frac{x_0}{x}\right)^{1/2}, \tag{9.51}$$

which can be integrated to give t as a function of x (see Problem 9.23). You can then (in principle) solve the equation to find x as a function of t and insert the result in (9.51) to find β as a function of t as well.

Equation (9.51) shows that

$$\beta_\infty = \sqrt{2}\,\beta_0. \tag{9.52}$$

For a terminal velocity of $0.2c$, this equation yields a value of 0.1414 for β_0, which, with an error of 4.3%, compares well with 0.147782 obtained relativistically. This agreement is actually misleading because the nonrelativistic limits of other important parameters going into the design of the nanocraft and light sail do not agree so well with the correct relativistic values.

One such parameter is the product LD, whose nonrelativistic limit, obtained from (9.48), is

$$LD \approx \frac{2.16\lambda Mc^3}{4NP_{\text{laser}}}\left(\frac{\beta_\infty^2}{2}\right) = \frac{2.16\lambda Mc^3\beta_\infty^2}{8NP_{\text{laser}}}. \tag{9.53}$$

For a nanocraft with a mass of 10 grams, driven by 100 ground-based 100-GW x-ray lasers with $\lambda = 1$ nm, to achieve a terminal speed of $0.2c$, we need an LD of

$$LD \approx \frac{2.16 \times 10^{-9} \times 0.01 \times (3 \times 10^8)^3 \times (0.2)^2}{8 \times 100 \times 10^{11}} = 0.2916\ \text{m}^2.$$

This is to be compared with 0.3931 m obtained in Equation (9.49). The error is substantial!

Other quantities are also affected non-negligibly. For example, consider the time required to cover the first stage of the trip. From (9.33),

$$x_0 = \frac{LD}{2.16\lambda} = \frac{2.916 \times 0.01}{2.16 \times 10^{-9}} = 1.35 \times 10^7\ \text{m},$$

and

$$t_0 = \sqrt{\frac{2x_0}{a}} = \sqrt{\frac{x_0 Mc}{NP_{\text{laser}}}} = \sqrt{\frac{1.35 \times 10^7 \times 0.01 \times 3 \times 10^8}{100 \times 10^{11}}} = 2.012\ \text{s},$$

which is to be compared with the 7.8 seconds obtained before. The conclusion is that while nonrelativistic equations of motion give a fairly accurate approximation to some parameters and a roughly good overall description of the journey, they are not accurate enough to be relied on when constructing a spacecraft and designing a laser. Relativity plays an important role in the project.

9.4 PROBLEMS

9.1. Assume that $M >> dm_g$ and consider the ejection of dm_g from the rocket as a decay. Then use the energy and 3-momentum components of the 4-momentum conservation $\mathbf{P} = \mathbf{p}_1 + \mathbf{p}_2$ to show that

$$d(M\gamma) + dm_g\gamma_g' = 0$$
$$d(M\gamma\vec{\beta}) + dm_g\gamma_g'\vec{\beta}_g' = 0.$$

9.2. Show that

$$d(M\gamma) = \gamma dM + M\beta\gamma^3 d\beta$$
$$d(M\gamma\beta) = \gamma\beta dM + M\gamma^3 d\beta.$$

9.3. Derive Equation (9.3) from the equations preceding it.

9.4. Consider the acceleration part of the photon propulsion (9.6) with initial speed zero. Show that you get Equation (9.4) with $\beta_g = 1$.

9.5. Suppose that a rocket accelerates from rest to β_0 upon launch and decelerates from β_0 to zero upon landing. If M_0 is the mass at launch and M_f is the mass at landing, show that (9.7) gives $M_f/M_0 = [\gamma_0(\beta_0 + 1)]^{-2}$, in agreement with (the first half of) Equation (9.5).

9.6. Show that both equations in (9.8) are consistent with $m(0) = 1$ for $\beta(0) = \beta_0$.

9.7. The first equation in (9.8) yields $\beta = 1$ when $m(t)$ is completely depleted. Is that a violation of relativity or a confirmation of it? Think about it for a while before you look up the answer in Note 5.2.3.

9.8. Let $-k$ be the rate of depletion of the fuel of a rocket.

(a) Solve $dM_{\text{fuel}}/dt = -k$, when k is a positive constant subject to the condition that $M_{\text{fuel}}(0) = M_{\text{fuel}}^{\text{beg}}$. Here $M_{\text{fuel}}^{\text{beg}}$ is the amount of fuel at the beginning (which is *not* necessarily the take-off) of the part of motion under consideration.

(b) Assuming that the fuel mass at take-off is M_0, determine k from the condition that $M_{\text{fuel}}(T_0) = 0$, where T_0 is the time it takes for the fuel to be completely depleted.

(c) Show that if time is measured in units of T_0, then

$$M_{\text{tot}}(t) = M_f + M_{\text{fuel}}(t) = M_f + M_{\text{fuel}}^{\text{beg}} - M_0 t,$$

where M_f is the mass left over at the end of the trip.

9.9. Derive Equation (9.11) and by taking the time derivative of (9.12) show that the latter is the integral of (9.11).

9.10. Show that the time required to accelerate the rocket from rest to β_0 is given by (9.13).

9.11. Show that $x(t_{acc})$ is given by Equation (9.14).

9.12. Differentiate Equation (9.16) to get (9.15).

9.13. Show that the time required to decelerate the rocket from β_0 to rest is given by (9.17) and the distance covered by (9.18).

9.14. The maximum possible speed attainable is when no fuel is left upon landing. Show that this speed is given by Equation (9.19), and for such a speed, t_{acc} and t_{dec} are given by Equation (9.20).

9.15. Derive Equations (9.21) and (9.22). Plot each separately as well as the sum $x_{acc} + x_{dec}$ and verify that the sum is a decreasing function with a maximum of 0.57 light second occurring when $m_f = 0$.

9.16. I discussed the case of maximum attainable speed when m_f is small. What about the case when $m_f \approx 1$? Write $m_f = 1 - m_0$ and assume that m_0 is very small.

(a) Show from Equation (9.19) that the maximum attainable speed is $\frac{1}{2}m_0$.
(b) From the expansion of (9.21) show that $x_{acc} \to \frac{1}{8}m_0$.
(c) From (9.22) conclude that $x_{dec} \to \frac{1}{8}m_0$.

9.17. Problem 9.16 is based on the assumption of uniform depletion of the fuel. The result, however, turns out to be general, at least for the case of a laser-propelled rocket as discussed on page 238.

(a) Suppose you use half the fuel (saving the other half for landing), i.e., the lasing photons, on a massless rocket ($M_f = 0$) to accelerate the rocket from rest to a speed β_f. Using the first equation in (9.7), show that β_f is given by

$$\frac{1+0.5\kappa}{1+\kappa} = \frac{1}{\gamma_f(\beta_f + 1)}, \qquad \kappa << 1.$$

(b) Solve this equation for β_f in terms of κ to show that $\beta_f \approx 0.5\kappa$.

9.18. Lorentz transform both the energy E of a photon and the sum $E_{sail} + P_{sail}$ of the energy and momentum of the light sail to another RF and show that you get the same equality (9.25) in the new RF.

9.19. In this problem you are going to verify Equation (9.27).

(a) Lorentz transform the event of the emission of the ith photon to the instantaneous reference frame of the spacecraft and show that

$$(t_i^{nano}, x_i^{nano}) = \gamma t_i(1, -\beta).$$

(b) How long does it take for this photon to reach the craft? Adding t_i^{nano} to this time, find $t_i^{\text{nano, ref}}$.

(c) From (a) and (b) and the fact that reflections occur at the spacecraft, conclude that the events of the reflection of the ith and $i+1$th photons in the spacecraft's RF are

$$\big(\gamma(1+\beta)t_i, 0\big), \quad \text{and} \quad \big(\gamma(1+\beta)t_{i+1}, 0\big).$$

(d) Now Lorentz transform this back to the laser RF and derive (9.27).

9.20. Why reflection coefficient R makes no sense in one-dimensional relativity! Write Equation (8.18) as

$$\begin{aligned} E - E_{\text{ref}} &= E'_{\text{sail}} - E_{\text{sail}} = \delta E_{\text{sail}} \\ p + p_{\text{ref}} &= P'_{\text{sail}} - P_{\text{sail}} = \delta P_{\text{sail}}, \end{aligned} \tag{9.54}$$

where no approximation is made regarding the mass of the spacecraft.

(a) Using R, show that these equations yield

$$\frac{1-R}{1+R} = \frac{\delta E_{\text{sail}}}{\delta P_{\text{sail}}}.$$

(b) Use $E_{\text{sail}} = \sqrt{P_{\text{sail}}^2 + M^2}$ to show that

$$\frac{1-R}{1+R} = \beta,$$

where β is the instantaneous speed of the craft.

9.21. Recall that the solid angle subtended at point P_0 by an infinitesimal area da located at point P is given by

$$d\Omega = \frac{\left|\hat{\mathbf{e}}_n \cdot (\vec{r} - \vec{r}_0)\right|}{|\vec{r} - \vec{r}_0|^3} da,$$

where $\hat{\mathbf{e}}_n$ is the unit vector normal to da and $\vec{r}$ and $\vec{r}_0$ are the position vectors of P and P_0, respectively. Now center a square of side L in the xy-plane and let P_0 have coordinates $(0, 0, z_0)$.

(a) Show that the total solid angle Ω subtended at P_0 by the square is

$$\Omega = 4\tan^{-1}\left(\frac{L^2}{2z_0\sqrt{2L^2 + 4z_0^2}}\right).$$

(b) Show that if $z_0 >> L$, then

$$\Omega \approx \frac{L^2}{z_0^2}.$$

9.22. This problem treats the first stage of the journey of a spacecraft non-relativistically. Assume perfect reflection of a beam of photons from the light sail.

(a) What is the momentum transfer to the light sail by a beam of photons carrying energy ΔE? Show that the force exerted on it is $2P/c$, where P is the fraction of laser power hitting the light sail. What is the acceleration if the light sail has a mass M?

(b) For the first leg of the trip, show that the acceleration is uniform. Find the value of this acceleration for a nanocraft with a mass of 10 grams driven by a 100-GW ground laser.

9.23. This problem treats the second stage of the journey of a spacecraft non-relativistically. Assume perfect reflection of a beam of photons from the light sail.

(a) For the second leg of the trip, show that the acceleration is

$$a = \frac{2P_{\text{laser}}}{Mc}\frac{x_0^2}{x^2}.$$

(b) Write dv/dt as $v dv/dx$ and integrate the equation in (a) to get

$$\tfrac{1}{2}v^2 - \tfrac{1}{2}v_0^2 = \frac{2P_{\text{laser}}}{Mc}x_0\left(1 - \frac{x_0}{x}\right)$$

assuming that the speed is v_0 at x_0.

(c) Show that the result in (b) can also be written as

$$v = v_0\sqrt{2 - \frac{x_0}{x}},$$

giving $v_\infty = \sqrt{2}\, v_0$. Hint: Find x_0 in terms of v_0 from Problem 9.22.

(d) Integrate (c) to find t as a function of x with the constant of integration determined by the condition that $x = x_0$ at $t = 0$.

CHAPTER 10

A Painless Introduction to Tensors

Electromagnetic theory started relativity. It is, therefore, natural to expect a feedback loop between the two disciplines. In fact, the most effective way of studying electromagnetism is by casting it in tones of relativity theory. However, with four dimensions and six components of the electric and magnetic fields, vector and matrix notations become extremely unwieldy. A tremendous amount of simplification ensues if indexed components of vectors and matrices are used. Although this chapter has the word "tensors" in its heading, it has no pretense of treating the fascinating subject of tensor analysis in any detail. The sole purpose of the chapter is to expose you to very simple, yet powerful, rules of manipulating vectors and matrices and use them to derive some important results in electrodynamics in the next chapter.

10.1 FROM BOLDFACE TO INDICES

Vector manipulations are greatly simplified if you write equations in terms of a *general* component. How do you accomplish this? Start with a generic vector equation, which can be written as

Boldfaced symbols represent 3-vectors here!

$$\mathbf{U} = \mathbf{V},$$

where $\mathbf{U}$ and $\mathbf{V}$ are, in general, vector *expressions*. The use of boldfaced letters for 3-vectors should not confuse you as we are dealing only with 3-vectors in this section. Examples of vector equations are

$$\mathbf{B} = \nabla \times \mathbf{A}, \qquad \mathbf{E} = -\nabla\Phi, \qquad \mathbf{A} = \int_a^b f(r)\hat{\mathbf{e}}_r\, dr. \tag{10.1}$$

You can also write each of these vector equations as three equations involving components. Thus, the middle term of (10.1) becomes

$$E_x = -\frac{\partial\Phi}{\partial x}, \qquad E_y = -\frac{\partial\Phi}{\partial y}, \qquad E_z = -\frac{\partial\Phi}{\partial z}.$$

It is very helpful to convert letter indices into number indices. Let $x \to 1$, $y \to 2$, and $z \to 3$, and write

$$U_1 = V_1, \qquad U_2 = V_2, \qquad U_3 = V_3.$$

Special Relativity. DOI:10.1016/B978-0-12-810411-8.00010-9

These equations are abbreviated as

$$U_i = V_i, \qquad i = 1, 2, 3. \tag{10.2}$$

This is what is meant by an equation in terms of a *general component*: The index i refers to any one of the components of the vectors on either side of the equation. It is called a **free** index because you are free to choose any symbol instead of i. An important property of a free index is that

Free index defined.

> **Note 10.1.1.** *A* **free index** *appears once and only once on both sides of a vector equation.*

Although the most common symbols used are i, j, k, l, m, and n, you can write Equation (10.2) in any one of the following alternative ways:

$$\begin{aligned} U_j &= V_j, \qquad j = 1, 2, 3, \\ U_p &= V_p, \qquad p = 1, 2, 3, \\ U_\heartsuit &= V_\heartsuit, \qquad \heartsuit = 1, 2, 3. \end{aligned}$$

An abbreviation that is used for derivatives with respect to Cartesian coordinates is given as follows. First $\partial/\partial x$ is replaced by $\partial/\partial x_1$, and the latter by the much shorter notation, ∂_1. Similarly, $\partial/\partial y$ becomes ∂_2, and $\partial/\partial z$ becomes ∂_3. A good example is the general component of the gradient of a function f:

Components of gradient.

$$\nabla f \quad \text{is totally equivalent to} \quad \partial_k f, \; k = 1, 2, 3. \tag{10.3}$$

With this notation, I can write the middle term of (10.1) as

$$E_i = \partial_i \Phi, \qquad i = 1, 2, 3.$$

All operations on vectors can be translated into the language of indexed relations. For example, $\mathbf{A} + \mathbf{B} = \mathbf{C}$ and $\mathbf{A} = \alpha \mathbf{B}$ are, respectively, equivalent to the following:

$$\begin{aligned} A_k + B_k = C_k, \quad & k = 1, 2, 3 \\ A_j = \alpha B_j \quad & j = 1, 2, 3. \end{aligned}$$

The two operations of vector multiplication are a little more involved and are treated separately in the following.

10.1.1 Dot and Cross Products

First consider the dot product. In terms of components, the dot product of **A** and **B** is

$$\mathbf{A} \cdot \mathbf{B} = A_x B_x + A_y B_y + A_z B_z.$$

Converting to number indices, you get

$$\mathbf{A}\cdot\mathbf{B} = A_1B_1 + A_2B_2 + A_3B_3 = \sum_{i=1}^{3} A_iB_i.$$

Now introduce a further simplification in notation due to Einstein, which gets rid of the clumsy summation sign:

Einstein summation convention.

Note 10.1.2. **(*Einstein Summation Convention*).** *Whenever an index is repeated it is summed from* 1 *to* 3.

Using this convention, the dot product becomes

Dot product.

$$\mathbf{A}\cdot\mathbf{B} = A_iB_i. \tag{10.4}$$

No summation sign is needed as long as you remember that the *repeated* index i is summed over. Since the repeated index is a *dummy index*, you can change it to any other symbol. Thus,

$$\mathbf{A}\cdot\mathbf{B} = A_kB_k = A_jB_j = A_nB_n = A_\heartsuit B_\heartsuit = \cdots.$$

The process of setting two indices equal and summing over them is called **contraction**.

Contraction of two indices.

To be able to continue, you should be prepared to leave the confines of single-indexed vectors and become familiar with multi-indexed quantities or **tensors**. In the simple spirit of this section, we will not discuss tensors that have more than two indices, with one exception that you'll see below. The first double-indexed tensor, δ_{ij}, is the **Kronecker delta**, which is defined as

Kronecker delta.

$$\delta_{ij} \equiv \begin{cases} 1 & \text{if } i = j, \\ 0 & \text{if } i \neq j. \end{cases} \tag{10.5}$$

For example,

$$\delta_{13} = \delta_{32} = 0,\ \delta_{11} = \delta_{22} = \delta_{33} = 1,\ \delta_{jj} = 3,$$

where the last equality incorporates the summation convention. The most important property of the Kronecker delta occurs when it shares a common repeated index with another tensor:

Note 10.1.3. *When an index of a tensor* **T** *is contracted with one of the indices of the Kronecker delta, the result is an expression in which the Kronecker delta is removed and the contracted index of* **T** *is replaced by the other index of the Kronecker delta.*

In symbols

$$T_{ijkpml}\delta_{p\heartsuit} \equiv \sum_{p=1}^{3} T_{ijkpml}\delta_{p\heartsuit} = T_{ijk\heartsuit ml}. \tag{10.6}$$

In practice, the other indices of **T** may also be contracted with other indexed quantities. The statement in Note 10.1.3 concentrates only on the sum involving **T** and the Kronecker delta, ignoring the other summations. For example,

$$A_j\delta_{j3} = A_3, \qquad A_i B_k \delta_{ik} = A_k B_k = A_i B_i = \mathbf{A}\cdot\mathbf{B}, \qquad A_k\delta_{ik}\delta_{ij} = A_k\delta_{jk} = A_j.$$

An important "application" of the Kronecker delta, which should be obvious, is the following:

Note 10.1.4. *If $(x_1, x_2, \ldots, x_n)$ are independent variables, then*

$$\frac{\partial x_i}{\partial x_j} = \delta_{ij}.$$

Levi-Civita Symbol ϵ_{ijk} defined.

The second multi-indexed tensor is described in

Note 10.1.5. *The* ***Levi-Civita Symbol*** ϵ_{ijk} *has the following two properties:*

1. *Convention:* $\epsilon_{123} = 1$.
2. *Antisymmetry: Interchanging any two indices changes the sign of* ϵ_{ijk}.

A consequence of the second property is a third property: When two indices of the Levi-Civita symbol are equal, the result is zero (show this!). Here are some examples illustrating the properties of the Levi-Civita symbol:

$$\epsilon_{11j} = 0 = \epsilon_{\heartsuit k \heartsuit},\ \epsilon_{213} = -1,\ \epsilon_{ijk}\delta_{ik} = 0,\ \epsilon_{ijk} = \epsilon_{kij} = -\epsilon_{kji}.$$

Let S_{ij} be a symmetric tensor, i.e., $S_{ij} = S_{ji}$. Then

$$\epsilon_{ijk}S_{jk} = -\epsilon_{ikj}S_{jk} = -\epsilon_{imn}S_{nm} = -\epsilon_{imn}S_{mn} = -\epsilon_{ijk}S_{jk}.$$

In the first equality, use the antisymmetry of the Levi-Civita symbol; in the second equality use the "dummyness" of the repeated indices; in the third equality, use the symmetry of S; and finally in the last equality again use the "dummyness" of the repeated indices. Now note that any quantity that is equal to its negative must be zero. This result is important enough to warrant a Note:

Note 10.1.6. *When the two indices of a symmetric tensor are contracted with the Levi-Civita symbol, the result is zero. In other words, if* $S_{ij} = S_{ji}$, *then* $\epsilon_{ijk}S_{jk} = 0$.

Table 10.1 Vector versus index notation of some familiar vector quantities.

Vector notation	Index notation
$\mathbf{E} = -\nabla\Phi$	$E_k = -\partial_k \Phi$
$\nabla \cdot \mathbf{A}$	$\partial_j A_j$
$\oiint_S \mathbf{A} \cdot d\mathbf{a} = \iiint_V \nabla \cdot \mathbf{A} \, dV$	$\oiint_S A_k \, da_k = \iiint_V \partial_j A_j \, dV$
$\nabla^2 \Phi$	$\partial_m \partial_m \Phi$
$\nabla \cdot (f\mathbf{A}) = \mathbf{A} \cdot \nabla f + f \nabla \cdot \mathbf{A}$	$\partial_i (f A_i) = A_i \partial_i f + f \partial_i A_i$
$\mathbf{A} \cdot (\mathbf{B} \times \mathbf{C})$	$\epsilon_{ijk} A_i B_j C_k$

The Levi-Civita symbol is conveniently used in the cross product:[1]

Cross product in the language of indices.

$$(\mathbf{A} \times \mathbf{B})_i = \epsilon_{ijk} A_j B_k, \qquad i = 1, 2, 3. \tag{10.7}$$

As a good practice in index manipulation, you should verify the above relation. If $\mathbf{A} = \mathbf{B}$ and you consider $A_j A_k$ as a symmetric double-indexed quantity, then the obvious vector equality $\mathbf{A} \times \mathbf{A} = 0$ follows from Equation (10.7) and Note 10.1.6. Table 10.1 shows both vector and index notations for some familiar vector quantities. You should verify all these relations, remembering the Einstein summation convention.

The Kronecker delta and the Levi-Civita symbol are mingled in one of the most important and frequently used tensor identities:

Note 10.1.7. *The following equation holds:*

$$\epsilon_{ijk} \epsilon_{imn} = \delta_{jm} \delta_{kn} - \delta_{jn} \delta_{km}. \tag{10.8}$$

You should verify this equation component by component, *remembering the sum over i* on the left.

Example 10.1.8. In this example, I'll derive the *bac cab* rule

$$\mathbf{A} \times (\mathbf{B} \times \mathbf{C}) = \mathbf{B}(\mathbf{A} \cdot \mathbf{C}) - \mathbf{C}(\mathbf{A} \cdot \mathbf{B}),$$

to demonstrate most of the properties of the Levi-Civita symbol discussed so far.

Start with a general component of the LHS and work through index manipulations until you reach the RHS:

$$\begin{aligned}
[\mathbf{A} \times (\mathbf{B} \times \mathbf{C})]_i &= \epsilon_{ijk} A_j (\mathbf{B} \times \mathbf{C})_k = \epsilon_{ijk} A_j \epsilon_{kmn} B_m C_n \\
&= \epsilon_{kij} \epsilon_{kmn} A_j B_m C_n = \left(\delta_{im} \delta_{jn} - \delta_{in} \delta_{jm}\right) A_j B_m C_n \\
&= \delta_{im} \delta_{jn} A_j B_m C_n - \delta_{in} \delta_{jm} A_j B_m C_n = A_j B_i C_j - A_j B_j C_i \\
&= B_i (A_j C_j) - C_i (A_j B_j) = B_i (\mathbf{A} \cdot \mathbf{C}) - C_i (\mathbf{A} \cdot \mathbf{B}).
\end{aligned} \tag{10.9}$$

But this last expression is the ith component of the RHS. ■

[1] It is common to leave the expression $i = 1, 2, 3$ out of vector equations. Since the range of the free index is understood from the context (in this section, it is always from 1 to 3), there is no need to attach it to all vector equations. From now on, I'll drop it out of all equations.

Notice how care was taken to use two new indices (m and n) in the second equality on the first line of (10.9). This is an extremely important practice worth emphasizing:

Note 10.1.9. *When inserting an indexed term containing dummy (repeated) indices in an existing indexed expression, always use a new set of dummy indices in the term being inserted.*

No index can occur more than twice!

As a rule, no term should contain an index that occurs more than twice. If you see such an index in any expression, it means you have broken the rule of Note 10.1.9.

Above, I asked you to verify the identity in Note 10.1.7 component by component. There is another way:

Note 10.1.10. *To show that two tensors are equal, contract them with a set of* arbitrary *vectors and prove that they yield the same result.*

And if the tensors contain vectors, it may help if you choose one of the axes of your coordinate system to lie along one of those vectors. I'll illustrate this in the following example.

Example 10.1.11. Let $\hat{\mathbf{e}}$ be a unit vector. In combination with the Levi-Civita symbol, the following identity sometimes turns out to be very useful:

$$\epsilon_{ijk}\hat{e}_i\hat{e}_m + \epsilon_{ikm}\hat{e}_i\hat{e}_j + \epsilon_{imj}\hat{e}_i\hat{e}_k = \epsilon_{jkm}. \tag{10.10}$$

To prove the identity, I contract both sides with $A_j B_k C_m$, where A_j, B_k, and C_m are components of arbitrary vectors **A**, **B**, and **C**. On the right-hand side, I get $\mathbf{A}\cdot(\mathbf{B}\times\mathbf{C})$. For the left-hand side, I further assume that $\hat{\mathbf{e}}$ lies along the first axis of my coordinate system. This does not reduce the generality of my proof, because I have total freedom in choosing the orientation of my coordinate system. So now I have $\hat{e}_1 = 1, \hat{e}_2 = 0, \hat{e}_3 = 0$. Then, the first term of the left-hand side gives

$$\begin{aligned}\text{first term} &= \epsilon_{ijk}\hat{e}_i\hat{e}_m A_j B_k C_m = A_j B_k(\epsilon_{ijk}\hat{e}_i)(\hat{e}_m C_m)\\ &= A_j B_k \underbrace{(\epsilon_{1jk}\hat{e}_1 + \epsilon_{2jk}\hat{e}_2 + \epsilon_{3jk}\hat{e}_3)}_{=\epsilon_{1jk}}\underbrace{(\hat{e}_1 C_1 + \hat{e}_2 C_2 + \hat{e}_3 C_3)}_{=C_1}\\ &= C_1\epsilon_{1jk}A_j B_k.\end{aligned}$$

Similarly

$$\begin{aligned}\text{second term} &= \epsilon_{imj}\hat{e}_i\hat{e}_k A_j B_k C_m = -B_1\epsilon_{1jm}A_j C_m\\ \text{third term} &= \epsilon_{ikm}\hat{e}_i\hat{e}_j A_j B_k C_m = A_1\epsilon_{1km}B_k C_m.\end{aligned}$$

I rewrite the sum of the first and second terms as

$$\text{sum 1 \& 2} = A_j(\epsilon_{1jk}B_k C_1 + \epsilon_{j1m}B_1 C_m) = A_j(\epsilon_{jk1}B_k C_1 + \epsilon_{j1k}B_1 C_k).$$

Now I sum over j and note that it cannot be 1 (because the Levi-Civita symbol already has a 1) to get

$$\begin{aligned}\text{sum } 1 \ \& \ 2 &= A_2(\epsilon_{2k1}B_kC_1 + \epsilon_{21k}B_1C_k) + A_3(\epsilon_{3k1}B_kC_1 + \epsilon_{31k}B_1C_k)\\ &= A_2\epsilon_{2km}B_kC_m + A_3\epsilon_{3km}B_kC_m.\end{aligned}$$

If I now add the third term, for the left-hand side of (10.10), I'll get

$$LHS = A_1\epsilon_{1km}B_kC_m + A_2\epsilon_{2km}B_kC_m + A_3\epsilon_{3km}B_kC_m = A_j\epsilon_{jkm}B_kC_m = \mathbf{A}\cdot(\mathbf{B}\times\mathbf{C}).$$

I have shown that contracting both sides of the identity (10.10) with an *arbitrary* set of vectors $A_jB_kC_m$ gives the same result. Therefore, the identity holds. ■

10.1.2 Proof of Note 7.6.2

As promised in Section 7.6.3, I'll now prove Note 7.6.2. To better understand this proof, you'll have to go back to Section 7.6.3 to see the premise of what I'm doing here. It'll also be helpful if you familiarize yourself with the content of Appendix C.2.

The equation of the contour in O' is given by (C.16) with all quantities primed. Denoting $\mathbf{r}-\mathbf{r}_0$ by $\mathbf{R}$ and $\mathbf{r}'-\mathbf{r}'_0$ by $\mathbf{R}'$ and using indices, I can write the left-hand side of that equation as

$$R'_i(\vec{\xi}'\times\vec{\eta}')_i = \epsilon_{ijk}R'_i\xi'_j\eta'_k.$$

But

$$\xi'_j = \frac{\partial x'_j}{\partial\xi} = \frac{\partial(\Lambda_{jl}x_l)}{\partial\xi} = \Lambda_{jl}\frac{\partial x_l}{\partial\xi} = \Lambda_{jl}\xi_l, \quad \text{and similarly} \quad \eta'_k = \Lambda_{km}\eta_m,$$

where Λ_{jl} is the jlth element of the Lorentz transformation submatrix (B.9). Therefore,

$$(\mathbf{r}'-\mathbf{r}'_0)\cdot(\vec{\xi}'\times\vec{\eta}') = \epsilon_{ijk}\Lambda_{jl}\Lambda_{km}R'_i\xi_l\eta_m. \tag{10.11}$$

If I can show that the Lorentz transform of $\mathbf{r}-\mathbf{r}_0$, i.e., $R'_i = \Lambda_{in}R_n$, makes (10.11) zero, I am done. So, I substitute $R'_i = \Lambda_{in}R_n$ in (10.11) and write

$$(\mathbf{r}'-\mathbf{r}'_0)\cdot(\vec{\xi}'\times\vec{\eta}') = \epsilon_{ijk}\Lambda_{jl}\Lambda_{km}\Lambda_{in}R_n\xi_l\eta_m. \tag{10.12}$$

Now, (B.9) gives

$$\begin{aligned}\epsilon_{ijk}\Lambda_{jl}\Lambda_{km}\Lambda_{in} &= \epsilon_{ijk}[\delta_{jl}+(\gamma-1)\hat{\beta}_j\hat{\beta}_l][\delta_{km}+(\gamma-1)\hat{\beta}_k\hat{\beta}_m][\delta_{in}+(\gamma-1)\hat{\beta}_i\hat{\beta}_n]\\ &= \epsilon_{ijk}[\delta_{jl}\delta_{km}+(\gamma-1)\delta_{jl}\hat{\beta}_k\hat{\beta}_m\\ &\quad +(\gamma-1)\delta_{km}\hat{\beta}_j\hat{\beta}_l][\delta_{in}+(\gamma-1)\hat{\beta}_i\hat{\beta}_n],\end{aligned}$$

where I used Note 10.1.6 and the fact that $\hat{\beta}_j\hat{\beta}_l\hat{\beta}_k\hat{\beta}_m$ is symmetric in indices j and k. I leave it for you to continue the multiplication and show that

$$\epsilon_{ijk}\Lambda_{jl}\Lambda_{km}\Lambda_{in} = \epsilon_{ijk}[\delta_{jl}\delta_{km}\delta_{in}+(\gamma-1)(\delta_{jl}\delta_{km}\hat{\beta}_i\hat{\beta}_n+\delta_{jl}\delta_{in}\hat{\beta}_k\hat{\beta}_m+\delta_{km}\delta_{in}\hat{\beta}_j\hat{\beta}_l)].$$

Substituting this in (10.12) yields

$$\begin{aligned}(\mathbf{r}'-\mathbf{r}'_0)\cdot(\vec{\xi}'\times\vec{\eta}') &= \epsilon_{ijk}[R_i\xi_j\eta_k+(\gamma-1)(R_n\xi_j\eta_k\hat{\beta}_i\hat{\beta}_n+R_i\xi_j\eta_m\hat{\beta}_k\hat{\beta}_m\\ &\quad +R_i\xi_l\eta_k\hat{\beta}_j\hat{\beta}_l)]\\ &= \epsilon_{ijk}R_i\xi_j\eta_k+(\gamma-1)R_n\xi_j\eta_k(\epsilon_{ijk}\hat{\beta}_i\hat{\beta}_n+\epsilon_{inj}\hat{\beta}_i\hat{\beta}_k+\epsilon_{ikn}\hat{\beta}_i\hat{\beta}_j),\end{aligned}$$

where in the last step I just renamed some of the dummy indices and rearranged the indices of the Levi-Civita tensor with due regard to its change of sign. Now note that $\hat{\beta}$ is a unit vector and you can use Equation (10.10) to finally obtain

$$\begin{aligned}(\mathbf{r}'-\mathbf{r}'_0)\cdot(\vec{\xi}'\times\vec{\eta}') &= \epsilon_{ijk}R_i\xi_j\eta_k+(\gamma-1)R_n\xi_j\eta_k\epsilon_{jkn}\\ &= \epsilon_{ijk}R_i\xi_j\eta_k+(\gamma-1)R_n\xi_j\eta_k\epsilon_{njk}\\ &= \gamma\epsilon_{ijk}R_i\xi_j\eta_k=\gamma(\mathbf{r}-\mathbf{r}_0)\cdot(\vec{\xi}\times\vec{\eta}).\end{aligned}$$

Therefore, the left-hand side is zero if and only if the right-hand side is zero. This means, in particular, that

Note 10.1.12. *The equation in ξ and η (C.17) from which you can find one in terms of the other is the same in O and O'.*

10.2 INDEX MANAGEMENT IN SPECIAL RELATIVITY

Much of the discussion of the previous section carries over smoothly to special relativity. But the carry-over is not, as you might expect, perfectly smooth. After all, the 4-vectors satisfy a dot product not all of whose summed members have the same sign (see Note 6.1.1). To handle this subtle difference I have to make some changes in the notation I have been using.

By the convention of Note 6.2.1, the **speed of light is set to 1 here.**

The first change is to adhere to the almost universal agreement of using lowercase Greek letters as indices. The second change is to use superscripts for coordinates: $\mathbf{r}\equiv x^\mu=(x^0,x^1,x^2,x^3)$. And since $\mathbf{r}$ is a typical 4-vector,[2] the same is true for *all* 4-vectors:

$$\begin{aligned}\text{The 4-velocity is}\quad & \mathbf{u}\equiv u^\mu=(u^0,u^1,u^2,u^3)\equiv(u^0,\vec{u})\\ \text{The 4-momentum is}\quad & \mathbf{p}\equiv p^\mu=(p^0,p^1,p^2,p^3)\equiv(p^0,\vec{p})\\ \text{The 4-force is}\quad & \mathbf{f}\equiv f^\mu=(f^0,f^1,f^2,f^3)\equiv(f^0,\vec{f}).\end{aligned}\tag{10.13}$$

The shared property of all 4-vectors is that under a LT, they transform in the same way that coordinates do. Sometimes, especially when summations are involved, the space components are indexed by Latin letters such as i, j, k, etc.

[2] From here on, I revert back to the convention that boldface letters represent 4-vectors.

In this vein, (10.13) becomes

$$\begin{aligned}
\mathbf{u} &\equiv u^\mu = (u^0, u^i) \equiv (u^0, \vec{u}), \quad i = 1, 2, 3, \\
\mathbf{p} &\equiv p^\mu = (p^0, p^j) \equiv (p^0, \vec{p}), \quad j = 1, 2, 3, \\
\mathbf{f} &\equiv f^\mu = (f^0, f^k) \equiv (f^0, \vec{f}), \quad k = 1, 2, 3.
\end{aligned} \tag{10.14}$$

Are there any *sub*scripts in special relativity? There must be, because otherwise introducing superscripts would not have been a smart thing to do! A prototype of a subscripted quantity is the gradient. The four-dimensional gradient of a function f has four components $\partial f/\partial x^\mu$, with μ taking values from 0 to 3. Logical consistency in notation dictates that the components of this gradient be *sub*scripted: the *super*script of the coordinates has come *down* to a denominator. But, as you'll see shortly there is more to this choice of notation than logic! For now, let's accept the implication of logic and denote the components of the gradient as $\partial_\mu f$. How do they transform?

Let $\mathbf{r}'$ be the transformed coordinates, so that $\mathbf{r}' = \Lambda\mathbf{r}$. Write this, and its inverse, in component form, and at this point don't worry about the location (whether they are subscripts or superscripts) of the matrix indices:

$$x'^\mu = \Lambda_{\mu\nu} x^\nu \qquad x^\mu = (\Lambda^{-1})_{\mu\nu} x'^\nu, \tag{10.15}$$

where as usual, the repeated indices are summed over. Since Λ is a function of (and only of) the velocity $\vec{\beta}$, write it as $\Lambda(\vec{\beta})$. Now note from the discussion leading to Equation (B.11) in Appendix B that $\Lambda^{-1} = \Lambda(-\vec{\beta})$. Therefore, you can write the preceding equation as

$$x'^\mu = \Lambda_{\mu\nu}(\vec{\beta}) x^\nu \qquad x^\mu = \Lambda_{\mu\nu}(-\vec{\beta}) x'^\nu. \tag{10.16}$$

This equation tells us how $\partial_\mu f$ transforms:

$$\begin{aligned}
\partial'_\mu f &\equiv \frac{\partial f}{\partial x'^\mu} = \frac{\partial f}{\partial x^\sigma}\frac{\partial x^\sigma}{\partial x'^\mu} = \partial_\sigma f \frac{\partial}{\partial x'^\mu}\left(\Lambda_{\mu\nu}(-\vec{\beta}) x'^\nu\right) \\
&= \partial_\sigma f \Lambda_{\mu\nu}(-\vec{\beta}) \frac{\partial x'^\sigma}{\partial x'^\nu} = \Lambda_{\mu\nu}(-\vec{\beta}) \partial_\sigma f \delta_{\nu\sigma} = \Lambda_{\mu\sigma}(-\vec{\beta}) \partial_\sigma f.
\end{aligned} \tag{10.17}$$

(Use the chain rule to get the second equality, and Note 10.1.4 to get the next to the last equality.) This shows that the gradient transforms via *inverse* LT, therefore it is not a 4-vector.

Is there a way to make a 4-vector out of the gradient, i.e., construct a quantity out of it which transforms according to $\Lambda(\vec{\beta})$ rather than $\Lambda(-\vec{\beta})$? If you write (10.17) in matrix form, you may find the way:

$$\boldsymbol{\partial}' f = \Lambda(-\vec{\beta}) \boldsymbol{\partial} f = \eta \Lambda(\vec{\beta}) \eta \boldsymbol{\partial} f,$$

where you have to use (B.12) to obtain the last equality. Now note that except for the factor of η at the beginning of the right-hand side, this equation is the transformation of a 4-vector. It is actually good that η is there, because if you left-multiply both sides of the equation by η, the unwanted factor on the right-hand side disappears ($\eta^2 = 1$), and you'll get a *needed* η to the left of $\boldsymbol{\partial}' f$. Conclusion: $\eta\boldsymbol{\partial} f$ transforms as a 4-vector! And like all 4-vectors, it has to be *super*scripted. However, instead of giving the entire expression $\eta\boldsymbol{\partial} f$ a superscript, it is common to assign the superscript to ∂, just as the *sub*script of the gradient was (naturally) assigned to ∂. So, write

$$\begin{aligned}(\boldsymbol{\partial} f)^\mu &\equiv \partial^\mu f = \eta^{\mu\nu}\partial_\nu f = \eta^{\mu 0}\partial_0 f + \eta^{\mu 1}\partial_1 f + \eta^{\mu 2}\partial_2 f + \eta^{\mu 3}\partial_3 f \\ &\equiv \eta^{\mu 0}\partial_0 f + \eta^{\mu i}\partial_i f, \end{aligned} \tag{10.18}$$

and note that since $\eta^{\mu\nu} = 0$ unless $\mu = \nu$, and since $\eta^{00} = 1$ and $\eta^{ii} = -1$, you should get

$$\begin{aligned}\partial^0 f &= \partial_0 f = \frac{\partial f}{\partial x^0} \\ \partial^i f &= -\partial_i f = -\frac{\partial f}{\partial x^i}, \quad i = 1, 2, 3. \end{aligned} \tag{10.19}$$

You can write this also as

$$\boldsymbol{\partial} f \equiv \partial^\mu f = (\partial^0 f, \partial^i f) \equiv (\partial^0 f, \vec{\partial} f) = (\partial f/\partial t, -\vec{\nabla} f), \tag{10.20}$$

where $\vec{\nabla} f$ is the ordinary three-dimensional gradient and, as usual, factors of c have been ignored.

Four-covectors.

Four-dimensional vectors that transform according to the inverse LT are called **four-covectors** and are *subscripted*. You just learned how to make 4-vectors out of 4-covectors. You can reverse the process as well. In fact, if $\mathbf{v}$ is a 4-vector, in matrix notation, it satisfies $\mathbf{v}' = \Lambda\mathbf{v}$. Use the first equation in (B.12) to write

$$\mathbf{v}' = \eta\Lambda(-\vec{\beta})\eta\mathbf{v} \;\Rightarrow\; \eta\mathbf{v}' = \Lambda(-\vec{\beta})\eta\mathbf{v}$$

The 3-vector part of a 4-vector is the negative of the 3-vector part of the 4-covector obtained from it.

and note that $\eta\mathbf{v}'$ transforms via the inverse LT. The components of $\eta\mathbf{v}$ are written as $v_\mu \equiv \eta_{\mu\nu}v^\nu$ or

$$\begin{aligned} v_\mu &\equiv (v_0, v_i) \equiv \eta_{\mu\nu}v^\nu \equiv (\eta_{0\nu}v^\nu, \eta_{i\nu}v^\nu) \\ &= (\eta_{00}v^0, \eta_{ij}v^j) = (v^0, -v^i). \end{aligned} \tag{10.21}$$

Equation (10.21) demonstrates a general rule in special relativity:

Note 10.2.1. *The 3-vector part of a 4-vector is the negative of the 3-vector part of the 4-covector obtained from it.*

You may wonder why η has two superscripts in (10.18) and two subscripts in (10.21). The simple answer is that in the first equation there is a superscript on the left-hand side, which has to be matched with one on the right-hand side. Furthermore, it is also more natural to sum a superscript with a subscript. To see this, look at the following manipulation of the expression $\partial^\mu f \partial_\mu f$:

$$\partial^\mu f \partial_\mu f \equiv \eta^{\mu\nu}\partial_\nu f \partial_\mu f = \eta^{00}\partial_0 f \partial_0 f + \eta^{ii}\partial_i f \partial_i f$$
$$= (\partial_0 f)^2 - (\partial_1 f)^2 - (\partial_2 f)^2 - (\partial_3 f)^2 \equiv (\boldsymbol{\partial} f) \bullet (\boldsymbol{\partial} f).$$

More generally, take one 4-vector a^μ and one 4-covector b_μ, and do the same:

$$a^\mu b_\mu \equiv a^\mu \eta_{\mu\nu} b^\nu = \eta_{\mu\nu} a^\mu b^\nu = \eta_{00} a^0 b^0 + \eta_{ii} a^i b^i$$
$$= a^0 b^0 - a^1 b^1 - a^2 b^2 - a^3 b^3 = a^0 b^0 - \vec{a} \cdot \vec{b} \equiv \mathbf{a} \bullet \mathbf{b}.$$

Thus, for the four-dimensional dot product, it is *necessary* to sum over repeated indices with different locations. This process of equating a subscript and a superscript and summing over them is called **contraction**. When you contract a subscript-superscript pair, you take away those two indices. In both of the equations above, the indices of a 4-vector and a 4-covector are taken away, leaving behind a dot product that has no index at all, because it is a scalar, namely a Lorentz invariant quantity, i.e., a quantity that, like the spacetime distance, is the same for all observers.

Contraction in relativity.

A more elaborate answer to the question of the placement of the indices of the matrix η, whose detail is beyond the scope of this book, replaces $\eta_{\mu\nu}$ with the more general **metric tensor** $g_{\mu\nu}$, whose inverse is chosen to have superscripts. A by-product of this choice of indices is that the inverse $g^{\mu\nu}$ of the metric raises a subscript to a superscript, as does our simple $\eta^{\mu\nu}$ [see (10.18)], while the metric itself has the opposite effect, as does our simple $\eta_{\mu\nu}$ [see (10.21)]. This raising or lowering of indices is not confined to 4-covectors and 4-vectors, but is applicable to tensors with more than one index.

What is the $\mu\nu$th element of the product of the matrices η^{-1} and η? It is $\eta^{\mu\sigma}\eta_{\sigma\nu}$, i.e., it is $\eta^{\mu\sigma}$ whose superscript σ has been lowered to a subscript ν (or it is $\eta_{\sigma\nu}$ whose subscript σ has been raised to a superscript μ). It is, therefore, natural to write $\eta^{\mu\sigma}\eta_{\sigma\nu} = \eta^\mu{}_\nu$. But you know that this matrix product should yield the 4×4 identity matrix, whose elements are expressed as the Kronecker delta. This argument shows that the Kronecker delta should be written as $\delta^\mu{}_\nu$.

The following Note summarizes what has been discussed above.

Note 10.2.2. *The number of free superscripts or subscripts on the left of an equation must match the number on the right. Repeated indices must include one superscript and one subscript;* $\eta^{\mu\nu}$ *raises an index like so:* $a^\mu = \eta^{\mu\nu} a_\nu$; $\eta_{\mu\nu}$ *lowers an index like so:* $b_\mu = \eta_{\mu\nu} b^\nu$. *Moreover,* $\eta^{\mu\sigma}\eta_{\sigma\nu} = \delta^\mu{}_\nu$, *where* $\delta^\mu{}_\nu$ *is the four dimensional Kronecker delta.*

According to this rule, you have to write the transformation of a 4-vector and a 4-covector respectively as

$$a'^\mu = \Lambda^\mu{}_\nu a^\nu \quad \text{and} \quad b'_\mu = \Lambda_\mu{}^\nu b_\nu. \tag{10.22}$$

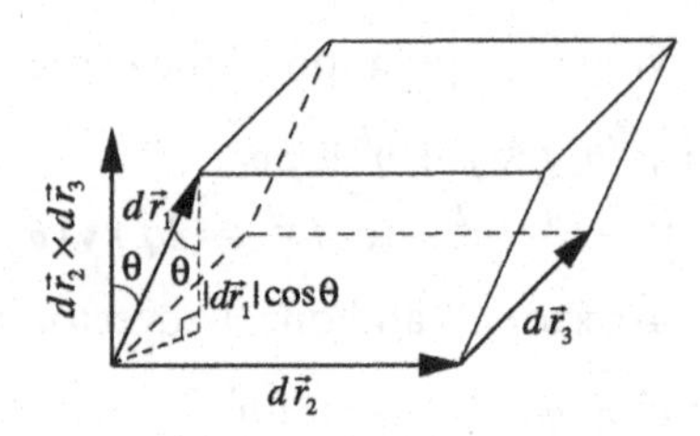

FIGURE 10.1 The details of calculating the volume of a parallelepiped. Note that the area of the base of the parallelepiped is $|d\vec{r}_2 \times d\vec{r}_3|$ and its height is $|d\vec{r}_1 \cos\theta|$, where θ is the angle between $d\vec{r}_1$ and $d\vec{r}_2 \times d\vec{r}_3$.

When the vectors are represented by columns (as is done normally), the dummy index of the matrix runs through the elements of a row, or, equivalently, it labels the columns. Therefore, in the first equation of (10.22), the lower index labels the *columns* and the top index labels the rows. The matrix of $\Lambda_\mu{}^\nu$ is the inverse of the matrix of $\Lambda^\mu{}_\nu$. Note that the superscripts are not directly on top of the subscripts so that it is clear which index has been raised or lowered.

Equation (10.22) illustrates, in a more general setting, the process of contraction described earlier. The matrix Λ is a two-indexed quantity. When one of its indices is contracted with the index of a 4-vector (or a 4-covector), that index, as well as the index of the 4-vector (or a 4-covector), will disappear, leaving behind a single index corresponding to a 4-vector in the first equation and a 4-covector in the second equation of (10.22). This holds for all multi-indexed quantities, not just Λ, as we'll see in the next chapter.

Example 10.2.3. TRANSFORMATION OF 3-VOLUME ELEMENT
Lorentz transformation of a 3-volume element is a good place to exercise your index manipulation. The reference frame O is moving with velocity $\vec{\beta}$ relative to O'. The most general volume element in O defined by three non-coplanar vectors $d\vec{r}_1$, $d\vec{r}_2$, and $d\vec{r}_3$ is $d\vec{r}_1 \cdot (d\vec{r}_2 \times d\vec{r}_3)$, as demonstrated in Figure 10.1. The same volume element in O' is

$$\begin{aligned} dV' &= d\vec{r}\,'_1 \cdot (d\vec{r}\,'_2 \times d\vec{r}\,'_3) = (d\vec{r}\,'_1)_i \, (d\vec{r}\,'_2 \times d\vec{r}\,'_3)_i \\ &= \epsilon_{ijk} \, (d\vec{r}\,'_1)_i \, (d\vec{r}\,'_2)_j \, (d\vec{r}\,'_3)_k \end{aligned}$$

where summation over repeated indices is implied and $d\vec{r}\,'_1$, $d\vec{r}\,'_2$, and $d\vec{r}\,'_3$ are given by the second equation in (6.44). You have to set $dt' = 0$ [or $dt = -\vec{\beta} \cdot d\vec{r}$ by the first equation in (6.44)] because length is the spacial distance between two *simultaneous* events (see Section 1.3). Stated differently, when you calculate the element of spatial volume, you have to set time equal to a constant, just as when you calculate the element of area $dxdy$, you set z equal to a constant. So, write

$$d\vec{r}\,'_1 = d\vec{r}_1 + \gamma\vec{\beta}\left(-\vec{\beta} \cdot d\vec{r}_1 + \frac{\gamma}{\gamma+1}\vec{\beta} \cdot d\vec{r}_1\right)$$

$$\equiv d\vec{r}_1 + \gamma\vec{\beta}ds_1, \quad ds_1 \equiv -\frac{1}{\gamma+1}\vec{\beta}\cdot d\vec{r}_1,$$

with similar expressions for $d\vec{r}\,'_2$ and $d\vec{r}\,'_3$. Now substitute the resulting expressions in the previous equation to obtain

$$\begin{aligned} dV' &= \epsilon_{ijk}[(d\vec{r}_1)_i + \gamma\beta_i ds_1][(d\vec{r}_2)_j + \gamma\beta_j ds_2][(d\vec{r}_3)_k + \gamma\beta_k ds_3] \\ &= \epsilon_{ijk}[(d\vec{r}_1)_i(d\vec{r}_2)_j + \gamma\beta_j(d\vec{r}_1)_i ds_2 + \gamma\beta_i(d\vec{r}_2)_j ds_1] \\ &\quad \times [(d\vec{r}_3)_k + \gamma\beta_k ds_3]. \end{aligned}$$

You should convince yourself why the $\beta_i\beta_j$ term is absent in the first square bracket of the last line of the equation. Multiply the two square brackets and note that for the same reason the $\beta_i\beta_k$ and $\beta_j\beta_k$ terms will not contribute. This leads to the following expression for dV':

$$\begin{aligned} dV' &= \epsilon_{ijk}[(d\vec{r}_1)_i(d\vec{r}_2)_j(d\vec{r}_3)_k + \gamma\beta_k(d\vec{r}_1)_i(d\vec{r}_2)_j ds_3 \\ &\quad + \gamma\beta_j(d\vec{r}_1)_i(d\vec{r}_3)_k ds_2 + \gamma\beta_i(d\vec{r}_2)_j(d\vec{r}_3)_k ds_1]. \end{aligned} \tag{10.23}$$

Call the sum of the last three terms in the square bracket $\mathcal{X}$ and substitute for ds_1, ds_2, and ds_3 to obtain

$$\begin{aligned} \mathcal{X} &= -\frac{\gamma}{\gamma+1}\epsilon_{ijk}[\beta_k(d\vec{r}_1)_i(d\vec{r}_2)_j\beta_m(d\vec{r}_3)_m \\ &\quad + \beta_j(d\vec{r}_1)_i(d\vec{r}_3)_k\beta_m(d\vec{r}_2)_m + \beta_i(d\vec{r}_2)_j(d\vec{r}_3)_k\beta_m(d\vec{r}_1)_m]. \end{aligned}$$

Now change the dummy indices so that you have the same set of indices for $d\vec{r}_1$, $d\vec{r}_2$, and $d\vec{r}_3$; for instance, change m to k and k to m in the first term. Then you will obtain

$$\begin{aligned} \mathcal{X} = -\frac{\gamma}{\gamma+1}[&\epsilon_{ijm}\beta_k\beta_m(d\vec{r}_1)_i(d\vec{r}_2)_j(d\vec{r}_3)_k \\ &+ \epsilon_{imk}\beta_j\beta_m(d\vec{r}_1)_i(d\vec{r}_3)_k(d\vec{r}_2)_j \\ &+ \epsilon_{mjk}\beta_i\beta_m(d\vec{r}_2)_j(d\vec{r}_3)_k(d\vec{r}_1)_i]. \end{aligned}$$

Using $\beta_n = \hat{\beta}_n\beta$ and $\gamma^2\beta^2 = \gamma^2 - 1$ reduces this equation to

$$\mathcal{X} = -\frac{\gamma-1}{\gamma}(d\vec{r}_1)_i(d\vec{r}_2)_j(d\vec{r}_3)_k[\epsilon_{ijm}\hat{\beta}_k\hat{\beta}_m + \epsilon_{imk}\hat{\beta}_j\hat{\beta}_m + \epsilon_{mjk}\hat{\beta}_i\hat{\beta}_m]$$

and Example 10.1.11 shows that (see also Problem 10.6)

$$\epsilon_{ijm}\hat{\beta}_k\hat{\beta}_m + \epsilon_{imk}\hat{\beta}_j\hat{\beta}_m + \epsilon_{mjk}\hat{\beta}_i\hat{\beta}_m = \epsilon_{ijk}.$$

Substituting this in the expression for $\mathcal{X}$ and the result back in (10.23), you should obtain

$$dV' = \frac{1}{\gamma}\epsilon_{ijk}(d\vec{r}_1)_i(d\vec{r}_2)_j(d\vec{r}_3)_k = \frac{dV}{\gamma}. \tag{10.24}$$

Problem 10.7 shows how you can find the same result using the parallel and perpendicular components of $d\vec{r}_1$, $d\vec{r}_2$, and $d\vec{r}_3$. It also shows that a volume can be thought of as an "area" perpendicular to $\vec{\beta}$ times a "height" parallel to $\vec{\beta}$. Since lengths perpendicular to the direction of motion don't change while lengths parallel to the direction of motion shrink by a factor of γ, the overall effect is for the volume to shrink by a factor of γ. ■

10.3 INVARIANCE OF THE ELEMENTAL 4-VOLUME

The mechanics of a relativistic fluid such as the one ejected by some compact astronomical objects requires the consideration of its elements of mass (more precisely, energy) and, therefore, volume. Since the manifold of relativity is space*time*, volume means *four*-volume. What is most important is how the 4-volume in one frame transforms into the 4-volume of another frame. To gain insight into the transformation of 4-volume elements, let's first look at the three-volume.

Any two infinitesimally close points define an element of three-volume. If Q, with coordinates $(x+dx, y+dy, z+dz)$, is a point close to P, with coordinates (x, y, z), then the vector $\vec{PQ}$ can be considered as the diagonal of an infinitesimal box with components (dx, dy, dz), whose product gives the volume of the box. If another Cartesian coordinate system is used in which Q and P have coordinates $(x' + dx', y' + dy', z' + dz')$ and (x', y', z'), respectively, then the vector $\vec{PQ}$ has components (dx', dy', dz'). The fact that $dx'dy'dz' = dxdydz$ is a peculiarity of the Cartesian coordinate systems as you may know from your experience with cylindrical and spherical coordinates.

To be completely general, consider a transformation of the form

$$x^1 = f(w^1, w^2, w^3), \qquad x^2 = g(w^1, w^2, w^3), \qquad x^3 = h(w^1, w^2, w^3), \quad (10.25)$$

where $x^1 = x$, $x^2 = y$, $x^3 = z$, and w^1, w^2, and w^3 are some other coordinates. Then

$$\begin{aligned} dx^1 &= \frac{\partial f}{\partial w^1} dw^1 + \frac{\partial f}{\partial w^2} dw^2 + \frac{\partial f}{\partial w^3} dw^3, \\ dx^2 &= \frac{\partial g}{\partial w^1} dw^1 + \frac{\partial g}{\partial w^2} dw^2 + \frac{\partial g}{\partial w^3} dw^3, \\ dx^3 &= \frac{\partial h}{\partial w^1} dw^1 + \frac{\partial h}{\partial w^2} dw^2 + \frac{\partial h}{\partial w^3} dw^3. \end{aligned}$$

By fixing w^2 and w^3 and allowing w^1 to vary by dw^1, you obtain $d\vec{l}_1$, an element of length in the w^1 direction. Similarly for the second and third elements of length. Therefore

$$\begin{aligned} d\vec{l}_1 &= \hat{\mathbf{e}}_1 \frac{\partial f}{\partial w^1} dw^1 + \hat{\mathbf{e}}_2 \frac{\partial g}{\partial w^1} dw^1 + \hat{\mathbf{e}}_3 \frac{\partial h}{\partial w^1} dw^1, \\ d\vec{l}_2 &= \hat{\mathbf{e}}_1 \frac{\partial f}{\partial w^2} dw^2 + \hat{\mathbf{e}}_2 \frac{\partial g}{\partial w^2} dw^2 + \hat{\mathbf{e}}_3 \frac{\partial h}{\partial w^2} dw^2, \\ d\vec{l}_3 &= \hat{\mathbf{e}}_1 \frac{\partial f}{\partial w^3} dw^3 + \hat{\mathbf{e}}_2 \frac{\partial g}{\partial w^3} dw^3 + \hat{\mathbf{e}}_3 \frac{\partial h}{\partial w^3} dw^3, \end{aligned}$$

where $\hat{\mathbf{e}}_1$, $\hat{\mathbf{e}}_2$, and $\hat{\mathbf{e}}_2$ are unit vectors in the direction of the three w-coordinates. From Figure 10.1, the volume element is the absolute value of

$$d\vec{l}_1 \cdot (d\vec{l}_2 \times d\vec{l}_3) = \det \begin{pmatrix} \frac{\partial f}{\partial w^1} dw^1 & \frac{\partial g}{\partial w^1} dw^1 & \frac{\partial h}{\partial w^1} dw^1 \\ \frac{\partial f}{\partial w^2} dw^2 & \frac{\partial g}{\partial w^2} dw^2 & \frac{\partial h}{\partial w^2} dw^2 \\ \frac{\partial f}{\partial w^3} dw^3 & \frac{\partial g}{\partial w^3} dw^3 & \frac{\partial h}{\partial w^3} dw^3 \end{pmatrix},$$

or

$$dV = \left| \det \begin{pmatrix} \frac{\partial x^1}{\partial w^1} & \frac{\partial x^2}{\partial w^1} & \frac{\partial x^3}{\partial w^1} \\ \frac{\partial x^1}{\partial w^2} & \frac{\partial x^2}{\partial w^2} & \frac{\partial x^3}{\partial w^2} \\ \frac{\partial x^1}{\partial w^3} & \frac{\partial x^2}{\partial w^3} & \frac{\partial x^3}{\partial w^3} \end{pmatrix} \right| dw^1 dw^2 dw^3,$$

which is more suggestively written as

$$d^3x = \left| \det \begin{pmatrix} \frac{\partial x^1}{\partial w^1} & \frac{\partial x^2}{\partial w^1} & \frac{\partial x^3}{\partial w^1} \\ \frac{\partial x^1}{\partial w^2} & \frac{\partial x^2}{\partial w^2} & \frac{\partial x^3}{\partial w^2} \\ \frac{\partial x^1}{\partial w^3} & \frac{\partial x^2}{\partial w^3} & \frac{\partial x^3}{\partial w^3} \end{pmatrix} \right| d^3w. \tag{10.26}$$

The matrix in (10.26) is called the **Jacobian matrix** and the absolute value of the determinant of that matrix is simply called the **Jacobian** of the coordinate transformation.

The Jacobian matrix and Jacobian.

Example 10.3.1. Apply Equation (10.26) to spherical coordinates with $r = w^1$, $\theta = w^2$, and $\varphi = w^3$. The transformation is

$$x^1 = w^1 \sin w^2 \cos w^3, \qquad x^2 = w^1 \sin w^2 \sin w^3, \qquad x^3 = w^1 \cos w^2.$$

This gives

$$\frac{\partial x^1}{\partial w^1} = \sin w^2 \cos w^3, \quad \frac{\partial x^1}{\partial w^2} = w^1 \cos w^2 \cos w^3, \quad \frac{\partial x^1}{\partial w^3} = -w^1 \sin w^2 \sin w^3$$
$$\frac{\partial x^2}{\partial w^1} = \sin w^2 \sin w^3, \quad \frac{\partial x^2}{\partial w^2} = w^1 \cos w^2 \sin w^3, \quad \frac{\partial x^2}{\partial w^3} = w^1 \sin w^2 \cos w^3$$
$$\frac{\partial x^3}{\partial w^1} = \cos w^2, \quad \frac{\partial x^3}{\partial w^2} = -w^1 \sin w^2, \quad \frac{\partial x^3}{\partial w^3} = 0$$

with the Jacobian matrix J given by

$$\mathsf{J} = \begin{pmatrix} \sin w^2 \cos w^3 & \sin w^2 \sin w^3 & \cos w^2 \\ w^1 \cos w^2 \cos w^3 & w^1 \cos w^2 \sin w^3 & -w^1 \sin w^2 \\ -w^1 \sin w^2 \sin w^3 & w^1 \sin w^2 \cos w^3 & 0 \end{pmatrix},$$

whose determinant is easily calculated to be $\det \mathsf{J} = (w^1)^2 \sin w^2$ leading to

$$d^3x = (w^1)^2 \sin w^2 d^3w = (w^1)^2 \sin w^2 dw^1 dw^2 dw^3$$
$$= r^2 \sin\theta dr d\theta d\varphi,$$

which is the familiar element of volume in spherical coordinates. ■

If the functions in (10.25) are linear in w^i, then the most general form of these functions is

$$x^i = \sum_{j=1}^{3} a^i_j w^j + b^i, \quad i = 1, 2, 3, \tag{10.27}$$

where a^i_j and b^i are all constants. Furthermore, $\partial x^i / \partial w^j = a^i_j$, and thus the Jacobian matrix becomes

$$\mathsf{J} = \begin{pmatrix} a^1_1 & a^2_1 & a^3_1 \\ a^1_2 & a^2_2 & a^3_2 \\ a^1_3 & a^2_3 & a^3_3 \end{pmatrix}. \tag{10.28}$$

The most general transformation considered in classical mechanics are the so-called rigid motions consisting of rotations and translations. With a slight change of notation of (10.27), these transformations are written as

$$x'^i = \sum_{j=1}^{3} R^i_j x^j + b^i, \quad i = 1, 2, 3, \tag{10.29}$$

with R^i_j the elements of the rotation matrix (e.g., in terms of Euler angles) and b^i the amount of translation of the ith coordinate. The Jacobian matrix is now a rotation matrix whose determinant is 1. Now we see why $dx'dy'dz' = dxdydz$.

The generalization to four dimensions is straightforward now that the procedure has been outlined for three dimensions. If E_2, with coordinates $(t + dt, x + dx, y + dy, z + dz)$ is an event[3] close to E_1, with coordinates (t, x, y, z), then the 4-vector $\mathbf{E_1E_2}$ can be considered as the diagonal of an infinitesimal four-dimensional box with components (dt, dx, dy, dz), whose product gives the 4-volume of the box. If another coordinate system is used in which the two events have coordinates $(t' + dt', x' + dx', y' + dy', z' + dz')$ and (t', x', y', z'), respectively, then the 4-vector $\mathbf{E_1E_2}$ has components (dt', dx', dy', dz'). You may be tempted to set $dt'dx'dy'dz' = dtdxdydz$, but you need to justify this temptation.

The relation between volume elements in any number of dimensions (not just 3 or 4) is given by the Jacobian. In fact, making the dependence of the Jacobian

[3] As usual, $c = 1$.

on the coordinate derivatives more explicit, you can generalize (10.26) to n dimensions without any major change:

Volume element in n dimensions.

$$d^n x = \left|\det \mathsf{J}(\partial x^i/\partial w^j)\right| d^n w, \tag{10.30}$$

where

$$\mathsf{J}(\partial x^i/\partial w^j) = \begin{pmatrix} \dfrac{\partial x^1}{\partial w^1} & \dfrac{\partial x^2}{\partial w^1} & \cdots & \dfrac{\partial x^n}{\partial w^1} \\ \dfrac{\partial x^1}{\partial w^2} & \dfrac{\partial x^2}{\partial w^2} & \cdots & \dfrac{\partial x^n}{\partial w^2} \\ \vdots & \vdots & & \vdots \\ \dfrac{\partial x^1}{\partial w^n} & \dfrac{\partial x^2}{\partial w^n} & \cdots & \dfrac{\partial x^n}{\partial w^n} \end{pmatrix} \tag{10.31}$$

In general relativity, where the functions relating coordinates are not assumed to be linear, this matrix (with $n = 4$) connects volume elements. In special relativity, LT is the most general coordinate transformation.[4] Since LTs are linear, the Jacobian matrix is simply the matrix of the transformation, which has been denoted by Λ. Hence, designating the coordinates as primes and unprimed, you get

$$d^4x' = |\det \Lambda| d^4 x.$$

Equation (6.4) gives the defining property of the Lorentz transformation. Take the determinant of that equation to obtain

$$(\det \widetilde{\Lambda})\,(\det \eta)\,(\det \Lambda) = \det \eta,$$

or, recalling that the determinant of the transpose of a matrix is the same as the determinant of the matrix,

$$(\det \Lambda)^2 = 1 \;\Rightarrow\; |\det \Lambda| = 1.$$

Therefore,

$$dt'dx'dy'dz' = dt dx dy dz. \tag{10.32}$$

Actually, (10.32) is more general. Since LT connects *all* 4-vectors, (10.32) applies to all such vectors. For example, it is often necessary to integrate over the 4-momenta of a particle or fluid. Then (10.32) yields

$$dE'dp'_x dp'_y dp'_z = dE dp_x dp_y dp_z. \tag{10.33}$$

---

[4] One can also include a translation as in (10.27). Then the relations are known as **Poincaré transformation**. We shall not discuss Poincaré transformations in this book.

10.4 PROBLEMS

10.1. Using indices, show that the divergence of the curl of any 3-vector is zero. Similarly, show that the curl of the gradient of any function is zero.

10.2. Using the elementary determinant way of calculating cross products, show that

$$(\mathbf{A}\times\mathbf{C})\cdot(\mathbf{B}\times\mathbf{D})=(\mathbf{A}\cdot\mathbf{B})(\mathbf{C}\cdot\mathbf{D})-(\mathbf{A}\cdot\mathbf{D})(\mathbf{B}\cdot\mathbf{C}).$$

Now use this vector identity and Note 10.1.10 to prove Equation (10.8).

10.3. Express $\mathbf{A}\cdot(\mathbf{B}\times\mathbf{C})$ in index form. Then using it show the cyclic property of this triple product:

$$\mathbf{A}\cdot(\mathbf{B}\times\mathbf{C})=\mathbf{C}\cdot(\mathbf{A}\times\mathbf{B})=\mathbf{B}\cdot(\mathbf{C}\times\mathbf{A}).$$

10.4. Using indices, prove the following vector identities:

$$\begin{aligned}\nabla\cdot(f\mathbf{A})&=(\nabla f)\cdot\mathbf{A}+f\nabla\cdot\mathbf{A}\\ \nabla\times(f\mathbf{A})&=(\nabla f)\times\mathbf{A}+f\nabla\times\mathbf{A}\\ \nabla\times(\nabla\times\mathbf{A})&=\nabla(\nabla\cdot\mathbf{A})-\nabla^2\mathbf{A}\end{aligned}$$

10.5. Show that the determinant of a 3×3 matrix A with elements a_{ij} is given by

$$\det\mathsf{A}=\epsilon_{ijk}a_{1i}a_{2j}a_{3k}.$$

10.6. Let $\hat{e}_i$, $i=1,2,3$ be the components of a unit vector. Let

$$a_{ijk}=\epsilon_{ijm}\hat{e}_k\hat{e}_m+\epsilon_{imk}\hat{e}_j\hat{e}_m+\epsilon_{mjk}\hat{e}_i\hat{e}_m.$$

Show that $a_{123}=1$, $a_{jik}=-a_{ijk}$, and $a_{ikj}=-a_{ijk}$. Therefore, $a_{ijk}=\epsilon_{ijk}$.

10.7. By substituting the components of $d\vec{r}_1$, $d\vec{r}_2$, and $d\vec{r}_3$ parallel and perpendicular to $\vec{\beta}$ in $dV=d\vec{r}_1\cdot(d\vec{r}_2\times d\vec{r}_3)$,

(a) show that

$$\begin{aligned}dV=&|(d\vec{r}_1)_\parallel||(d\vec{r}_2)_\perp\times(d\vec{r}_3)_\perp|+|(d\vec{r}_2)_\parallel||(d\vec{r}_3)_\perp\times(d\vec{r}_1)_\perp|\\ &+|(d\vec{r}_3)_\parallel||(d\vec{r}_1)_\perp\times(d\vec{r}_2)_\perp|.\end{aligned}\tag{10.34}$$

This shows that the volume element is the product of an area perpendicular to the direction of motions [terms like $|(d\vec{r}_2)_\perp\times(d\vec{r}_3)_\perp|$] times a height in the direction of motion [terms like $|(d\vec{r}_1)_\parallel|$].

(b) From (6.17) and $dt'=0$, show that $|(d\vec{r}\,'_1)_\parallel|=|(d\vec{r}_1)_\parallel|/\gamma$, with similar relations for $|(d\vec{r}\,'_2)_\parallel|$ and $|(d\vec{r}\,'_3)_\parallel|$.

(c) Insert these results in the expression for dV' as given by (10.34) and derive the formula $dV'=dV/\gamma$.

CHAPTER 11

Relativistic Electrodynamics

The previous chapter worked out the details of the manipulation of indices in tensors sufficient for our purposes. The main application of that chapter will be in the study of the electromagnetic theory. As the first application of tensor calculation, you'll learn how electric and magnetic fields transform from one RF to another.

11.1 TRANSFORMATION OF $\vec{E}$ AND $\vec{B}$

The second postulate of relativity is a consequence of the assumption that Maxwell's equations are universal, that is, that they hold in all reference frames. This assumption leads to the transformation properties of the electric and magnetic fields, the main subject of this section.

Start with the second and third Maxwell's equations, the so-called **homogeneous** equations (because they don't include any charges or currents). Concentrate on the second equation, $\nabla \cdot \vec{B} = 0$. Although it is always true that the divergence of the curl of any 3-vector is zero (you should have shown this in Problem 10.1 using indices), its converse is not always true: the vanishing of the divergence of a vector field does not generally mean that the vector is the curl of another vector field. Nevertheless, under some very mild assumptions the converse is also true and we can state that the divergence of a vector field vanishes if and only if that vector field is the curl of another vector field.

Homogeneous Maxwell's equations.

Thus, the second Maxwell equation implies the existence of a vector field $\vec{A}$ such that $\vec{B} = \nabla \times \vec{A}$, or

$$B_i = \epsilon_{ijk}\partial_j A_k \equiv \epsilon_{ijk}\frac{\partial A_k}{\partial x^j} \quad \text{or} \quad B_i = \partial_j A_k - \partial_k A_j, \tag{11.1}$$

where in the last relation the indices i, j, and k are all different and we have to cyclically permute them to get all the components of $\vec{B}$. You should verify that the following useful relation also holds:

$$\partial_j A_k - \partial_k A_j = \epsilon_{jki} B_i. \tag{11.2}$$

Special Relativity. DOI:10.1016/B978-0-12-810411-8.00011-0

The vector $\vec{A}$ is called the **vector potential**. Substitute this in the third Maxwell equation to get

$$\nabla \times \vec{E} = -\frac{\partial}{\partial t}(\nabla \times \vec{A}) = -\nabla \times \left(\frac{\partial \vec{A}}{\partial t}\right),$$

or

$$\nabla \times \left(\vec{E} + \frac{\partial \vec{A}}{\partial t}\right) = 0.$$

Just as in (11.1), the vanishing of the curl implies the existence of a gradient of a function (because the curl of the gradient of any function is zero). This function is historically denoted by $-\Phi$ (so that in the static case it reduces to the familiar electrostatic potential). Thus,

$$\vec{E} = -\nabla\Phi - \frac{\partial \vec{A}}{\partial t} \quad \text{or} \quad E_i = -\partial_i \Phi - \partial_0 A_i. \tag{11.3}$$

Comparison of (11.3) with the last equation in (11.1) suggests identifying $-\Phi$ as the zeroth component of a 4-vector A^μ whose 3-vector part is $\vec{A}$. This 4-vector is called the **four-potential**:

Four-potential.

$$A^\mu = (-\Phi, \vec{A}). \tag{11.4}$$

With this identification, the electric field can be expressed as[1]

$$E_i = \partial_i A_0 - \partial_0 A_i. \tag{11.5}$$

11.1.1 Transformation of $\vec{E}$

We are now ready to see how the electric and magnetic fields transform under a LT. Start with the electric field and use (10.22) on the 4-covectors ∂_μ and A_μ:

$$\begin{aligned} E'_i &= \partial'_i A'_0 - \partial'_0 A'_i = \left(\Lambda_i{}^\nu \partial_\nu\right)\left(\Lambda_0{}^\sigma A_\sigma\right) - \left(\Lambda_0{}^\sigma \partial_\sigma\right)\left(\Lambda_i{}^\nu A_\nu\right) \\ &= \Lambda_0{}^\sigma \Lambda_i{}^\nu (\partial_\nu A_\sigma - \partial_\sigma A_\nu) \\ &= \Lambda_0{}^0 \Lambda_i{}^\nu (\partial_\nu A_0 - \partial_0 A_\nu) + \Lambda_0{}^j \Lambda_i{}^\nu (\partial_\nu A_j - \partial_j A_\nu) \\ &= \gamma \underbrace{\Lambda_i{}^\nu (\partial_\nu A_0 - \partial_0 A_\nu)}_{\equiv \mathcal{X}} - \gamma\beta_j \underbrace{\Lambda_i{}^\nu (\partial_\nu A_j - \partial_j A_\nu)}_{\equiv \mathcal{Y}}. \end{aligned} \tag{11.6}$$

The last line follows because, $\Lambda_0{}^0 = \gamma$ and $\Lambda_0{}^j = -\gamma\beta_j$ by Equation (B.11). To continue, calculate the designated underbraced terms separately, keeping in mind that Latin letters represent the space indices:

$$\mathcal{X} = \Lambda_i{}^0(\partial_0 A_0 - \partial_0 A_0) + \Lambda_i{}^j(\partial_j A_0 - \partial_0 A_j) = \Lambda_i{}^j E_j$$

---

[1] You should be warned that definition (11.4) of the 4-potential is not unique. Some authors give the space part of the 4-potential a negative sign and define their electric field as $E_i = \partial_0 A_i - \partial_i A_0$, with a similar change in their magnetic field.

and

$$\begin{aligned}\mathcal{Y} &= \Lambda_i{}^0(\partial_0 A_j - \partial_j A_0) + \Lambda_i{}^k(\partial_k A_j - \partial_j A_k)\\ &= -\Lambda_i{}^0 E_j + \Lambda_i{}^k(\epsilon_{kjm} B_m) = \gamma\beta_i E_j + \epsilon_{kjm}\Lambda_i{}^k B_m.\end{aligned}$$

Put the results for $\mathcal{X}$ and $\mathcal{Y}$ back in Equation (11.6) to obtain

$$E'_i = \gamma\Lambda_i{}^j E_j - \gamma^2\beta_i\beta_j E_j - \gamma\epsilon_{kjm}\Lambda_i{}^k\beta_j B_m.$$

Now use (10.4) and (10.7) to rewrite the last equation as

$$E'_i = \gamma\Lambda_i{}^j E_j - \gamma^2\beta_i(\vec{\beta}\cdot\vec{E}) - \gamma\Lambda_i{}^k(\vec{\beta}\times\vec{B})_k. \tag{11.7}$$

The only thing left to do is to substitute $\Lambda_i{}^j$ in terms of the components of $\vec{\beta}$. For this see (B.9) and note that the 3×3 submatrix of $\Lambda(\vec{\beta})$ and its inverse $\Lambda(-\vec{\beta})$ are the same because they both involve the product of two factors of the components of $\hat{\beta}$. Hence, substituting (B.9) in (11.7), you get

$$\begin{aligned}E'_i &= \gamma\left[\delta_{ij} + (\gamma-1)\hat{\beta}_i\hat{\beta}_j\right]E_j - \gamma^2\beta_i(\vec{\beta}\cdot\vec{E})\\ &\quad - \gamma\left[\delta_{ik} + (\gamma-1)\hat{\beta}_i\hat{\beta}_k\right](\vec{\beta}\times\vec{B})_k\\ &= \gamma E_i + \gamma(\gamma-1)\hat{\beta}_i\hat{\beta}\cdot\vec{E} - \gamma^2\beta^2\hat{\beta}_i(\hat{\beta}\cdot\vec{E})\\ &\quad - \gamma(\vec{\beta}\times\vec{B})_i - \gamma(\gamma-1)\hat{\beta}_i\underbrace{\hat{\beta}\cdot(\vec{\beta}\times\vec{B})}_{=0}.\end{aligned}$$

Using $\gamma^2\beta^2 = \gamma^2 - 1$, you should convince yourself that the last expression simplifies to

$$E'_i = \gamma\left[E_i - (\vec{\beta}\times\vec{B})_i\right] - (\gamma-1)\hat{\beta}_i\hat{\beta}\cdot\vec{E}$$

or

$$\vec{E}' = \gamma(\vec{E} - \vec{\beta}\times\vec{B}) - (\gamma-1)\hat{\beta}(\hat{\beta}\cdot\vec{E}). \tag{11.8}$$

Transformation rule for electric field.

11.1.2 Transformation of $\vec{B}$

Now it's the magnetic field's turn.

$$\begin{aligned}B'_i &= \epsilon_{ijk}\partial'_j A'_k = \epsilon_{ijk}\left(\Lambda_j{}^\nu\partial_\nu\right)\left(\Lambda_k{}^\sigma A_\sigma\right) = \epsilon_{ijk}\Lambda_j{}^\nu\Lambda_k{}^\sigma\partial_\nu A_\sigma\\ &= \epsilon_{ijk}\Lambda_j{}^\nu\left(\Lambda_k{}^0\partial_\nu A_0 + \Lambda_k{}^m\partial_\nu A_m\right) = -\gamma\epsilon_{ijk}\beta_k\Lambda_j{}^\nu\partial_\nu A_0 + \epsilon_{ijk}\Lambda_k{}^m\left(\Lambda_j{}^\nu\partial_\nu A_m\right) \qquad (11.9)\\ &= -\gamma\underbrace{\epsilon_{ijk}\beta_k\Lambda_j{}^\nu\partial_\nu A_0}_{\equiv\mathcal{U}} + \underbrace{\epsilon_{ijk}\Lambda_k{}^m\Lambda_j{}^0\partial_0 A_m}_{\equiv\mathcal{V}} + \underbrace{\epsilon_{ijk}\Lambda_k{}^m\Lambda_j{}^l\partial_l A_m}_{\equiv\mathcal{W}}.\end{aligned}$$

The designated terms can be calculated as before:

$$\mathcal{U} = \epsilon_{ijk}\beta_k\Lambda_j{}^0\partial_0 A_0 + \epsilon_{ijk}\beta_k\Lambda_j{}^m\partial_m A_0$$

$$= -\underbrace{\epsilon_{ijk}\beta_k\beta_j}_{=0}\partial_0 A_0 + \epsilon_{ijk}\beta_k[\delta_{jm} + (\gamma - 1)\hat{\beta}_j\hat{\beta}_m]\partial_m A_0$$
$$= \epsilon_{imk}\beta_k\partial_m A_0 + (\gamma - 1)\underbrace{\epsilon_{ijk}\beta_k\hat{\beta}_j}_{=0}\hat{\beta}_m\partial_m A_0$$

and

$$\mathcal{V} = -\gamma\epsilon_{ijk}\beta_j[\delta_{km} + (\gamma - 1)\hat{\beta}_k\hat{\beta}_m]\partial_0 A_m$$
$$= -\gamma\epsilon_{ijk}\beta_j\partial_0 A_k - \gamma(\gamma - 1)\underbrace{\epsilon_{ijk}\beta_j\hat{\beta}_k}_{=0}\hat{\beta}_m\partial_0 A_m.$$

For $\mathcal{W}$, you have to use a trick! Employ the antisymmetry of the Levi-Civita symbol to write $\mathcal{W}$ as

$$\mathcal{W} = \tfrac{1}{2}\epsilon_{ijk}\Lambda_k{}^m\Lambda_j{}^l\partial_l A_m - \tfrac{1}{2}\epsilon_{ikj}\Lambda_k{}^m\Lambda_j{}^l\partial_l A_m.$$

In the second term change the dummy index k to j and the dummy index j to k:

$$\mathcal{W} = \tfrac{1}{2}\epsilon_{ijk}\Lambda_k{}^m\Lambda_j{}^l\partial_l A_m - \tfrac{1}{2}\epsilon_{ijk}\Lambda_j{}^m\Lambda_k{}^l\partial_l A_m.$$

Now in the second term change the dummy index l to m and the dummy index m to l:

$$\mathcal{W} = \tfrac{1}{2}\epsilon_{ijk}\Lambda_k{}^m\Lambda_j{}^l\partial_l A_m - \tfrac{1}{2}\epsilon_{ijk}\Lambda_j{}^l\Lambda_k{}^m\partial_m A_l$$
$$= \tfrac{1}{2}\epsilon_{ijk}\Lambda_k{}^m\Lambda_j{}^l(\partial_l A_m - \partial_m A_l) = \tfrac{1}{2}\epsilon_{ijk}\epsilon_{lmn}\Lambda_k{}^m\Lambda_j{}^l B_n \tag{11.10}$$
$$= \tfrac{1}{2}\epsilon_{ijk}\epsilon_{lmn}[\delta_{km} + (\gamma - 1)\hat{\beta}_k\hat{\beta}_m][\delta_{jl} + (\gamma - 1)\hat{\beta}_j\hat{\beta}_l]B_n.$$

It is a good exercise (and a good test of your patience and perseverance in handling multiple indices) to go through the details of the manipulation of the last line above and show that

$$\mathcal{W} = \gamma B_i - (\gamma - 1)\hat{\beta}_i(\hat{\beta}\cdot\vec{B}). \tag{11.11}$$

Now that you have $\mathcal{U}$, $\mathcal{V}$, and $\mathcal{W}$, substitute them in (11.9) to obtain

$$B_i' = -\gamma\epsilon_{imk}\beta_k\partial_m A_0 - \gamma\epsilon_{ijk}\beta_j\partial_0 A_k$$
$$+ \gamma B_i - (\gamma - 1)\hat{\beta}_i(\hat{\beta}\cdot\vec{B}). \tag{11.12}$$

You should show for yourself that the sum of the first two terms gives $\gamma\epsilon_{ijk}\beta_j E_k$. Therefore,

$$B_i' = \gamma\epsilon_{ijk}\beta_j E_k + \gamma B_i - (\gamma - 1)\hat{\beta}_i(\hat{\beta}\cdot\vec{B})$$

or

Transformation rule for magnetic field.

$$\vec{B}' = \gamma(\vec{B} + \vec{\beta}\times\vec{E}) - (\gamma - 1)\hat{\beta}(\hat{\beta}\cdot\vec{B}). \tag{11.13}$$

Note the striking similarity between the transformation rules of the electric [Equation (11.8)] and magnetic fields!

It is instructive to separate the transformation rules for the components of the fields parallel and perpendicular to the relative velocity $\vec{\beta}$. Write each field on either side of (11.8) and (11.13) as the sum of their parallel and perpendicular components and equate each component on the left to the corresponding component on the right to obtain

$$\vec{E}_{\parallel}' = \vec{E}_{\parallel}, \qquad \vec{E}_{\perp}' = \gamma(\vec{E}_{\perp} - \vec{\beta} \times \vec{B}_{\perp})$$
$$\vec{B}_{\parallel}' = \vec{B}_{\parallel}, \qquad \vec{B}_{\perp}' = \gamma(\vec{B}_{\perp} + \vec{\beta} \times \vec{E}_{\perp}). \tag{11.14}$$

11.2 FIELD OF A UNIFORMLY MOVING CHARGE

When an electric charge moves, it creates both an electric and a magnetic field. You can calculate both of these fields if you use Equations (11.8) and (11.13), as well as (6.16). Put a charge q at the origin of the coordinate system O and give it a boost $\vec{\beta}$ relative to the coordinate system O'. At a certain time t, an observer in O measures the electric and magnetic fields at a point P with position vector $\vec{r}$. This constitutes an event which could be transformed to O' to see what the measurement looks like there. The electric and magnetic fields of this point charge at P in O are

$$\vec{E} = \frac{k_e q}{r^3}\vec{r}, \qquad \vec{B} = 0. \tag{11.15}$$

Substituting this in (11.8) and (11.13) yields the electric and magnetic fields of the charge in O'. First the electric field:

$$\vec{E}' = \frac{k_e q}{r^3} \underbrace{[\gamma\vec{r} - (\gamma - 1)\hat{\beta}\hat{\beta} \cdot \vec{r}\,]}_{\equiv \vec{\mathbb{S}}}. \tag{11.16}$$

To find $\vec{E}'$ entirely in terms of variables in O', you have to write $\vec{r}$ in terms of $\vec{r}\,'$. This is done by the inverse of (6.16), which differs from it by the sign of $\vec{\beta}$. Thus,

$$\vec{r} = \vec{r}\,' - \gamma\vec{\beta}\left(t' - \frac{\gamma}{\gamma+1}\vec{\beta} \cdot \vec{r}\,'\right), \tag{11.17}$$

and multiplying both sides of this by $\hat{\beta}$ you get (after some algebra that you should go through)

$$\hat{\beta} \cdot \vec{r} = \gamma(\hat{\beta} \cdot \vec{r}\,' - \beta t'). \tag{11.18}$$

Insert (11.17) and (11.18) in the square bracket of (11.16) to obtain

$$\begin{aligned}\vec{\mathbb{S}} &= \gamma\vec{r}\,' - \gamma^2\vec{\beta}\left(t' - \frac{\gamma}{\gamma+1}\vec{\beta} \cdot \vec{r}\,'\right) - (\gamma - 1)\hat{\beta}\gamma(\hat{\beta} \cdot \vec{r}\,' - \beta t') \\ &= \gamma\vec{r}\,' - \gamma^2\vec{\beta}t' + \frac{\gamma^3\beta^2}{\gamma+1}\hat{\beta}\hat{\beta} \cdot \vec{r}\,' - (\gamma - 1)\hat{\beta}\gamma\hat{\beta} \cdot \vec{r}\,' + (\gamma - 1)\vec{\beta}\gamma t'.\end{aligned}$$

Using $\gamma^2\beta^2 = \gamma^2 - 1$ and simplifying, you get

$$\vec{s} = \gamma(\vec{r}\,' - \vec{\beta}t') \equiv \gamma\vec{r}\,'_{\text{inst}}, \tag{11.19}$$

where $\vec{r}\,'_{\text{inst}}$ is the instantaneous vector connecting the charge to the field point (observation point) in O'. Therefore, you can now write $\vec{E}'$ as

$$\vec{E}' = \frac{k_e q\gamma}{r^3}\vec{r}\,'_{\text{inst}}. \tag{11.20}$$

It is convenient to write $r = |\vec{r}\,|$ in terms of $r'_{\text{inst}} \equiv |\vec{r}\,'_{\text{inst}}|$, rather than $|\vec{r}\,'|$. You can do this by substituting $\vec{r}\,'$ in terms of $\vec{r}\,'_{\text{inst}}$ in (11.17). Then, after some algebra, you get

$$\vec{r} = \vec{r}\,'_{\text{inst}} + (\gamma - 1)\hat{\beta}\hat{\beta}\cdot\vec{r}\,'_{\text{inst}}, \tag{11.21}$$

from which you should be able to obtain

$$r^2 = r'^2_{\text{inst}} + \gamma^2(\vec{\beta}\cdot\vec{r}\,'_{\text{inst}})^2 \tag{11.22}$$

and

$$\vec{E}' = \frac{k_e q\gamma}{\left[r'^2_{\text{inst}} + \gamma^2(\vec{\beta}\cdot\vec{r}\,'_{\text{inst}})^2\right]^{3/2}}\vec{r}\,'_{\text{inst}}. \tag{11.23}$$

If θ_{inst} is the angle between the direction of motion of the charge and $\vec{r}\,'_{\text{inst}}$, then (11.23) can also be written as

Instantaneous electric field of a uniformly moving charge.

$$\vec{E}' = \frac{k_e q\gamma}{\left(\gamma^2\cos^2\theta_{\text{inst}} + \sin^2\theta_{\text{inst}}\right)^{3/2}}\frac{\vec{r}\,'_{\text{inst}}}{|\vec{r}\,'_{\text{inst}}|^3}. \tag{11.24}$$

For the magnetic field of a moving charge, use (11.13) and note that $\vec{B} = 0$. Then, you get

$$\vec{B}' = \gamma\vec{\beta}\times\left(\frac{k_e q}{r^3}\vec{r}\right) = \frac{k_e q\gamma}{r^3}\vec{\beta}\times\vec{r}.$$

Instantaneous magnetic field of a uniformly moving charge.

But by (11.21), $\vec{\beta}\times\vec{r} = \vec{\beta}\times\vec{r}\,'_{\text{inst}}$. Therefore, using (11.22), you get

$$\vec{B}' = \frac{k_e q\gamma}{\left(\gamma^2\cos^2\theta_{\text{inst}} + \sin^2\theta_{\text{inst}}\right)^{3/2}}\frac{\vec{\beta}\times\vec{r}\,'_{\text{inst}}}{|\vec{r}\,'_{\text{inst}}|^3} = \vec{\beta}\times\vec{E}'. \tag{11.25}$$

This indicates that the electric and magnetic fields of an electric charge in uniform motion are perpendicular to each other.

It is interesting to note that the last equality in (11.25) is identical to the classical case and is the content of the Biot-Savart law in magnetism. In fact, when β is small, (11.25) becomes (after inserting the missing factors of c)

$$\vec{B}' = \frac{1}{c}\frac{k_e q}{r'^3_{\text{inst}}}(\vec{v}/c)\times\vec{r}\,'_{\text{inst}} = \frac{\vec{v}}{c^2}\times\vec{E}' = \frac{k_m q\,\vec{v}\times\vec{r}\,'_{\text{inst}}}{r'^3_{\text{inst}}}, \tag{11.26}$$

which is the Biot-Savart law for a moving point charge. In Equation (11.26), $k_m \equiv \mu_0/4\pi$ is the magnetic constant related to the electric constant $k_e \equiv 1/4\pi\epsilon_0$ via $c^2 = k_e/k_m$.

11.3 ELECTROMAGNETIC FIELD TENSOR

The electric and magnetic fields, as given by (11.1) and (11.5) in terms of the derivatives of the 4-potential, carry *two indices*. Therefore, it is natural to introduce a two-indexed tensor $F_{\mu\nu} \equiv \partial_\mu A_\nu - \partial_\nu A_\mu$, called the **electromagnetic field tensor**, which is *antisymmetric* under the interchange of its indices. Any two-indexed tensor can be represented by a matrix, and the matrix of $F_{\mu\nu}$ has the added property that it is antisymmetric; in particular, its diagonal elements are necessarily zero:

$$\partial_\mu A_\nu - \partial_\nu A_\mu \equiv F_{\mu\nu} \equiv \begin{pmatrix} 0 & F_{01} & F_{02} & F_{03} \\ F_{10} & 0 & F_{12} & F_{13} \\ F_{20} & F_{21} & 0 & F_{23} \\ F_{30} & F_{31} & F_{32} & 0 \end{pmatrix} = \begin{pmatrix} 0 & -E_1 & -E_2 & -E_3 \\ E_1 & 0 & B_3 & -B_2 \\ E_2 & -B_3 & 0 & B_1 \\ E_3 & B_2 & -B_1 & 0 \end{pmatrix}. \tag{11.27}$$

You should verify that the elements of the matrix in terms of the electric and magnetic fields are indeed as given above. Pay particular attention to the signs!

Sometimes the matrix is written in terms of the tensor with *superscripts*. And there is a good reason for this: superscripts transform straightforwardly as 4-vectors with the LT matrix Λ given by (B.5). Recall that to raise an index you have to multiply by η. To raise both indices, you have to multiply by two factors of η:

$$F^{\mu\nu} = \eta^{\mu\sigma}\eta^{\nu\xi}F_{\sigma\xi}. \tag{11.28}$$

Write this in matrix form, paying attention to the order of matrices (see Problem 11.11), and multiply the three matrices to get the matrix whose elements are $F^{\mu\nu}$. Let's denote this matrix by F from now on:

$$\mathsf{F} = \begin{pmatrix} 0 & E_1 & E_2 & E_3 \\ -E_1 & 0 & B_3 & -B_2 \\ -E_2 & -B_3 & 0 & B_1 \\ -E_3 & B_2 & -B_1 & 0 \end{pmatrix} \tag{11.29}$$

Electromagnetic field tensor $F^{\mu\nu}$.

The transformation of $F^{\mu\nu}$ under a boost is the product of two Lorentz boosts. This can be seen by noting that $F^{\mu\nu} \equiv \partial^\mu A^\nu - \partial^\nu A^\mu$, where ∂^μ is the 4-vector obtained from the gradient operator ∂_μ by raising its subscript. Since both ∂^μ and A^ν are 4-vectors, each transforms according to the first equation of (10.22). Therefore, the transformation rule for $F^{\mu\nu}$ is

$$F'^{\mu\nu} = \Lambda^\mu{}_\sigma \Lambda^\nu{}_\xi F^{\sigma\xi} = \Lambda^\nu{}_\xi \left(\Lambda^\mu{}_\sigma F^{\sigma\xi}\right) = \Lambda^\nu{}_\xi \left(\Lambda\mathsf{F}\right)^{\mu\xi}$$

$$
\begin{aligned}
&= \Lambda^{\nu}{}_{\xi}\left(\widetilde{\Lambda \mathsf{F}}\right)^{\xi\mu} = \left(\Lambda\widetilde{\Lambda \mathsf{F}}\right)^{\nu\mu} = \left(\Lambda\tilde{\mathsf{F}}\tilde{\Lambda}\right)^{\nu\mu} \\
&= -(\Lambda \mathsf{F} \Lambda)^{\nu\mu} = (\Lambda \mathsf{F} \Lambda)^{\mu\nu}.
\end{aligned} \tag{11.30}
$$

You should go through each step of the derivation above as an illuminating exercise in index and matrix manipulation, especially the transposition properties of Λ and F. In matrix form, this equation becomes

$$\mathsf{F}' = \Lambda \mathsf{F} \Lambda, \tag{11.31}$$

where F is given by (11.29) and Λ by (B.5). As a *must* exercise, you should verify that Equation (11.31) gives exactly the same transformation rules as (11.8) and (11.13) for electric and magnetic fields (Problem 11.13).

11.3.1 Lorentz Force Law

You know that a particle of charge q and velocity $\vec{v}$ in a magnetic field $\vec{B}$ experiences a magnetic force given by $q\vec{v} \times \vec{B}$. You also just learned that $F_{\mu\nu}$ incorporates the magnetic (as well as the electric) field. It is therefore instructive to investigate the properties of the product of $F_{\mu\nu}$ and the 4-velocity of a charged particle. More specifically, let us find the space and time components of $qF_{\mu\nu}u^{\mu}$, where u^{μ} is the 4-velocity of the charge q. Since there is only one free subscript in this expression, introduce a 4-covector $f_{\nu} \equiv qF_{\mu\nu}u^{\mu}$, and proceed to calculate the components of f_{ν} using Equation (11.27).

$$
\begin{aligned}
f_0 &= q\left(F_{10}u^1 + F_{20}u^2 + F_{30}u^3\right) \\
&= q\left(E_1u^1 + E_2u^2 + E_3u^3\right) \\
&= q\vec{E}\cdot\vec{u} = q\gamma_\alpha \vec{E}\cdot\vec{\alpha},
\end{aligned}
$$

where (6.42) was used in the last equality. Now find f_1:

$$
\begin{aligned}
f_1 &= q\left(F_{01}u^0 + F_{21}u^2 + F_{31}u^3\right) = q\left(-E_1u^0 - B_3u^2 + B_2u^3\right) \\
&= q\left[-E_1u^0 - (\vec{u}\times\vec{B})_1\right] = -q\gamma_\alpha\left[E_1 + (\vec{\alpha}\times\vec{B})_1\right],
\end{aligned}
$$

with a similar expression for the other two space components. Combining all the results obtained above, you get

$$
\begin{aligned}
f_0 &= q\gamma_\alpha \vec{E}\cdot\vec{\alpha} \\
f_i &= -q\gamma_\alpha\left[E_i + (\vec{\alpha}\times\vec{B})_i\right], \quad i = 1, 2, 3.
\end{aligned}
$$

The 4-vector associated with this 4-covector has components $f^{\mu} = \eta^{\mu\nu}f_{\nu}$, with

$$
\begin{aligned}
f^0 &= \eta^{00}f_0 = f_0 = q\gamma_\alpha \vec{E}\cdot\vec{\alpha} \\
f^i &= \eta^{ij}f_j = -f_i = q\gamma_\alpha\left[E_i + (\vec{\alpha}\times\vec{B})_i\right], \quad i = 1, 2, 3.
\end{aligned} \tag{11.32}
$$

This is a 4-force whose spacial part, by (6.62), is $\gamma_\alpha\vec{F}$, with

$$\vec{F} = q(\vec{E} + \vec{\alpha}\times\vec{B}), \tag{11.33}$$

Lorentz 4-force.

the Lorentz force law. Thus, you can call $qF_{\mu\nu}u^{\mu}$ the **Lorentz 4-force.**

11.3.2 The Cyclotron

Our knowledge of the relativistic second law of motion can help us analyze the motion of a charged particle in an electromagnetic field. Equations (6.65) and (11.33) lead immediately to[2]

$$\frac{dE}{dt} = q\vec{\alpha} \cdot \vec{E},$$

which indicates the known fact that only an electric field can change the (kinetic) energy of the particle. In particular, if there is no electric field in the RF of interest, then $E = m\gamma_\alpha$ is constant, implying that γ_α, and therefore $|\vec{\alpha}|$, is constant. This is the same classical result that states that a magnetic field does not accelerate a charged particle.

The case of a constant magnetic field is of practical importance. Let $\vec{B}$ be constant and in the positive z-direction. Then with $\vec{E} = 0$, (11.33) yields

$$\vec{F} = q\vec{\alpha} \times \vec{B} = q \det \begin{pmatrix} \hat{\mathbf{e}}_x & \hat{\mathbf{e}}_y & \hat{\mathbf{e}}_z \\ \alpha_x & \alpha_y & \alpha_z \\ 0 & 0 & B \end{pmatrix} = qB\alpha_y\hat{\mathbf{e}}_x - qB\alpha_x\hat{\mathbf{e}}_y,$$

and (6.63) with $\vec{p} = m\vec{u} = m\gamma_\alpha\vec{\alpha}$ gives

$$\frac{d\alpha_x}{dt} = \frac{qB}{m\gamma_\alpha}\alpha_y, \qquad \frac{d\alpha_y}{dt} = -\frac{qB}{m\gamma_\alpha}\alpha_x, \qquad \frac{d\alpha_z}{dt} = 0, \tag{11.34}$$

because γ_α is constant. You can ignore the motion along the z-axis because it is uniform. Differentiate the first equation with respect to t and substitute from the second equation to obtain

$$\frac{d^2\alpha_x}{dt^2} = \frac{qB}{m\gamma_\alpha}\frac{d\alpha_y}{dt} = \frac{qB}{m\gamma_\alpha}\left(-\frac{qB}{m\gamma_\alpha}\alpha_x\right) = -\left(\frac{qB}{m\gamma_\alpha}\right)^2 \alpha_x,$$

whose most general solution is

$$\alpha_x = A_1 \cos\omega t + A_2 \sin\omega t, \qquad \omega = \frac{qB}{m\gamma_\alpha}.$$

The first equation in (11.34) gives the y-component:

$$\alpha_y = \frac{1}{\omega}\frac{d\alpha_x}{dt} = -A_1 \sin\omega t + A_2 \cos\omega t.$$

Now choose the orientation of the axes in such a way that at $t = 0$, the particle moves with speed α_0 entirely in the positive y-direction. This means that at $t = 0$, $\alpha_x = 0$ and $\alpha_y = \alpha_0$.[3] For this initial condition to be true, we must have $A_1 = 0$ and $A_2 = \alpha_0$. Therefore, the final solution for $\vec{\alpha}$ is

$$\alpha_x = \alpha_0 \sin\omega t, \qquad \alpha_y = \alpha_0 \cos\omega t, \qquad \omega = \frac{qB}{m\gamma_\alpha}. \tag{11.35}$$

---

[2] Don't confuse the energy with the (magnitude of the) electric field!

[3] Do not confuse the subscript "0" with the "zeroth" component of a 4-vector. Here the zero indicates the initial value.

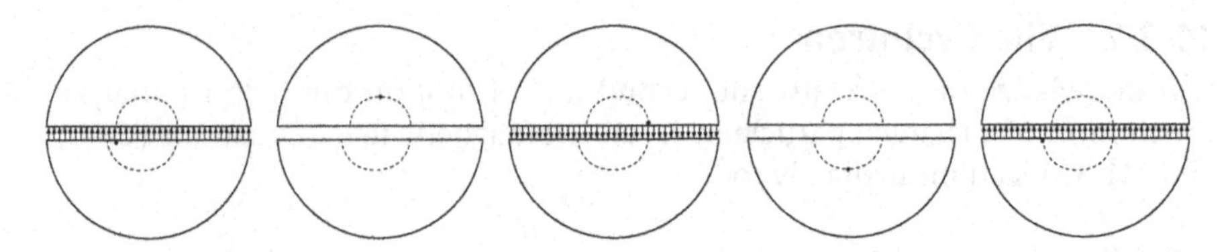

FIGURE 11.1 The little arrows show the direction of the electric field. The frequency of the electric field is the same as the frequency of the rotation of the charged particle.

Integrating these equations yields

$$x = -\frac{\alpha_0 c}{\omega}\cos\omega t + x_0 = -\frac{v_0}{\omega}\cos\omega t + x_0$$
$$y = \frac{\alpha_0 c}{\omega}\sin\omega t + y_0 = \frac{v_0}{\omega}\sin\omega t + y_0 \tag{11.36}$$

if you restore the factors of c. This is a clockwise motion of the particle in the xy-plane on a circle centered at (x_0, y_0) with radius

$$R = \frac{v_0}{\omega} = \frac{mv_0\gamma_{\alpha_0}}{qB} = \frac{mv_0}{qB\sqrt{1-(v_0/c)^2}} = \frac{p_0}{qB}, \tag{11.37}$$

Momentum measurement in particle chambers.

where p_0 is the magnitude of the momentum of the particle. It is interesting to note that the last equality holds both relativistically and non-relativistically. It also shows that if you measure the radius of the path of the particle in a known magnetic field, you can determine its momentum regardless of how fast it is moving. This is indeed how the momenta of high-energy particles are measured in cloud chambers, spark chambers, and various other chambers of particle detectors.

Cyclotron works only for low speeds.

A **cyclotron** is a large pill-box shaped device cut in half in the shape of two "dee"s with a gap between them as shown in Figure 11.1. A constant magnetic field perpendicular to the flat areas of the dees keeps the charged particle on a circle while an alternating electric field in the gap between the two dees accelerates the particle. If the speed of the charged particle is much smaller than the speed of light, then $\gamma_\alpha \approx 1$ and (11.35) shows that ω is independent of the speed of the particle. In that case, you can choose the electric field to have the same frequency as that of the rotation of the particle on the circle. A positively charged particle (represented by the heavy dot in Figure 11.1) enters the gap of the cyclotron in the first diagram on the left of the figure and is accelerated by the electric field. The magnetic field guides the particle on a circle until it reaches the gap (the third diagram) where the electric field has now changed direction and thus accelerates the particle once more. As the particle enters the lower dee, it has a larger speed, and therefore moves on a larger circle until it reaches the lower edge of the gap (the last diagram on the right) where the electric field has once again changed direction, ready to accelerate the particle for the third time. Note that although the radius of the particle path is larger in the lower dee, the time it takes the particle to cover the lower semi-circle is the same as the time it takes it to cover the upper semi-circle:

$$t = \frac{\pi R}{v_0} = \frac{\pi m}{qB} = \frac{\pi}{\omega} = \frac{T}{2},$$

independent of v_0 and R. Here T is the period of the revolution of the particle in the dees, which is equal to the period of oscillation of the electric field.

The first cyclotron was developed by the American physicist Ernest Lawrence and his graduate student Stanley Livingston in 1932. It had a diameter of 2.5 inches and accelerated protons to about 1.3% the speed of light, for which the approximation $\gamma_\alpha \approx 1$ is good. However, as the need for higher and higher speeds grew, so did the size of the cyclotron, and when relativistic effects became noticeable and the cyclotron frequency (11.35) became speed-dependent, the design of the cyclotron had to be altered and the frequency of the electric field had to be *synchronized* with the speed of the particle. The second generation of cyclotrons was called synchrocyclotrons or **synchrotrons**.

Synchrotrons adjust the frequency of the electric field to account for the speed of the particle.

Example 11.3.1. You want to accelerate a proton of mass 1.67×10^{-27} kg or 938.27 MeV (remember that $c = 1$; therefore, $m = mc^2$) and charge 1.6×10^{-19} C to an energy of 7 TeV (1 TeV is 10^{12} eV) in a magnetic field $B = 5.5$ Tesla in a circular accelerator. What is the minimum size of the circumference of this accelerator?

From $E = m\gamma_{\alpha_0}$ you find

$$\gamma_{\alpha_0} = \frac{E}{m} = \frac{7 \times 10^{12}}{938.27 \times 10^6} = 7461,$$

corresponding to $v_0 = 0.999999991c$. Substitute these values in Equation (11.37) to obtain

$$R = \frac{mv_0\gamma_{\alpha_0}}{qB} = \frac{(1.67 \times 10^{-27})(3 \times 10^8)(7461)}{(1.6 \times 10^{-19})(5.5)} = 4248 \text{ m}.$$

The circumference of this circle is 26,690 m, or about 27 km. These are roughly the parameters used in the construction of the Large Hadron Collider (LHC), which discovered the elusive Higgs boson on July 4th, 2012. ■

Circumference of LHC and discovery of Higgs boson.

11.3.3 Maxwell's Equations

The electromagnetic field tensor encompasses both the electric and magnetic fields. Maxwell's equations tie together the derivatives of electric and magnetic fields as well as charges. Therefore, it is instructive to examine the derivative of the electromagnetic field tensor. The derivative has a subscript [see the discussion after Equation (10.14)]. The electromagnetic field tensor of Equation (11.28) has two superscripts. What if you contracted one of the superscripts of $F^{\mu\nu}$ with the subscript of the derivative? Then these two indices disappear and one superscript of $F^{\mu\nu}$ is left, indicating a resulting 4-vector. What is this 4-vector? Let's find out.

First write

$$\partial_\mu F^{\mu\nu} = \partial_0 F^{0\nu} + \partial_1 F^{1\nu} + \partial_2 F^{2\nu} + \partial_3 F^{3\nu}. \tag{11.38}$$

Now evaluate this equation for $\nu = 0, 1, 2, 3$. For $\nu = 0$, you get

$$\begin{aligned}\partial_0 F^{00} &+ \partial_1 F^{10} + \partial_2 F^{20} + \partial_3 F^{30} \\ &= -\partial_1 E_1 - \partial_2 E_2 - \partial_3 E_3 = -\vec{\nabla} \cdot \vec{E}.\end{aligned} \tag{11.39}$$

For $\nu = 1$, you get

$$\begin{aligned}\partial_0 F^{01} &+ \partial_1 F^{11} + \partial_2 F^{21} + \partial_3 F^{31} \\ &= \partial_0 E_1 + \underbrace{\partial_2(-B_3) + \partial_3 B_2}_{=(-\vec{\nabla}\times\vec{B})_1} \\ &= \frac{\partial E_1}{\partial t} - (\vec{\nabla} \times \vec{B})_1,\end{aligned}$$

with similar results for the other two space indices, so that

$$\partial_\mu F^{\mu i} = \frac{\partial E_i}{\partial t} - (\vec{\nabla} \times \vec{B})_i, \quad i = 1, 2, 3. \tag{11.40}$$

4-current density and inhomogeneous Maxwell's equations.

Equations (11.39) and (11.40) suggest introducing a **4-current density** J^μ with $J^0 = J_0 = -\rho$ and $J^i = -J_i$, where ρ is the charge density and J_i is the ith component of the current density $\vec{J}$. Then Maxwell's first and last equations, the so-called **inhomogeneous Maxwell's equations**, can be combined into the single equation:

$$\partial_\mu F^{\mu\nu} = \mu_0 J^\nu, \quad \nu = 0, 1, 2, 3, \tag{11.41}$$

where μ_0 is the permeability of vacuum, which is equal to $1/\epsilon_0$, because we have set the speed of light equal to 1.

The inhomogeneous Maxwell's equations were obtained by contracting the subscript of the derivative with one of the superscripts of the electromagnetic field tensor. Contraction is the generalization of the dot product. So, loosely speaking, we obtained two of the Maxwell's equations by taking dot products. Could it be that the other two equations emerge by taking some sort of four-dimensional cross product? Recall that the three-dimensional cross product is obtained by contracting two indices of the Levi-Civita symbol with the indices of two vectors [see Equation (10.7)]. Now we have three indices, one coming from the derivative and two from the electromagnetic field tensor. And the obviously generalized four-dimensional Levi-Civita tensor has four indices. That's perfect! Because contraction of three indices of the two sets of indices leaves only one index free, which could be the index of a 4-vector (or covector).

Since the derivative carries a natural *subscript*, if we want 3 indices, we should let it act on $F_{\mu\nu}$, in which case we end up with three subscripts. Now, contract these with the four-dimensional Levi-Civita symbol, whose indices are all superscripts. So, here is the 4-vector we are after:

$$V^{\nu} \equiv \epsilon^{\sigma\xi\mu\nu}\partial_\sigma F_{\xi\mu}, \quad \nu = 0, 1, 2, 3. \tag{11.42}$$

To get the zeroth component of this vector, first sum over all possible values of σ. Then for each term so obtained, sum over ξ, and finally over μ, keeping in mind that all indices of the Levi-Civita symbol must be different:

$$\begin{aligned} V^0 &= \epsilon^{\sigma\xi\mu 0}\partial_\sigma F_{\xi\mu} = \epsilon^{1\xi\mu 0}\partial_1 F_{\xi\mu} + \epsilon^{2\xi\mu 0}\partial_2 F_{\xi\mu} + \epsilon^{3\xi\mu 0}\partial_3 F_{\xi\mu} \\ &= \epsilon^{12\mu 0}\partial_1 F_{2\mu} + \epsilon^{13\mu 0}\partial_1 F_{3\mu} + \epsilon^{21\mu 0}\partial_2 F_{1\mu} \\ &\quad + \epsilon^{23\mu 0}\partial_2 F_{3\mu} + \epsilon^{31\mu 0}\partial_3 F_{1\mu} + \epsilon^{32\mu 0}\partial_3 F_{2\mu}. \end{aligned}$$

The only index left is μ, and as you can see, in each term there is only one possibility for it:

$$\begin{aligned} V^0 &= \epsilon^{1230}\partial_1 F_{23} + \epsilon^{1320}\partial_1 F_{32} + \epsilon^{2130}\partial_2 F_{13} \\ &\quad + \epsilon^{2310}\partial_2 F_{31} + \epsilon^{3120}\partial_3 F_{12} + \epsilon^{3210}\partial_3 F_{21}. \end{aligned}$$

Noting that $\epsilon^{0123} = 1$ and that $\epsilon^{\sigma\xi\mu\nu}$ is completely antisymmetric, you should be able to obtain the following:

$$\begin{aligned} V^0 &= -\partial_1 F_{23} + \partial_1 F_{32} + \partial_2 F_{13} - \partial_2 F_{31} - \partial_3 F_{12} + \partial_3 F_{21} \\ &= 2\partial_1 F_{32} + 2\partial_2 F_{13} + 2\partial_3 F_{21}. \end{aligned}$$

The last line follows because $F_{\xi\mu} = -F_{\mu\xi}$. Now consult Equation (11.27) and note that $F_{32} = -B_1$, $F_{13} = -B_2$, and $F_{21} = -B_3$. Then the equation above yields $V^0 = -2\vec{\nabla}\cdot\vec{B}$, or

$$\epsilon^{\sigma\xi\mu 0}\partial_\sigma F_{\xi\mu} = -2\vec{\nabla}\cdot\vec{B}.$$

You should verify that

$$V^i \equiv \epsilon^{\sigma\xi\mu i}\partial_\sigma F_{\xi\mu} = 2[\partial_0 B_i + (\vec{\nabla}\times\vec{E})_i] = 2\left[\frac{\partial B_i}{\partial t} + (\vec{\nabla}\times\vec{E})_i\right].$$

The last two equations tell us that the equation

$$\epsilon^{\sigma\xi\mu\nu}\partial_\sigma F_{\xi\mu} = 0, \quad \nu = 0, 1, 2, 3$$

is equivalent to the two homogeneous Maxwell's equations. Therefore,

Note 11.3.2. *Maxwell's equations are succinctly written as*

$$\begin{aligned} \partial_\mu F^{\mu\nu} &= \mu_0 J^\nu, \quad \nu = 0, 1, 2, 3, \\ \epsilon^{\sigma\xi\mu\nu}\partial_\sigma F_{\xi\mu} &= 0, \quad \nu = 0, 1, 2, 3. \end{aligned}$$

Because of the antisymmetry of the electromagnetic field tensor, Maxwell's equations *imply* the conservation of the electric charge, which is contained in the continuity equation of Note A.1.2 (Problem 11.21).

What is interesting is that the 4-vector nature of J^μ leads to the invariance of the electric charge under a LT. Let O be the rest frame of some electric charge

Invariance of the electric charge.

whose density is ρ. Let O' see O move with velocity $\vec{\beta}$. Since $\vec{J}=0$ (the charges are not moving in O, so there is no current density there), the transformation rule for 4-vectors (6.16) yields

$$\rho'=\gamma\left(\rho+\vec{\beta}\cdot\vec{J}\right)=\gamma\rho$$

$$\vec{J}'=\vec{J}+\gamma\vec{\beta}\left(\rho+\frac{\gamma}{\gamma+1}\vec{\beta}\cdot\vec{J}\right)=\gamma\vec{\beta}\rho.$$

Therefore the amount of charge dq enclosed in a volume dV in O, which is obviously $\rho\, dV$, appears as

$$dq'=\rho' dV'=\gamma\rho\frac{dV}{\gamma}=dq,$$

in O' because of Equation (10.24). That is,

Note 11.3.3. *Electric charge is an invariant quantity, i.e., it is the same for all observers.*

11.4 PROBLEMS

11.1. From $B_i=\epsilon_{ijk}\partial_j A_k$ show that $\partial_j A_k-\partial_k A_j=\epsilon_{jkm}B_m$.

11.2. Go through the details of the manipulation of the last line of (11.10) and derive (11.11).

11.3. Show that the sum of the first two terms of (11.12) gives $\gamma\epsilon_{ijk}\beta_j E_k$.

11.4. Derive Equation (11.14) from (11.8) and (11.13).

11.5. Derive Equation (11.18).

11.6. Derive Equation (11.19).

11.7. Derive Equation (11.21). Now take the dot product of the equation with itself to derive (11.22).

11.8. Derive Equation (11.25).

11.9. Consider an infinite plate charged uniformly with surface density σ. The plate is moving uniformly with speed β perpendicular to its surface. There are two ways to calculate the electric and magnetic fields of this charge distribution.

a. Calculate the fields $\vec{E}$ and $\vec{B}$ in the rest frame of the plate using the elementary method of Gauss's law. Now transform those fields to the frame in which the plate is moving.
b. Use (11.24) and (11.25) to write the contribution from each element of charge on the surface and integrate over the plate to get the fields.

11.10. Verify that the elements of the matrix (11.27) are indeed the electric and magnetic fields as given there.

11.11. Let $\mathsf{F}^{\mathrm{sup}}$ and $\mathsf{F}_{\mathrm{sub}}$ denote the matrices of $F^{\mu\nu}$ and $F_{\mu\nu}$, respectively. Show that the matrix form of Equation (11.28) is $\mathsf{F}^{\mathrm{sup}} = \eta \mathsf{F}_{\mathrm{sub}} \eta$. Now carry out the multiplication of the three matrices to come up with Equation (11.29).

11.12. Verify Equation (11.30) by going through each step of its derivation.

11.13. Noting that F' is given by the same matrix as (11.29) except that its elements carry a prime, use Equation (11.31) to derive the transformation rules of (11.8) and (11.13) for electric and magnetic fields.

11.14. Show that $F^{\mu\nu}F_{\mu\nu} = 2(|\vec{B}|^2 - |\vec{E}|^2)$. Hint: Prove that $F^{\mu\nu}F_{\mu\nu}$ is the negative of the trace (the sum of the diagonal elements) of the product of the two matrices in (11.29) and (11.27). Then find that trace.

11.15. Show that $\det \mathsf{F} = (\vec{E} \cdot \vec{B})^2$, where F is as given by (11.29).

11.16. The second cyclotron that Lawrence's group built could accelerate a proton to a kinetic energy of 1 MeV in a magnetic field of 1.26 Tesla.

a. What is γ_{α_0} and v_0 for such a proton?
b. What was the diameter of the cyclotron?

11.17. A K^0 meson at rest decays into two charged pions π^+ and π^- in a bubble chamber in which a magnetic field $B = 1.25$ T is present. The pions have equal mass $m_\pi = 139.57$ MeV. If the radius of curvature of the pions is 55 cm, determine the momenta and speeds of the pions and the mass of the K^0.

11.18. A Λ particle at rest decays into a proton and a π^- in a bubble chamber in which a magnetic field $B = 0.5$ T is present. The mass of the pion is $m_\pi = 139.57$ MeV and that of the proton is $m_p = 938.27$ MeV. If the radius of curvature of the pion is 67 cm, determine its speed, the radius of curvature of the proton, the speed of the proton, and the mass of Λ.

11.19. Verify Equation (11.40) for $i = 2$ and $i = 3$.

11.20. Show that $\epsilon^{\sigma\xi\mu i}\partial_\sigma F_{\xi\mu} = 2[\partial_0 B_i + (\vec{\nabla} \times \vec{E})_i]$.

11.21. In this problem, you'll show that Maxwell's equations imply the conservation of the electric charge.

(a) Write the continuity equation (see Note A.1.2) in terms of indices in four-dimensional spacetime.
(b) Suppose that $S^{\mu\nu}$ and $A^{\mu\nu}$ are, respectively, symmetric and antisymmetric under the interchange of their indices. Show that $S^{\mu\nu}A_{\mu\nu} = 0$.
(c) Differentiate the first equation in Note 11.3.2 and use (b) to prove that Maxwell's equations imply charge conservation.

CHAPTER 12

Early Universe

Einstein's general theory of relativity (GTR) predicts that the universe is expanding on a large scale, meaning that in the past, the universe was smaller, and if you go far enough in the past, the universe collapses into a single spacetime event (a singularity) now called the big bang. The "bang" refers to the colossal explosion caused by the enormous amount of energy confined in the spacetime point that was the universe. GTR even gives an expression for the expansion of the universe from which one can estimate the age of the universe, which has been measured to be 13.82 billion years.

When the entire universe is condensed into a single point, you get infinite energy and mass density, infinite temperature, infinite pressure, etc. This is a sign of the incompleteness of our theoretical description. GTR is very accurate when applied to large structures. However, the infinities indicate that it fails at microscopic scales, where quantum theory is applicable. It is hoped that a quantum theory of gravity, which unifies GTR and quantum theory, will remove the infinities and give us a handle on the big bang.

Although the big bang itself is an enigma, the behavior of the universe slightly after the big bang is within the realm of modern physics. In fact, with some simple results borrowed from GTR, we can follow the history of the early universe using our knowledge of Newtonian gravity and special relativity accumulated in this book.

12.1 PHOTON GAS

A major component of the universe is electromagnetic radiation. When the universe was very young, this radiation was in thermal equilibrium with other constituents of the universe. The distribution of the frequencies of the EM radiation and the intensity of each frequency depends on the temperature of the container of the radiation. The change of the color of an object (say a metal) as it gets hotter is a familiar phenomenon. In this section I'll discuss the behavior of a photon gas in terms of the frequency of its constituents and the temperature of its surrounding.

Special Relativity. DOI:10.1016/B978-0-12-810411-8.00012-2

In Section A.4, I have examined EM waves in a cubic cavity, which has frequencies given by a triplet of integers or *modes*. Now I want to hold the cavity at temperature T and find the total thermal energy of the photon gas in it. First I find the average energy of each mode. A photon in mode $\mathbf{n}$ has an energy $\varepsilon_n = \hbar\omega_n$, and if there are m photons in that mode, then their total energy is $m\hbar\omega_n$. What is the average number of photons in mode $\mathbf{n}$? The fundamental principle of statistical mechanics gives the probability for m photons to be in mode $\mathbf{n}$ as $P(m, \mathbf{n}) = Ce^{-m\hbar\omega_n/k_BT}$, with k_B the Boltzmann constant. Since $\sum_m P(m, \mathbf{n}) = 1$, I can find C—which, as is common, I write as $1/Z$—and get

Fundamental principle of statistical mechanics.

$$P(m, \mathbf{n}) = \frac{e^{-m\hbar\omega_n/k_BT}}{Z}, \qquad Z = \sum_{m=0}^{\infty} e^{-m\hbar\omega_n/k_BT}. \tag{12.1}$$

Partition function.

The sum in Z (the *partition function*) is a geometric series and can be evaluated in closed form:

$$Z = \sum_{m=0}^{\infty} \left(e^{-\hbar\omega_n/k_BT}\right)^m = \frac{1}{1 - e^{-\hbar\omega_n/k_BT}}.$$

Therefore, $\langle m_n \rangle$, the average number of photons in mode $\mathbf{n}$, is

$$\langle m_n \rangle = \sum_{m=0}^{\infty} mP(m, \mathbf{n}) = \frac{1}{Z}\sum_{m=0}^{\infty} me^{-m\hbar\omega_n/k_BT}.$$

Let $x = \hbar\omega_n/k_BT$ and note that

$$\sum_{m=0}^{\infty} me^{-mx} = -\frac{d}{dx}\sum_{m=0}^{\infty} e^{-mx} = -\frac{dZ}{dx} = \frac{e^{-x}}{\left(1 - e^{-x}\right)^2}.$$

Hence,

$$\langle m_n \rangle = \frac{e^{-x}}{1 - e^{-x}} = \frac{1}{e^x - 1} = \frac{1}{e^{\hbar\omega_n/k_BT} - 1}, \tag{12.2}$$

and the average energy $\langle \varepsilon_n \rangle$ of mode $\mathbf{n}$ is

$$\langle \varepsilon_n \rangle = \langle m_n \rangle \hbar\omega_n = \frac{\hbar\omega_n}{e^{\hbar\omega_n/k_BT} - 1}. \tag{12.3}$$

The total energy U_γ of the radiation is the sum of the average energies:[1]

$$U_\gamma = \sum_{n_x, n_y, n_z} \langle \varepsilon_n \rangle = \sum_{n_x, n_y, n_z} \frac{\hbar\omega_n}{e^{\hbar\omega_n/k_BT} - 1}.$$

I'm going to let the size of the cubic cavity of side L be large. This means that as n_x, n_y, and n_z are increased by a few integers, ω_n varies very little. Therefore

[1] γ is a universal symbol for photon. I'll use the subscript to indicate that the quantity describes photons.

I can treat n_x, n_y, and n_z as continuous variables and turn the sum above into an integral over $dn_x dn_y dn_z$:

$$\iiint \frac{\hbar\omega_n}{e^{\hbar\omega_n/k_BT}-1} dn_x dn_y dn_z.$$

And since by (A.24) ω_n depends only on the magnitude $|\mathbf{n}|$, I can switch to spherical coordinates and write

$$U_\gamma = (2)\left(\frac{1}{8}\right)\int_0^\infty 4\pi n^2 \frac{\hbar\omega_n}{e^{\hbar\omega_n/k_BT}-1} dn.$$

The factor 2 comes from two independent polarizations and the factor $\frac{1}{8}$ comes because I'm integrating only over positive values of n_x, n_y, and n_z. Using Equation (A.24), I finally obtain

$$U_\gamma = \pi\hbar\left(\frac{L}{\pi c}\right)^3 \int_0^\infty \frac{\omega_n^3}{e^{\hbar\omega_n/k_BT}-1} d\omega_n,$$

or, getting rid of the superfluous subscript n,

$$u_\gamma = \frac{U_\gamma}{V} \equiv \int_0^\infty u_\gamma(\omega, T) d\omega = \frac{\hbar}{\pi^2 c^3}\int_0^\infty \frac{\omega^3}{e^{\hbar\omega/k_BT}-1} d\omega \qquad (12.4)$$

where $V = L^3$ and u_γ is the energy density. The quantity

$$u_\gamma(\omega, T) = \frac{\hbar}{\pi^2 c^3}\frac{\omega^3}{e^{\hbar\omega/k_BT}-1} \qquad (12.5)$$

Black body radiation law.

is the **Planck radiation law**, or the **black body radiation law**. It is the *spectral energy density*, i.e., the energy density per unit frequency: when multiplied by $\Delta\omega$, it gives the energy density of photons having frequencies in the range ω to $\omega + \Delta\omega$. Planck found this formula empirically and worked backward to arrive at the quantization of electromagnetic energy $\varepsilon = \hbar\omega$.

The radiation law is oftentimes written in terms of the wavelength. Using $\omega = 2\pi c/\lambda$, you can change the integration variable in (12.4) and write

$$u_\gamma = \int_0^\infty u_\gamma(\lambda, T) d\lambda, \quad u_\gamma(\lambda, T) = \frac{8\pi hc}{\lambda^5}\frac{1}{\exp\left(\frac{hc}{k_BT\lambda}\right)-1}, \qquad (12.6)$$

where $h = 2\pi\hbar$ is the Planck constant.

The total number N_γ of photons can also be found. It is given by

$$N_\gamma = \sum_{n_x,n_y,n_z} \langle m_n \rangle = \sum_{n_x,n_y,n_z} \frac{1}{e^{\hbar\omega_n/k_BT}-1}.$$

The only difference between this and the total energy is a factor of $\hbar\omega$. So, taking out that factor from (12.4), we obtain the photon number density:

$$n_\gamma = \frac{N_\gamma}{V} \equiv \int_0^\infty n_\gamma(\omega, T) d\omega = \frac{1}{\pi^2 c^3}\int_0^\infty \frac{\omega^2}{e^{\hbar\omega/k_BT}-1} d\omega, \qquad (12.7)$$

where

$$n_\gamma(\omega, T) \equiv \frac{1}{\pi^2 c^3} \frac{\omega^2}{e^{\hbar\omega/k_B T} - 1} \tag{12.8}$$

is the **spectral number density** of photons. It can also be written in terms of wavelength:

$$n_\gamma(\lambda, T) \equiv \frac{8\pi}{\lambda^4} \frac{1}{\exp\left(\frac{hc}{k_B T\lambda}\right) - 1}, \quad \text{with} \quad n_\gamma = \int_0^\infty n_\gamma(\lambda, T) d\lambda. \tag{12.9}$$

Change the variable of integration of (12.4) to $x = \hbar\omega/k_B T$, then $\omega = k_B T x/\hbar$ and $d\omega = k_B T dx/\hbar$, so that

$$u_\gamma = \frac{\hbar}{\pi^2 c^3} \left(\frac{k_B T}{\hbar}\right)^4 \int_0^\infty \frac{x^3}{e^x - 1} dx.$$

The definite integral has the value $\pi^4/15$. Therefore,

Stefan-Boltzmann radiation law.

$$u_\gamma = \frac{\pi^2 k_B^4}{15\hbar^3 c^3} T^4, \tag{12.10}$$

which is the **Stefan-Boltzmann radiation law**. Do the same with the integral of (12.7) and obtain

$$n_\gamma = \frac{1}{\pi^2 c^3} \left(\frac{k_B T}{\hbar}\right)^3 \underbrace{\int_0^\infty \frac{x^2}{e^x - 1} dx}_{=2.40411} = \frac{2.40411 k_B^3}{\pi^2 c^3 \hbar^3} T^3 = 2 \times 10^7 T^3 \text{ photons/m}^3. \tag{12.11}$$

The average energy of a photon is the energy density divided by the number density,

$$\langle \varepsilon_\gamma \rangle = \frac{u_\gamma}{n_\gamma} = \frac{\pi^4 k_B}{15 \times 2.40411} T = 2.7 k_B T. \tag{12.12}$$

This is of the same form as the familiar average kinetic energy of a non-relativistic particle: $\langle KE \rangle = \frac{3}{2} k_B T$.

Example 12.1.1. THRESHOLD TEMPERATURE

The photons at the beginning of the creation of the universe were very energetic, so much so that they could create a particle-antiparticle pair. The two-particle collision was studied in Section 8.2. The balanced reaction

$$\gamma + \gamma \longleftrightarrow p + \bar{p}, \tag{12.13}$$

where p is any particle and $\bar{p}$ its antiparticle, occurs when the photon gas is hot enough to create a particle-antiparticle pair. In Problem 12.2, you are asked to show that for the reaction (12.13) to occur, we must have

$$E_1 E_2 = \frac{m_p^2}{\sin^2(\theta/2)},$$

where m_p is the mass of p (or $\bar{p}$), E_1 and E_2 are the energies of the photons, and θ is the angle between the direction of motion of the initial photons. The minimum photon energy $E_{\min}$ required turns out to be $m_p c^2$.

When the photon gas is hot enough so that the average photon energy is larger than $E_{\min}$, the process goes both ways as in (12.13) and the photon gas is in equilibrium with $p\bar{p}$ gas. The temperature corresponding to $E_{\min}$ is called the **threshold temperature** for p. For example, the electron threshold temperature is

Threshold temperature.

$$T_e \equiv \frac{m_e c^2}{2.7 k_B} = \frac{511000 \text{ ev}}{2.7 \times 8.62 \times 10^{-5} \text{ ev K}^{-1}} = 2.2 \times 10^9 \text{ K}.$$

No place in the current universe is this hot. Even the hottest stars have an interior temperature of only several million degrees. ■

Example 12.1.2. Sometimes we are interested in the fraction of photons whose wavelengths lie between two given values, say λ_1 and λ_2. This is obtained by integrating the spectral number density $n_\gamma(\lambda, T)$ from λ_1 to λ_2. Denoting the corresponding number density by $n_\gamma(\lambda_1, \lambda_2, T)$, I get

$$n_\gamma(\lambda_1, \lambda_2, T) = 8\pi \int_{\lambda_1}^{\lambda_2} \frac{d\lambda}{\lambda^4 \left(e^{hc/\lambda k_B T} - 1\right)} = 8\pi \left(\frac{k_B T}{hc}\right)^3 \int_{y_1}^{y_2} \frac{dy}{y^4 (e^{1/y} - 1)},$$

from which I obtain

$$\frac{n_\gamma(\lambda_1, \lambda_2, T)}{n_\gamma} = \frac{1}{2.40411} \int_{y_1}^{y_2} \frac{dy}{y^4 (e^{1/y} - 1)}, \quad y_1 \equiv \frac{k_B T \lambda_1}{hc}, \quad y_2 \equiv \frac{k_B T \lambda_2}{hc}, \tag{12.14}$$

where I used Equation (12.11), noting that $\hbar = h/2\pi$. As an example, I can ask "At a temperature of 6000 K (Sun's surface temperature), what fraction of photons are visible?" The visible range is from $\lambda_1 = 0.4$ μm (violet) to $\lambda_2 = 0.7$ μm (red), and the question wants to know the ratio $n_\gamma(\lambda_1, \lambda_2, T)/n_\gamma$. With

$$y_1 = \frac{k_B T \lambda_1}{hc} = \frac{(1.38 \times 10^{-23})(6000)(0.4 \times 10^{-6})}{(6.63 \times 10^{-34})(3 \times 10^8)} = 0.1665$$

and $y_2 = 0.291375$, which is found similarly to y_1, I find

$$\frac{n_\gamma(\lambda_1, \lambda_2, T)}{n_\gamma} = \frac{1}{2.40411} \int_{0.1665}^{0.291375} \frac{dy}{y^4 (e^{1/y} - 1)} = 0.23.$$

Thus, assuming that the Sun is a black body radiator, only about 23% of the photons it emits are visible.

A related question is "What fraction of the total energy do the visible photons carry?" The answer comes in a relation similar to (12.14):

$$\begin{aligned} u_\gamma(\lambda_1, \lambda_2, T) &= 8\pi hc \int_{\lambda_1}^{\lambda_2} \frac{d\lambda}{\lambda^5 \left(e^{hc/\lambda k_B T} - 1\right)} = 8\pi hc \left(\frac{k_B T}{hc}\right)^4 \int_{y_1}^{y_2} \frac{dy}{y^4 (e^{1/y} - 1)} \\ &= \frac{u_\gamma}{(\pi^4/15)} \int_{y_1}^{y_2} \frac{dy}{y^5 (e^{1/y} - 1)}, \quad y_1 \equiv \frac{k_B T \lambda_1}{hc}, \quad y_2 \equiv \frac{k_B T \lambda_2}{hc}, \end{aligned} \tag{12.15}$$

which, with y_1 and y_2 as before, gives

$$\frac{u_\gamma(\lambda_1, \lambda_2, T)}{u_\gamma} = \frac{15}{\pi^4} \int_{0.1665}^{0.291375} \frac{dy}{y^5 (e^{1/y} - 1)} = 0.375.$$

So, although only 23% of the photons emitted by the Sun are visible, they carry 37.5% of the energy. This is because $u_\gamma(\lambda, T)$ peaks at about 0.5 μm, the middle of the visible spectrum (see Example 12.1.3).

Another question I can ask and answer is "At what temperature is the number of photons having a wavelength of λ_2 or shorter $1/1.6 \times 10^9$ the total number?" This may seem a random question, but it is relevant to the calculation of certain temperatures discussed later. I am looking for a temperature at which $n_\gamma(0, \lambda_2, T)/n_\gamma$ is 6.25×10^{-10}. So I have to solve the equation

$$\frac{n_\gamma(0, \lambda_2, T)}{n_\gamma(T)} = 6.25 \times 10^{-10} \quad \text{or} \quad \int_0^{y_2} \frac{dy}{y^4(e^{1/y} - 1)} = 2.404(6.25 \times 10^{-10}) = 1.5 \times 10^{-9}.$$

The equation can be solved numerically, with the solution $y_2 \approx 0.037$ and

$$\lambda_2 T = \frac{hcy_2}{k_B} = 5.33 \times 10^{-4} \text{ m K}. \tag{12.16}$$

I'll use this equation a couple of times later. ■

Intensity or brightness.

An important quantity is the **intensity** I_γ (sometimes called **brightness**) of the radiation coming out of a small hole on the surface of a cavity held at temperature T. Let the hole have area Δa and an outward normal $\hat{\mathbf{e}}_n$. Then, Equation (A.4) gives

$$I_\gamma = \frac{\Delta\phi}{\Delta a} = \mathbf{J} \cdot \hat{\mathbf{e}}_n = u_\gamma \mathbf{v} \cdot \hat{\mathbf{e}}_n = u_\gamma c\, \hat{\mathbf{e}}_v(\theta, \varphi) \cdot \hat{\mathbf{e}}_n,$$

where $\hat{\mathbf{e}}_v(\theta, \varphi)$ is the unit vector in the direction of motion of the photon impinging on the area, and its dependence on both spherical angles is explicitly shown. Let's orient our coordinate system so that $\hat{\mathbf{e}}_n$ is along the z-axis. Then, for each bundle of photons impinging on the hole in the solid angle $d\Omega$,

$$dI_\gamma = u_\gamma c \cos\theta\, d\Omega, \quad 0 \le \theta \le \pi/2.$$

The restriction for θ is because I'm interested in the radiation that goes *out of* the cavity. The photons move in random directions, so I'll have to average the above expression over all solid angles. Thus,

$$\langle I_\gamma \rangle \equiv \frac{u_\gamma c \int_0^{\pi/2} d\theta \int_0^{2\pi} d\varphi \cos\theta \sin\theta\, d\theta\, d\varphi}{4\pi} = \tfrac{1}{4} u_\gamma c.$$

It is common to ignore the average sign and write

$$I_\gamma = \frac{c}{4}\left(\frac{\pi^2 k_B^4}{15\hbar^3 c^3} T^4\right) = \frac{\pi^2 k_B^4}{60\hbar^3 c^2} T^4 \equiv \sigma_B T^4, \quad \sigma_B = \frac{\pi^2 k_B^4}{60\hbar^3 c^2}, \tag{12.17}$$

Stefan-Boltzmann constant.

where σ_B is the **Stefan-Boltzmann constant** and in SI units has the numerical value of

$$\sigma_B = 5.67 \times 10^{-8} \text{ W m}^{-2} \text{ K}^{-4}. \tag{12.18}$$

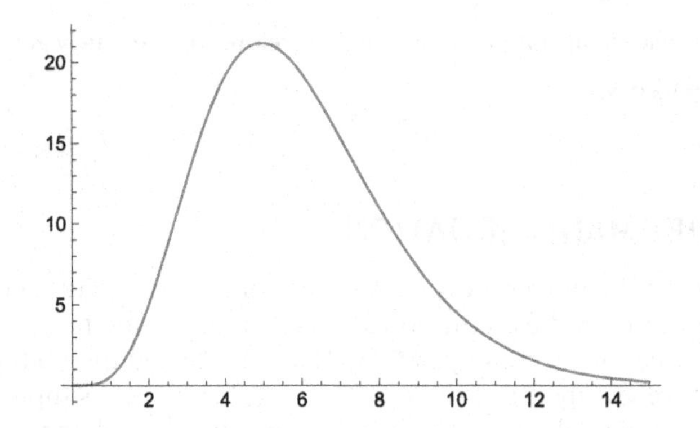

FIGURE 12.1 Plot of $x^5/(e^x - 1)$. This is a typical plot of the black body radiation curve. The maximum occurs at $x_{\max} = 4.965$.

Sometimes *spectral intensity* is used instead of spectral energy density. The difference is just the factor of $\frac{1}{4}c$. So, in terms of frequency,

$$I_\gamma(\omega, T) = \tfrac{1}{4}cu_\gamma(\omega, T) = \frac{\hbar}{4\pi^2 c^2}\frac{\omega^3}{e^{\hbar\omega/k_B T} - 1}, \tag{12.19}$$

and in terms of wavelength

$$I_\gamma(\lambda, T) = \tfrac{1}{4}cu_\gamma(\lambda, T) = \frac{2\pi hc^2}{\lambda^5}\frac{1}{\exp\left(\frac{hc}{k_B T\lambda}\right) - 1}. \tag{12.20}$$

The intensity of radiation, i.e., the amount of energy that a black body held at temperature T radiates per unit area per unit time in the range of frequencies from ω to $\omega + \Delta\omega$ is $I_\gamma(\omega, T)\Delta\omega$. Similarly for wavelength.

Let $x = \dfrac{hc}{k_B T\lambda}$ and write $I_\gamma(\lambda, T)$ in terms of x:

$$I_\gamma(x, T) = \frac{2\pi(k_B T)^5}{ch^4}\frac{x^5}{e^x - 1}.$$

The plot of $I_\gamma(x, T)$ as a function of x is shown in Figure 12.1. This is the black body radiation curve in terms of (inverse) wavelength. The peak occurs at $x_{\max} = 4.965$, or

Wien displacement law.

$$\lambda_{\max} T = \frac{hc}{4.965 k_B} = 0.0029 \text{ m K}, \tag{12.21}$$

which is called the **Wien displacement law**. It determines the surface temperature of a black body radiator once the wavelength of the peak of its curve is determined.

Example 12.1.3. The Sun is not a perfect black body, but it approximates it pretty well. It is known that the peak wavelength of the Sun occurs in the green range at about

500 nm. The surface temperature of the Sun is therefore approximately given by

$500 \times 10^{-9} T = 0.0029$,

or $T = 5800$ K. ■

12.2 FRIEDMANN EQUATION

The dynamics of the universe depends on its constituents. With gravity being the dominant force and extreme conditions holding at the beginning of the universe, we need the machinery of GTR. However, a (slightly modified) Newtonian theory of gravity can also describe the behavior of a simple model of the universe surprisingly well. The modification involves giving mass to the energy of EM radiation and making it a source of gravity.

I borrow certain results of GTR that are well known. One is that the universe is expanding, in the sense that typical distances between (distant) galaxies are increasing. Here is an elaboration of the notion of expansion.

Note 12.2.1. *Meaning of expansion: Look at two distant galaxies at some time and measure their distance* R_1. *Look at them at a later time and measure their distance* R_2. *Expansion of the universe means that* $R_2 > R_1$.

It turns out that the rate of expansion is proportional to the "scale" of the universe. Denoting the *scale factor* at time t by $a(t)$ and the distance between two distant galaxies at the same time by $R(t)$, I can write

$$\frac{R(t_2)}{R(t_1)} = \frac{a(t_2)}{a(t_1)}. \tag{12.22}$$

Note that $a(t)$ has no numerical value and only its ratio at different times has any meaning.

An important astronomical observation that verified the expansion of the universe took place in the late 1920s and early 1930s. This observation demonstrated that the relative velocity of two distant galaxies is proportional to their separation:

$$v = H_0 R \quad \text{or} \quad \frac{dR}{dt} = H_0 R$$

Hubble constant, Hubble parameter, and Hubble law.

where H_0 is called the **Hubble constant**. GTR had already predicted this relation, not only for now, but for all time. In terms of the scale factor, this prediction can be written as

$$\dot{a}(t) \equiv \frac{da}{dt} = H(t)a(t), \tag{12.23}$$

where $H(t)$ is called the **Hubble parameter** and (12.23) is called the **Hubble law**. In the following discussion, I'll set the time of the big bang to $t = 0$ and

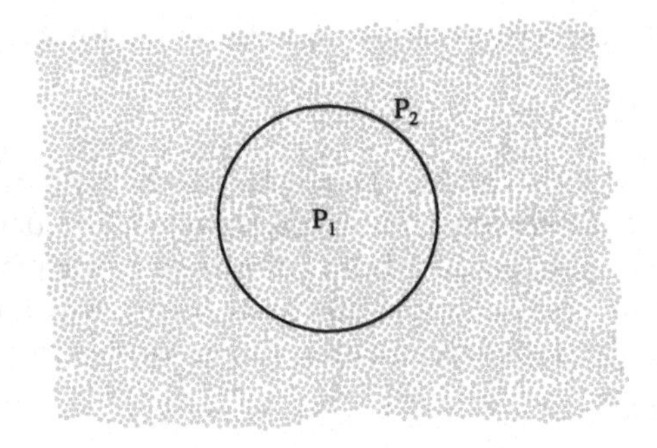

FIGURE 12.2 Two points P_1 and P_2 and the sphere filled with matter and radiation between them.

denote "now" by t_0. The present value of H is given in terms of a dimensionless parameter h:

$$H_0 \equiv H(t_0) = 100h \text{ (km/s)/Mpc},$$

where Mpc stands for mega parsec and each parsec is 3.26156 light years. The latest value of h is 0.673, giving the latest value of H_0 as

$$H_0 = 67.3 \text{ (km/s)/Mpc} = 20.63 \text{ (km/s)/Mly} = 2.18 \times 10^{-18} \text{ s}^{-1}. \quad (12.24)$$

I keep talking about *distant* galaxies because Hubble law does not apply locally. Clearly the Earth-Moon distance is not expanding! Neither is Earth-Sun or stellar distances, or even galactic distances for nearby galaxies: Andromeda galaxy is *approaching* our Milky Way even though it is two million light years away. So, even two million light years is not "distant" enough. When talking about expansion, I'm considering galaxies that are hundreds of million light years away. At such distances, the universe is both **isotropic** and **homogeneous**. The first means that if you look at the (distant) universe at different observation angles, you'll see the same structure. The second means that there are no preferred points in the universe. The universe looks the same from all its (distant) points.

Isotropic and homogeneous universe.

With these assumptions, I can derive the equation that governs the evolution of the universe. Two points P_1 and P_2 of our universe are separated by a distance R. Construct a sphere with its center at P_1 and its radius equal to $\overline{P_1P_2}$ as shown in Figure 12.2, and concentrate on the motion of an object of mass m at P_2. The sphere is filled with matter and radiation exerting a force of gravity on m. The rest of the universe has no effect on m, because m happens to be in a spherical cavity as far as the rest of the universe is concerned, and no gravitational force exists in such a cavity.[2] The mass m moves away from the center of

[2] This should be familiar from your experience with Gauss's law in electrostatics for a charge distribution that is spherically symmetric. Gravity also obeys Gauss's law, and an isotropic universe is spherically symmetric.

the sphere due to the expansion of the universe with a total energy of

$$E = \tfrac{1}{2}mv^2 - \frac{GMm}{R},$$

where M is the mass of the sphere. This mass can be written in terms of the *uniform* density ρ of the material filling the sphere.[3] The density ρ includes both matter and radiation, because both contribute to the gravitational force: matter due to its mass, and radiation, due to its energy, which by $E = mc^2$, has an equivalent mass. Then the last equation can be written as

$$E = \tfrac{1}{2}mv^2 - \tfrac{4}{3}Gm\pi R^2\rho$$

or equivalently as

$$\frac{2e}{R^2} = \left(\frac{v}{R}\right)^2 - \frac{8\pi G}{3}\rho$$

where $e = E/m$.

When the general theory of relativity is applied to the universe, the same equation is obtained except that the numerator of the left-hand side is replaced by $-kc^2$, where $k = 0, \pm 1$. Borrowing this piece of information from GTR, replacing R with a, $v = dR/dt \equiv \dot{R}$ with $\dot{a}$, and rearranging the last equation slightly, we get the **Friedmann equation**:

Friedmann equation.

$$\left(\frac{\dot{a}}{a}\right)^2 = \frac{8\pi G}{3}\rho - \frac{kc^2}{a^2}. \tag{12.25}$$

When $k = 1$, the universe is called **closed**, when it is -1, the universe is said to be **open**, and a **flat** universe corresponds to $k = 0$. Which one of the three our universe corresponds to depends on the density ρ. All observations now point to a flat universe, and I'll restrict my discussion to that case and rewrite (12.25) as

$$\left(\frac{\dot{a}}{a}\right)^2 = \frac{8\pi G}{3}\rho \quad \text{or} \quad \dot{a}^2 = \frac{8\pi G\rho}{3}a^2, \tag{12.26}$$

Vacuum density, dark energy density, cosmological constant.

where $\rho = \rho_m + \rho_\gamma + \rho_\nu + \rho_\Lambda$, each subscript representing one constituent of the universe: ρ_m is the matter (both visible and dark) density, and it is sometimes written as $\rho_m = \rho_b + \rho_{\text{cdm}}$, where b stands for baryon (protons and neutrons) and cdm for *cold dark matter*; ρ_γ is the EM radiation density; ρ_ν is the neutrino density; and ρ_Λ is the so-called **vacuum density** or **dark energy density** and is related to the cosmological constant Λ.

Equations (12.23) and (12.26) give

$$\rho(t) = \frac{3H^2(t)}{8\pi G}.$$

[3] Since the universe is assumed homogeneous and isotropic, its density cannot change from point to point, or from one angle to another.

The present value of ρ is called the **critical density** and, by (12.24), is equal to

Critical density.

$$\rho_{\text{crit}} = \frac{3H_0^2}{8\pi G} = \frac{3 \times (2.18 \times 10^{-18})^2}{8\pi \times 6.674 \times 10^{-11}} = 8.5 \times 10^{-27}\ \text{kg/m}^3. \tag{12.27}$$

The ratios of the present values of various densities to the critical density are widely used and are denoted by the symbol Ω:

$$\Omega_m \equiv \frac{\rho_m(t_0)}{\rho_{\text{crit}}} = \Omega_b + \Omega_{\text{cdm}}, \quad \Omega_\gamma \equiv \frac{\rho_\gamma(t_0)}{\rho_{\text{crit}}}, \quad \Omega_\nu \equiv \frac{\rho_\nu(t_0)}{\rho_{\text{crit}}}, \quad \Omega_\Lambda \equiv \frac{\rho_\Lambda(t_0)}{\rho_{\text{crit}}}. \tag{12.28}$$

For a flat universe that I am considering, $\Omega_m + \Omega_\gamma + \Omega_\nu + \Omega_\Lambda = 1$. The latest measured values of these Ωs are[4]

$$\Omega_m = 0.308 \pm 0.012,\ \Omega_b = 0.0484 \pm 0.001,\ \Omega_{\text{cdm}} = 0.258 \pm 0.011,$$
$$\Omega_\gamma = (5.38 \pm 0.15) \times 10^{-5},\ \Omega_\nu < 0.016,\ \Omega_\Lambda = 0.692 \pm 0.012. \tag{12.29}$$

So, as I mentioned before, all the indications are that the universe is indeed flat. Furthermore, the flatness is not caused by visible matter (baryons), which constitutes less than 5% of the universe, but, for the most part, by the invisible dark matter constituting 26.5%, and the invisible (and mysterious) dark energy making up a whopping 68.5% of the universe.

12.3 THE DENSITIES

To solve (12.26), we need to know the densities as a function of time [or $a(t)$]. For ρ_m, refer to Figure 12.2 and note that the amount of matter in the sphere does not change as the universe expands (or shrinks in reverse time). Therefore,

$$M = \tfrac{4}{3}\pi R^3(t)\rho_m(t) = \tfrac{4}{3}\pi R^3(t_0)\rho_m(t_0)$$

or, replacing $R(t)$ with $a(t)$,

$$\rho_m(t) = \rho_m(t_0)\left[\frac{a(t_0)}{a(t)}\right]^3 = \rho_{\text{crit}}\Omega_m\left[\frac{a(t_0)}{a(t)}\right]^3. \tag{12.30}$$

The radiation density $\rho_\gamma \equiv u_\gamma/c^2$ may be small now, but in the past when the universe was much smaller and hotter, it was much larger as Equation (12.10) indicates. To find ρ_γ in terms of the scale factor, note that the *number* of photons in a given volume of the universe remains constant. Therefore

$$N_\gamma = \tfrac{4}{3}\pi R^3(t)n_\gamma(t) = \tfrac{4}{3}\pi R^3(t_0)n_\gamma(t_0)$$

or

$$\frac{n_\gamma(t)}{n_\gamma(t_0)} = \left[\frac{a(t_0)}{a(t)}\right]^3. \tag{12.31}$$

[4] Taken from Particle Data Group at http://bit.ly/2eVASuo, which was last revised March 2016.

Since $u_\gamma(t) = n_\gamma(t)\langle\varepsilon_\gamma(t)\rangle$ by (12.12), I need to find $\langle\varepsilon_\gamma(t)\rangle$. Photon energy is related to its wavelength. So, I look for how λ changes with time or with the scale of the universe.

Consider two nearby points P_1 and P_2 in a perfectly homogeneous and isotropic universe separated by a distance R. Let $\lambda(t)$ denote the wavelength of the EM wave as it passes by P_1 at time t after the big bang. This same wave reaches P_2 at time $t + \Delta t$ after the big bang. This means that the wave travels the distance R in Δt seconds, so that $R = c\Delta t$. Let v be the relative speed of the two points. Now recall from our discussion of Doppler effect in Section 4.3.3 that the fractional change in the wavelength of an EM wave is simply v/c [see Equation (4.2)]. Thus, I can write

$$\frac{\Delta\lambda}{\lambda} = \frac{\dot{R}}{c} = \frac{\Delta R}{c\Delta t} = \frac{\Delta R}{R}$$

or

$$\frac{d\lambda}{\lambda} = \frac{dR}{R} = \frac{da}{a}.$$

Integrating both sides gives $\ln\lambda = \ln a + \ln C$, or $\lambda = Ca$, where $\ln C$ is the integration constant. Therefore,

$$\frac{\lambda(t_0)}{\lambda(t)} = \frac{a(t_0)}{a(t)}. \qquad (12.32)$$

Since a photon's energy is inversely proportional to its wavelength, I can rewrite the above as

$$\varepsilon_\gamma(t) = \varepsilon_\gamma(t_0)\frac{a(t_0)}{a(t)}.$$

Taking the average of both sides over the photons and noting that the ratio on the right side is constant in the process of averaging, I obtain

$$\frac{\langle\varepsilon_\gamma(t)\rangle}{\langle\varepsilon_\gamma(t_0)\rangle} = \frac{a(t_0)}{a(t)}, \qquad (12.33)$$

Temperature of the universe is inversely proportional to the scale of the universe.

from which I get the very important result,

$$\frac{T(t)}{T(t_0)} = \frac{a(t_0)}{a(t)}, \qquad (12.34)$$

because of (12.12). Although I derived (12.34) for EM radiation, T is actually the temperature of the universe, because all constituents are assumed to be in thermal contact with one another. Multiplying both sides of (12.33) and (12.31), and noting that $u_\gamma = n_\gamma\langle\varepsilon_\gamma\rangle$ and that $\rho_\gamma \equiv u_\gamma/c^2$, yields

$$\frac{\rho_\gamma(t)}{\rho_\gamma(t_0)} = \left[\frac{a(t_0)}{a(t)}\right]^4 \quad \text{or} \quad \rho_\gamma(t) = \rho_{\text{crit}}\Omega_\gamma\left[\frac{a(t_0)}{a(t)}\right]^4. \qquad (12.35)$$

It is convenient to have a relation between densities and temperature. We already know how radiation density varies with temperature. Equation (12.10) was derived without any resort to cosmology, and it gives

$$\rho_\gamma \equiv \frac{u_\gamma}{c^2} = \frac{\pi^2 k_B^4}{15\hbar^3 c^5} T^4 = 8.38 \times 10^{-33} T^4 \text{ kg/m}^3. \tag{12.36}$$

Radiation density actually applies to more particles than just photons. Any particle whose kinetic energy is much larger than its mass (times c^2), and therefore moves close to the speed of light (and many particles did when the universe was extremely hot), acts like a photon. Its density turns out to be *proportional* to ρ_γ. For neutrinos, which have a very small mass, the proportionality constant α_ν is 0.681 (actually, α_ν has not always been 0.681; see Table 12.1). Thus,

$$\rho_\nu = \alpha_\nu \rho_\gamma = 0.681 \frac{\pi^2 k_B^4}{15\hbar^3 c^5} T^4 = 5.7 \times 10^{-33} T^4 \text{ kg/m}^3. \tag{12.37}$$

Electrons and other fundamental particles have their own proportionality constants and

$$\begin{aligned} \rho_{\text{rel}} &\equiv \rho_\gamma + \rho_\nu + \rho_e + \cdots = (1 + \alpha_\nu + \alpha_e + \cdots)\rho_\gamma \equiv \alpha \rho_\gamma \\ \Omega_{\text{rel}} &\equiv \Omega_\gamma + \Omega_\nu + \Omega_e + \cdots = (1 + \alpha_\nu + \alpha_e + \cdots)\Omega_\gamma = \alpha \Omega_\gamma, \end{aligned} \tag{12.38}$$

where "rel" stands for relativistic and α is the sum of the proportionality constants for all particles whose masses could be ignored compared to their KE.

For matter, we don't have an equation similar to (12.10). So, we have to resort to the empirical relation (12.30). Inserting (12.34) in that relation gives

$$\rho_m(T) = \rho_{\text{crit}} \Omega_m \left[\frac{T(t)}{T(t_0)}\right]^3 = 2.68 \times 10^{-27} \left(\frac{T}{T_0}\right)^3 \text{ kg/m}^3, \tag{12.39}$$

where $T_0 \equiv T(t_0)$. Once T_0 is known, I can write an equation for matter similar to (12.36).

The dark energy density is written as

Dark energy density and cosmological constant.

$$\rho_\Lambda = \frac{c^2 \Lambda}{8\pi G}, \tag{12.40}$$

where Λ has the dimension of inverse squared length and is called the **cosmological constant**. Since Λ is a constant, ρ_Λ is constant, i.e., it does not change with time in the evolution of the universe. As the other densities decay with the increasing scale factor $a(t)$, ρ_Λ stays the same, so that it is the dominant density in the present universe.

Substitute all the densities in (12.26) and get

$$\dot{a}^2 = H_0^2 \left[\Omega_m \left(\frac{a_0}{a}\right)^3 + \Omega_{\text{rel}} \left(\frac{a_0}{a}\right)^4 + \Omega_\Lambda\right] a^2,$$

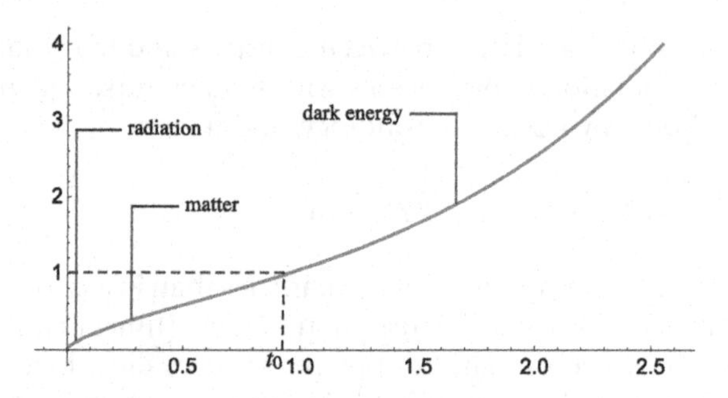

FIGURE 12.3 Plot of $a(t)$ as a function of t with $a(t_0) = 1$. The time is in units of Hubble time $t_H = 1/H_0$. The labels of various regions designate periods in which the density of one constituent of the universe dominated the other two.

where $a_0 \equiv a(t_0)$. Since all scales are compared to a_0, I'll set that equal to 1 and write the master equation of the cosmology of a flat universe as

$$\dot{a}^2 = H_0^2 \left(\frac{\Omega_m}{a} + \frac{\Omega_{\text{rel}}}{a^2} + \Omega_\Lambda a^2 \right). \tag{12.41}$$

This equation can be integrated if you note that it can be written as

$$\frac{a\,da}{\sqrt{\Omega_{\text{rel}} + \Omega_m a + \Omega_\Lambda a^4}} = H_0 dt \equiv \frac{dt}{t_H}$$

Hubble time.

where $t_H \equiv 1/H_0$ is called the **Hubble time** and is approximately 4.6×10^{17} s or 14.6 billion years. Assuming that $a = 0$ at $t = 0$, the solution to this equation is

$$\int_0^a \frac{u\,du}{\sqrt{\Omega_{\text{rel}} + \Omega_m u + \Omega_\Lambda u^4}} = \frac{t}{t_H}. \tag{12.42}$$

12.4 EVOLUTION OF THE UNIVERSE

Ideally, we want to have a as a function of t in closed form. This involves integrating the left-hand side of (12.42) and solving the resulting integrand for a in terms of t, both of which are very messy, if not impossible! However, if we treat t as a function of a, then the left-hand side can be evaluated numerically for any given value of a, and the resulting table displayed as a graph, with t as the horizontal axis. Figure 12.3 shows $a(t)$ as a function of t for the values of the Ωs given in (12.29).

Some general features of the plot are worth mentioning. First note the labels given to various regions of the plot. When one of the densities of the constituents of the universe is much larger than the other two, we say that that constituent is *dominant*. For example, a radiation-dominant universe has

$\rho_\gamma >> \rho_m$ and $\rho_\gamma >> \rho_\Lambda$. The first inequality occurs when

$$\rho_{\text{crit}}\Omega_\gamma \frac{1}{a^4} >> \rho_{\text{crit}}\Omega_m \frac{1}{a^3}$$

by (12.30) and (12.35) and our assumption that $a(t_0) = 1$. This inequality leads to

$$a << \frac{\Omega_\gamma}{\Omega_m} = 2.5 \times 10^{-4},$$

at which scale $\rho_\gamma >> \rho_\Lambda$ as well. So before the scale size of the universe reached 0.00025 its current scale, radiation was the dominant constituent of the universe.

The second feature I want to discuss is how the slope $\dot{a}$ of the graph changes with time. You can see that at the beginning, the slope decreases, meaning that $\ddot{a} < 0$. We have a *decelerating* universe. Then, sometime before the current time t_0, the slope starts to increase, meaning that $\ddot{a} > 0$. So, we are now in an *accelerating* universe. Up until 1998, all the observations pointed to a decelerating universe, consistent with the assumption that the universe was made up of matter and radiation. Then in 1998, two groups of astronomers discovered that not only is the universe not slowing down, but it is actually speeding up.

Since Equation (12.42) cannot be solved in closed form, what is usually done is to solve it assuming that the universe consisted of just the dominant constituent. This means that the density in Equation (12.26) is just the density of the dominant constituent, and only the dominant Ω in (12.42) is retained; the other two are set equal to zero.

Matter Dominance

With $\Omega_{\text{rel}} = 0 = \Omega_\Lambda$, Equation (12.42) becomes

$$\int_0^{a_m} \sqrt{u}\, du = \sqrt{\Omega_m}\, \frac{t}{t_H} \quad \text{or} \quad \tfrac{2}{3} a_m^{3/2} = \sqrt{\Omega_m}\,(t/t_H),$$

or

$$a_m(t) = \left(\tfrac{3}{2}\sqrt{\Omega_m}\right)^{2/3} (t/t_H)^{2/3} = 0.885\,(t/t_H)^{2/3}. \tag{12.43}$$

Radiation Dominance

With $\Omega_m = 0 = \Omega_\Lambda$, Equation (12.42) becomes

$$\int_0^{a_{\text{rel}}} u\, du = \sqrt{\Omega_{\text{rel}}}\, \frac{t}{t_H} \quad \text{or} \quad \tfrac{1}{2} a_{\text{rel}}^2 = \sqrt{\Omega_{\text{rel}}}\, \frac{t}{t_H},$$

or, using a more suggestive subscript

$$a_\alpha(t) = \left(4\alpha\Omega_\gamma\right)^{1/4} (t/t_H)^{1/2}. \tag{12.44}$$

Only those particles whose threshold temperatures (see Example 12.1.1) are below the ambient temperature of the universe at the scale size of a_α contribute to α.

Dark Energy Dominance

The initial condition for the solution to a dark energy dominated universe is different. Instead of having $a = 0$ at $t = 0$, I'll have $a = 1$ at $t = t_0$. With $\Omega_m = 0 = \Omega_{\text{rel}}$, Equation (12.42) yields

$$\int_1^{a_\Lambda} \frac{du}{u} = \sqrt{\Omega_\Lambda}\, H_0(t - t_0) \quad \text{or} \quad \ln a_\Lambda = \sqrt{\Omega_\Lambda}\, H_0(t - t_0),$$

or

$$a_\Lambda(t) = e^{H_\Lambda(t-t_0)}, \quad H_\Lambda^2 \equiv \Omega_\Lambda H_0^2 = \frac{c^2\Lambda}{3}, \tag{12.45}$$

where H_Λ is the Hubble constant for the dark energy-dominant universe.

The dark energy stage of the universe occurred fairly recently, toward the end of the matter dominance, when the universe was roughly $(\Omega_m/\Omega_\Lambda)^{1/3} = 0.77$ of its present size, or, by (12.43), when t was $(0.77)^{3/2} t_0$, i.e., about 4.5 billion years ago. Since I'm interested in the early universe, I'm going to stop any further discussion of the accelerating universe.

12.5 COSMIC BACKGROUND RADIATION

Temperature plays a key role in the development of the early universe, because it determines which particle species were in existence. Equation (12.34) relates temperature at time t to the size of the universe at the same time, and we just found the size of the universe at time t for various dominances. The only thing left in finding the temperature as a function of time is determining T_0, the present temperature of the universe.

In 1964, two radio astronomers, Arno Penzias and Robert Wilson, at Bell Laboratories were using a horn antenna, built earlier for communication via the *Echo* satellite, to measure the intensity of the radio waves coming from the Milky Way. The radio signals from all astronomical objects come in as "noise," much like the statics picked up by radios during a thunderstorm. Distinguishing the signal noise from other spurious noises is not trivial, although it is much easier if the source is small, such as a star. In this case, one can switch the antenna beam back and forth between the source and the empty sky. If there is a detectable difference, it must be due to the source. The enormous size of our galaxy makes such a directional distinction difficult. In order to observe any signal from the galaxy, the antenna had better be as "noise free" as possible.

Role of pigeons in the discovery of the big bang!

By a technique using liquid helium, Penzias and Wilson could reduce the expected spurious noise considerably. They started their observation in the spring of 1964 using a relatively short wavelength of 7.35 cm, where the radio noise from the Milky Way should have been negligible. To their surprise, they detected a strong signal. They changed the antenna direction; the noise was still there. It appeared that the noise was coming from practically every direction. To make sure that the fault was not of the antenna, they dismantled the 20-foot horn, and discovered that some pigeons had nested in the antenna and deposited a "white dielectric material" there! After cleaning the mess and pointing the antenna to the sky in early 1965, they observed very little difference in the level of the noise. The noise did not want to quit.

Theoretical argument points to the existence of a background radiation in the universe.

Puzzled by the persistent noise, Penzias contacted some colleagues, who eventually directed him to Princeton University. It turned out that a group of physicists there had been working on the formation of nuclei at the early universe. James Peebles, the theorist of the group, had argued that the observed structure of the visible universe, indicating a composition of about 99% hydrogen and helium, was a strong evidence for an intense radiation at the early universe. Peebles' calculation revealed that during the first few minutes of the evolution of the universe, the nuclear processes would take place at such a rapid pace that a large fraction of the nucleons would "cook" into the nuclei of heavy elements. The present absence of such elements must be the result of a mechanism that prevented their formation. The only candidate for such a prevention is a very dense and hot background radiation. Peebles estimated the present temperature of this radiation to be around 10 K.

In a subsequent meeting, the Princeton group and Penzias and Wilson decided to publish a pair of companion letters in the *Astrophysical Journal*, in which Penzias and Wilson would announce their observation, and the Princeton group would explain the cosmological implications. The radiation has come to be known as the **cosmic background radiation** (CBR).

If there is a radiation blanketing the universe, what kind of properties does it have? Is it a black body radiation? A black body radiator, by definition, contains radiation and matter in thermal equilibrium at some temperature T. In its early history, the universe consisted of protons, electrons, neutrons, and photons (plus neutrinos). Since photons interact very strongly with charged particles, these early photons were confined within the plasma of the charged particles and were in thermal equilibrium with them. Eventually, the electrons, protons, and neutrons combined to form neutral atoms of hydrogen and helium. Once the plasma was gone, the universe became transparent to photons and they were no longer in thermal equilibrium with matter; they started to roam the universe freely.

When radiation was interacting with the hot plasma, its spectral energy density was described by the Planck radiation law $u_\gamma(\lambda, T)$ in (12.6). Let λ and T be the wavelength and temperature of the radiation at time t when it was confined in the plasma. Let λ' and T' be its wavelength and temperature at time t' after it was decoupled from the plasma. Equations (12.32) and (12.34) give

$$\frac{\lambda}{\lambda'} = \frac{a(t)}{a(t')}, \quad \frac{T}{T'} = \frac{a(t')}{a(t)}, \quad \text{and} \quad \lambda T = \lambda' T'.$$

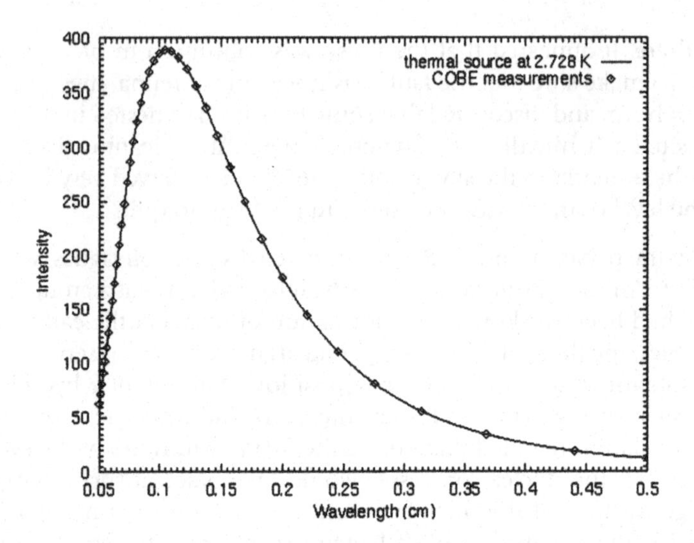

FIGURE 12.4 The black body radiation curve as detected by COBE. The squares are the data points, and the solid curve is the BBR curve corresponding to a temperature of 2.725 K.

Substituting in (12.6), it is trivial to show that

$$u_\gamma(\lambda', T') = \left[\frac{a(t)}{a(t')}\right]^5 u_\gamma(\lambda, T). \tag{12.46}$$

So, except for a scale factor, $u_\gamma(\lambda', T')$ is identical to $u_\gamma(\lambda, T)$. The radiation roaming the universe is indeed a black body radiation.

Penzias and Wilson could detect only a small portion of the curve characteristic of a black body radiator. As it reaches the Earth's surface, CBR loses most of its content to the atmosphere. Nevertheless, Penzias and Wilson could estimate the temperature of CBR to be about 3 K. To see the entire spectrum, and to determine the temperature more accurately, the antenna had to be lifted above the Earth's atmosphere. NASA's COsmic Background Explorer (COBE) satellite was launched on November 18, 1989.

CBR *is* a black body radiator, and its temperature is 2.725 K.

Less than two months later, COBE had sent enough information that the investigators could construct the shape of the radiation curve. Was the curve that of a black body radiation as predicted by Equation (12.46)? In the winter meeting of the American Astronomical Society held outside Washington, DC, on January 13, 1990, one of the principal investigators put up an image (reproduced in Figure 12.4) of the plotted data points on the screen. An eerie silence fell over the audience, which immediately turned into frenzied applause and a standing ovation. The data points fell exactly on the theoretical curve! The CBR *was* a black body radiator, and its temperature was 2.725 ± 0.001 K.

With the present temperature determined, I can now write (12.39) completely in terms of temperature

$$\rho_m(T) = \frac{2.68 \times 10^{-27}}{T_0^3} T^3 \text{ kg/m}^3 = 1.32 \times 10^{-28}\, T^3 \text{ kg/m}^3. \tag{12.47}$$

I can also give a relation between T and t for each stage of the early universe. From (12.34), with $a(t_0) = 1$, I get $T(t) = T_0/a(t)$; and from (12.43) and (12.44), I get

$$\begin{aligned} &\text{matter dominance:} \quad a_m(t) = 0.885(t/t_H)^{2/3}, \qquad T_m(t) = 1.13\, T_0 \left(\frac{t_H}{t}\right)^{2/3} \\ &\text{radiation dominance:} \quad a_\alpha(t) = \sqrt[4]{4\alpha\Omega_\gamma}\,(t/t_H)^{1/2}, \quad T_\alpha(t) = \frac{T_0}{\sqrt[4]{4\alpha\Omega_\gamma}} \left(\frac{t_H}{t}\right)^{1/2}, \end{aligned} \tag{12.48}$$

where $T_0 = 2.725$ K, $t_H = 1/H_0 = 4.59 \times 10^{17}$ s $= 14.5$ Gyr, and α includes the contribution from all the particles (and their antiparticles) present in the universe at time t. A convenient relation that gives t in terms of temperature in the radiation-dominant universe is obtained by inserting all the numerical values in the equation for T_α. Then, you can easily obtain

$$t_\alpha = \frac{2.32 \times 10^{20}}{\sqrt{\alpha}\, T_\alpha^2} \text{ s K}^2. \tag{12.49}$$

The temperature T_{eq} that separates matter and radiation dominance is called the **equilibrium temperature** and is given roughly by setting the two densities equal. This temperature was reached far after all relativistic particles were gone except for photons and neutrinos. Therefore, I'll use $\alpha = 1.681$ for ρ_{rel} in $\rho_m = \rho_{\text{rel}}$. Then (12.30), (12.35), (12.38), and our convention $a(t_0) = 1$ lead to

Equilibrium temperature.

$$\frac{\Omega_m}{[a(t)]^3} = \frac{\alpha\Omega_\gamma}{[a(t)]^4}, \quad \text{or} \quad a_{\text{eq}} = \frac{\alpha\Omega_\gamma}{\Omega_m}, \quad \text{and} \quad T_{\text{eq}} = \frac{T_0\Omega_m}{\alpha\Omega_\gamma}.$$

Substituting the values of Ω_m and Ω_γ from (12.29) yields $T_{\text{eq}} \approx 9300$ K and $a_{\text{eq}} \approx 2.9 \times 10^{-4}$. Equation (12.49) now gives a t of about 65800 years as the time at which matter took over. Thus, for almost 66 millennia, radiation dominated the evolution of the universe. Although a minute fraction of its age, the radiation period was crucial in determining the fate of the universe, because during this period the elements from which every subsequent object was created were formed.

12.6 COSMIC GENESIS

The standard cosmological model places all the known fundamental particles and their antiparticles in equal numbers right after the creation of the universe

Table 12.1 The fundamental particles so far known and discovered. All masses are in MeV and all temperatures are in K. Because quarks cannot be free, their masses—especially the light quarks—are hard to measure, so the values given for u and d are not as accurate as the other quarks.

Name	Symbol	Mass	α	T_{th}
Higgs boson	H^0	125700	0.5	5.4×10^{14}
Neutral weak boson	Z^0	91188	1.5	3.9×10^{14}
Charged weak boson	$W^{\pm}$	80385	3	3.5×10^{14}
Photon	γ	0	1	
Gluon	G	0	8	
Neutrino before e^-e^+ annihilation	ν	≈ 0	2.625	
Neutrino after e^-e^+ annihilation	ν	≈ 0	0.681	
Electron	e	0.511	1.75	2.2×10^9
Muon	μ	105.7	1.75	4.5×10^{11}
Tauon	τ	1777	1.75	7.6×10^{12}
Up quark	u	2.3	1.75	9.9×10^9
Down quark	d	4.8	1.75	2.1×10^{10}
Strange quark	s	95	1.75	4.1×10^{11}
Charm quark	c	1275	1.75	5.5×10^{12}
Bottom quark	b	4500	1.75	1.9×10^{13}
Top quark	t	173000	1.75	7.4×10^{14}

and lets the laws of physics evolve it and its content. Table 12.1 summarizes all the known particles and their relevant properties. The first entry is the famous Higgs boson, the last missing piece of the standard model puzzle discovered in 2012. The next four entries are gauge bosons, particles intermediating interactions among the so-called matter particles, filling up the rest of the table. Z^0, $W^{\pm}$, and γ are responsible for the electroweak force.[5] The eight gluons (4 particles and 4 antiparticles) are responsible for the strong nuclear force, in which only quarks participate. The strong nuclear force, unlike the familiar forces of gravity and electricity, *increases* with the distance between quarks. This leads to the confinement of quarks inside the particles that they make up, like protons and neutrons. The next four particles are leptons. There are three neutrinos (and three antineutrinos). Each of these neutrinos is associated with the remaining leptons: ν_e is associated with the electron, ν_μ with the muon, and ν_τ with the tauon. The two α's reported for neutrinos incorporate this "threeness." The remaining six entries of the table are quarks, which make up hundreds of particles called **hadrons**, only two of which, protons and neutrons, exist naturally, the remaining having been produced artificially in accelerators.

The strong nuclear force *increases* with the distance between quarks.

[5] The unification of the electromagnetic and weak nuclear force was proposed by Weinberg, Salam, and Glashow in 1968; was indirectly confirmed in various observations by 1979, securing a Nobel Prize for the discoverers; and was directly confirmed in 1983 when Z^0 and $W^{\pm}$ were produced in the laboratory at CERN.

The value of α for neutrino needs some explanation. For any spin-$\frac{1}{2}$ particle (all the entries in the first column of Table 12.1 below gluons) $\alpha = 1.75$; however, while other particles have two spin states, neutrinos have only one. Therefore, each neutrino has an α of 0.875, and since there are three kinds of ν, the total α for ν is $3 \times 0.875 = 2.625$. This is all before electron-positron annihilation (see the fourth epoch below). Since neutrinos are extremely weakly interacting, they fell out of equilibrium with the rest of the universe just before electrons and positrons annihilated themselves. Therefore, when the annihilation occurred, the released energy heated the photon gas but not the neutrino gas, causing a reduction of the neutrino temperature relative to the photon temperature, and, therefore, a reduction in α_ν from 2.625 to 0.681.

The last column, the threshold temperature T_{th} of a particle, is the ambient temperature of the universe when the average energy of a photon (12.12) is equal to the mass of that particle (see Example 12.1.1). For a particle of mass m, it is given by

$$T_{\text{th}} = \frac{mc^2}{2.7k_B} = 4.3 \times 10^9\, mc^2 \text{ MeV}^{-1} \text{ K}.$$

When the universe is hotter than this temperature, the particle and its antiparticle coexist with the CBR. Once the universe cools to a lower temperature, the reaction (12.13) goes from right to left, but not from left to right. So another interpretation of T_{th} is the temperature below which all the antiparticles (and an equal number of particles) of a given species disappear.

Below the threshold temperature all antiparticles of a given species disappear.

The history of the early universe cannot be told chronologically. The reason is that events proceed with extraordinary speed at the beginning. The number of things that happen in the first second may be far more than what happens in the next second. Our analysis so far has shown that the important parameter is the temperature T, as it determines the content of the universe, and, therefore, the processes that take place in it.

It is convenient to divide the course of the development of the universe into epochs, with each epoch specified by a range of values of the temperature.[6] Starting with a very simple and symmetric universe, you will see how the laws of physics can predict the content of each epoch and ultimately of the present universe. It is remarkable that the predictions of these laws so closely match the data collected in numerous observations of the universe. There is a caveat! The first three epochs are based on more speculative and less precise physical ideas than the rest of the history of the universe. Our understanding of the universe is based on firmer grounds starting with the fourth epoch in which electron-positron annihilation takes place.

---

[6] I have to warn you that the ranges of the temperature and therefore the epochs themselves are rather arbitrary.

The First Epoch 10^{15} K $< T < \infty$

The present laws of physics cannot explain how the big bang occurred. Such an explanation requires a quantum theory of gravity, which as of this writing is lacking. Ultimately, such a theory combines the general theory of relativity—which works so well in the large scale structures—with the quantum theory, which is the theory of the microscopic phenomena. So we don't know what happened at the *start* of the first epoch, but it is reasonable to assume that sometime during this epoch, the universe consisted of all the known fundamental particles and their antiparticles. All particles are assumed to populate the universe in equal numbers, because when the temperature is sufficiently high, every process takes place in both directions, and if the particles on the left of the process are more numerous, the left-right reaction occurs more frequently and more of the particles on the right are produced than on the left. This balances out the abundance of particles.

As the universe expands and cools down, particles whose threshold temperatures lie above the prevailing temperature disappear. This kind of annihilation raises the question: If all particles are accompanied by exactly the same number of their antiparticles, why did they not annihilate each other completely? After all, our existence indicates that matter survived!

Up until 1964, all fundamental processes seemed to obey the **matter-antimatter symmetry** principle: for every process involving particles (and antiparticles), there is a process in which all the particles are changed to their antiparticles and antiparticles to particles. For example, for the decay of μ^- (see Note 8.4.1 for notation),

$$\mu^- \longrightarrow e^- + \bar{\nu}_e + \nu_\mu,$$

where μ^- is considered the particle, there is the decay

$$\mu^+ \longrightarrow e^+ + \nu_e + \bar{\nu}_\mu.$$

The matter-antimatter symmetry is closely related to the so-called **CP symmetry**, where "C" stands for **charge conjugation** and "P" for **parity**. These are abstract mathematical symmetries: charge conjugation changes the sign of various charges of the particle and parity reflects the position of the particle about the origin. It was assumed that all fundamental processes obey these symmetries. In 1964, however, for reasons that we do not fully comprehend theoretically, certain interactions, ever so slightly, violated the CP symmetry. This important phenomenon, which has come to be known as **CP violation** and secured a Nobel Prize for its discoverers, explains the abundance of matter over antimatter in the present universe.

CP violation is the same as matter-antimatter asymmetry.

In Problem 12.5, I have asked you to show that for every baryon (protons and neutrons) there are approximately 1.6 billion photons. This ratio has persisted ever since the last matter-antimatter annihilation took place. It is also a good estimate of the CP violation in the early universe. Assuming that for every 1.6 billion processes that created antimatter in the very early universe there were 1.6 billion plus one processes that created matter, you can explain the dominance of matter over antimatter in the early universe. Furthermore, the equality of the number of all particles, including photons, and the fact that photons survive, because a photon doesn't have an antiparticle (or that it is its own antiparticle), explain the present ratio of the photon and baryon number densities.

The Second Epoch 10^{12} K < T < 10^{15} K

The second epoch starts about 40 picoseconds after the big bang. At the beginning of this epoch, all the particles in Table 12.1 are present in the universe. By the end of this epoch, all the very heavy particles, whose threshold temperatures are larger than 10^{12}, disappear, so that by the end of the epoch only the lighter particles survive.

Due to the small scale of the universe, the quarks are so close to each other at the beginning of the second epoch that they do not feel any strong force.[7] The quarks are the "free" roamers of the universe. But as the universe expands, the inter-quark distance increases and the quarks begin to feel the strong force and bag themselves and the gluons inside hadrons. The following example gives a very crude estimate of the **quark confinement temperature**.

Quark confinement temperature.

Example 12.6.1. QUARK CONFINEMENT TEMPERATURE
Quarks are more or less free inside a (spherical) proton, which has a mass of 1.67×10^{-27} kg and a radius of about 0.85×10^{-15} m. These two numbers give a density of

Estimating quark confinement temperature.

$$\rho = \frac{1.67 \times 10^{-27}\ \text{kg}}{\frac{4}{3}\pi(0.85 \times 10^{-15}\ \text{m})^3} = 6.5 \times 10^{17}\ \text{kg/m}^3$$

for the proton. As long as the quarks are separated by distances smaller than a typical proton radius, they are free. In other words, as long as the density of the universe is larger than the density calculated above, the quarks are free. However, once the density of the universe falls below the above density, they become bound. Now the question is "At what temperature is the density of the universe equal to the ρ given above?"

Assume that all particles are relativistic and we are close to the end of the second epoch so that all particles with a threshold temperature larger than 5×10^{12} are gone. Then using Table 12.1, I can set

$$\begin{aligned}\alpha &= \alpha_\gamma + \alpha_G + \alpha_\nu + \alpha_e + \alpha_\mu + \alpha_u + \alpha_d + \alpha_s \\ &= 1 + 8 + 2.625 + 5 \times 1.75 = 20.375.\end{aligned}$$

With this α, Equations (12.38) and (12.36) yield

$$6.5 \times 10^{17} = \alpha\rho_\gamma = 20.375 \times 8.38 \times 10^{-33} T^4 \quad \text{or} \quad T = 1.4 \times 10^{12}\ K.$$

This is surprisingly close to 2×10^{12} K obtained by a far more elaborate theoretical calculation involving supercomputers. ■

---

[7] Recall that the strong force between two quarks *increases* with increasing distance.

Confinement time.

Thus, the most significant event of the second epoch is the confinement of quarks inside hadrons, which occurred when the universe was about 13 microseconds old.

The Third Epoch 10^{10} K $< T <$ 10^{12} K

The universe is about 87 microseconds old now and consists almost entirely of electrons, muons, neutrinos and their antiparticles, and, of course, photons. Hadrons, which were formed out of quarks in the second epoch, disappear by the beginning of the third epoch for either (or both) of the following two reasons: decay or particle-antiparticle annihilation.

Recall that (see Example 2.1.3) if N_0 decaying particles are present at time $t = 0$, then at time t, the number is reduced to $N(t)$ given by

$$N(t) = N_0 e^{-t/\tau} = N_0 2^{-t/t_{\text{half}}}, \quad t_{\text{half}} = \tau \ln 2, \tag{12.50}$$

where τ is the **mean life** and t_{half} is the **half-life** of the particle.

The lightest hadrons, the pions, even though their threshold temperature is below 10^{12} K, have such a short mean life that the majority of them decay by the end of the second epoch (see Problem 12.9). The heavier hadrons, whose threshold temperatures are above 10^{12} K, disappear before the start of the third epoch by decay or by annihilating their antiparticles. A small number of protons and neutrons (about 1 for every 1.6 billion of the other constituents) are immersed in the sea of other particles. These protons and neutrons are the bound states of the minute "extra" quarks (without their corresponding antiquarks) produced in the first epoch.

In the middle of this epoch, when the temperature is 10^{11} K, the muons disappear (μ^+ and μ^- annihilate each other), and the only survivors will be electrons, positrons, neutrinos and antineutrinos, the photons, and a small "contamination" of protons and neutrons. The small "extra" number of muons (created in the first epoch due to matter-antimatter asymmetry) decay very rapidly due to their very short half-life. The density of the universe right after μ^+-μ^- annihilation is approximately 4.5×10^{12} kg/m^3. This is so large[8] that even neutrinos, which can easily pass through layers of lead that are light years thick, are trapped.

Equality of proton and neutron populations at the beginning of the third epoch.

Reactions such as $p + e^- \leftrightarrow n + \nu_e$ and $n + e^+ \leftrightarrow p + \bar{\nu}_e$ keep the number of protons and neutrons equal. Here is why. Suppose there are twice as many protons as neutrons. Then—as there is no shortage of the electrons, positrons, and neutrinos—it is twice as likely for the first reaction to go from left to right (than right to left) and the second reaction to go from right to left (than left to right), causing more protons to turn into neutrons than vice versa. Ultimately, the numbers become equal. Therefore, the third epoch starts with equal populations of protons and neutrons.

---

[8] A grain of sand made of this material would weigh several tons!

However, as the end of the third epoch approaches, the p–n conversion becomes asymmetric. The reason is in Equation (8.20). In the rest frame of the proton, for *neutron production* reactions $p + e^- \to n + \nu_e$ and $n + e^+ \leftarrow p + \bar{\nu}_e$, (8.20) becomes

$$m_p^2 + 2m_p\mathcal{E}_1 = m_n^2 + 2\mathcal{E}_3(\mathcal{E}_1 + m_p) - 2\mathcal{E}_1\mathcal{E}_3\cos\theta,$$

where $\mathcal{E}_1$ is the energy of the incident particle (electron or antineutrino), and $\mathcal{E}_3$ is the energy of the final *light* particle (positron or neutrino). In the rest frame of the neutron, for *proton production* reactions $p + e^- \leftarrow n + \nu_e$ and $n + e^+ \to p + \bar{\nu}_e$, (8.20) becomes

$$m_n^2 + 2m_n\mathcal{E}_1 = m_p^2 + 2\mathcal{E}_3(\mathcal{E}_1 + m_n) - 2\mathcal{E}_1\mathcal{E}_3\cos\theta,$$

where $\mathcal{E}_1$ is the energy of the incident positron or neutrino, and $\mathcal{E}_3$ is the energy of the final electron or antineutrino. In both equations, I have ignored the electron and neutrino masses compared to their KEs and to the masses of protons and neutrons.

Let m denote m_p or m_n (they are very nearly equal) and Δm their (very small) mass difference. Then, with obvious superscript notations, the last two equations can be expressed as

$$\mathcal{E}_1^n \approx \frac{\mathcal{E}_3 + \Delta m}{1 - (\mathcal{E}_3/m)(1 - \cos\theta)} \quad \text{for neutron production}$$
$$\mathcal{E}_1^p \approx \frac{\mathcal{E}_3 - \Delta m}{1 - (\mathcal{E}_3/m)(1 - \cos\theta)} \quad \text{for proton production.} \tag{12.51}$$

Since $\mathcal{E}_1^n > \mathcal{E}_1^p$, it takes more energy to produce neutrons than protons, and as the universe cools, neutron production becomes less and less likely. There is also the decay of neutrons into protons, $n \to p + e^- + \bar{\nu}_e$, which decreases the number of neutrons and increases the number of protons. However, this decay is too slow to have any effect on the p–n imbalance, at least in the present epoch. Detailed calculations beyond the scope of this book show that by the end of the third epoch, the protons outnumber the neutrons by a factor of three, i.e., out of all the baryons, 25% are neutrons and 75% protons.

Protons outnumber neutrons 3 to 1 at the end of the third epoch.

The Fourth Epoch 10^9 K $< T <$ 10^{10} K

About one second has passed since the big bang. The universe is still composed of electrons, positrons, three kinds of neutrino, photons, and a small "contamination" of protons and neutrons. The density has fallen down to the point that the neutrinos are no longer trapped. They *decouple* from the rest of the universe.

A little after the neutrino decoupling, at a temperature of about 2.2×10^9 K, when the universe is about 20 seconds old, the electrons and positrons annihilate each other, and the energy released as a result of the annihilation heats up the photons, making the photon gas hotter than the decoupled neutrino gas, thus reducing α_ν. The end of this epoch coincides with the end of the antimatter era. From now on, the matter of the universe is composed of a small number of protons and neutrons (formed in the second and third epochs out of the extra u and d quarks produced in the first epoch via CP violation), and a small number of electrons (also produced in the first epoch) that survived the e^+-e^- annihilation in the present epoch, immersed in a sea of photons and neutrinos.

Number of protons is equal to the number of electrons . . . always!

Since the electric charge is strictly conserved, and no physical process can create a positive charge without an equal negative charge, the number of electrons equals the number of protons, and this equality will hold for eternity. Any process that changes the number of protons will also change the number of electrons by the same amount. One such process is the neutron decay, which becomes more important as the age of the universe becomes comparable to the half-life of a neutron, 610 seconds.

The Fifth Epoch 10^8 K $< T <$ 10^9 K

The dominant constituents of the universe are now photons and the neutrinos. The universe is about three minutes old, and the time span is long enough that neutron decay plays a significant role in the abundance of p and n. Example 12.6.2 shows that during this time, about 20% of the neutrons turn into protons via beta decay. If this were the only process converting neutrons to protons, the fraction of neutrons would go from 0.25 to $0.25 - 0.25 \times 0.2 = 0.2$, or 20% neutrons, and the remaining 80% protons. However, the other processes of the third epoch also occur and reduce neutrons by another 28%. So, by the beginning of the fifth epoch 48% of the neutron population turns into protons, bringing the fraction of neutrons to $0.25 - 0.25 \times 0.48 = 0.13$, or 13%. The nuclear particles now consist of approximately 87% protons and 13% neutrons.

Shortly after the start of this epoch, a little over 3 minutes after the big bang, something remarkable happens. Every two neutrons pair up with two protons to form a helium nucleus. This means that 26% of the nucleons end up inside the helium nuclei, with free protons constituting the remaining 74%. Since nucleons have almost equal masses, the mass of all nuclear matter in the universe is now 26% helium nuclei and 74% protons.

Neutron decay in 5th epoch.

Example 12.6.2. The neutron beta decay $n \to p + e^- + \nu$, like all other decays, obeys the exponentially decaying law (12.50) with $t_{\text{half}} = 610$ s. The helium formation starts when the temperature reaches 950 million K, corresponding to 198.3 seconds after the big bang and 197.3 seconds after the third epoch, in which protons outnumbered neutrons three to one. The contribution to further reduction of the neutrons coming from their decay can be calculated using (12.50):

$$N(t) = N_0 2^{-197.3/610} = 0.80 N_0.$$

This shows that 20% of the neutrons present at the end of the third epoch decay by the beginning of the fifth epoch. ■

Helium nuclei are formed a little over 3 minutes after the big bang.

The formation of helium nuclei, although quick, is not an instantaneous process. There are intermediate steps that need to be taken before helium can be created. Even though the protons and neutrons had a lot of opportunity to fuse at earlier times, they did not form helium. Why? The answer is the **deuteron bottleneck**.

In order to make helium, a proton and a neutron must fuse together to form a deuteron: $p+n \to D+\gamma$. Then two deuterons fuse into a helium nucleus $D+D \to ^4\text{He}+\gamma$. And while ^{4}He is very tightly bound, a deuteron is not. It takes only 2.224 MeV to break up the deuteron into a proton and a neutron.[9] At earlier times, although deuterons *were* formed, they were immediately broken apart by the energetic photons, i.e., deuteron production went both ways: $p+n \leftrightarrow D+\gamma$. What exactly is the temperature at which deuterons are made and not broken apart? What is the threshold temperature for D-formation?

Based on Equation (8.33), you might think that the photons in the reaction $\gamma + D \to p+n$ need to have an *average* energy equal to the binding energy of deuterons, which is 2.224 MeV. If you make such an assumption, then $2.7k_BT = 2.224$ MeV gives a threshold temperature of about 9.5 billion K, corresponding to the beginning of the *fourth* epoch! So, why have I waited until now to talk about deuteron formation? The reason is the following.

Recall that matter-antimatter asymmetry increases the relative number densities of photons and baryons by a factor of 1.6 billion. When the *average* photon energy is 2.224 MeV, there are far more photons with this energy (and higher) than I need. Just one photon per baryon will do. So, I have to find the temperature at which I have a sufficient number of photons energetic enough to break up the deuterons that may be formed. "Energetic enough" means having an energy larger than the binding energy of the deuteron. So if just $1/(1.6 \times 10^9)$ or 6.25×10^{-10} of the population of photons has energies 2.224 MeV or higher, I have one energetic photon for every nucleon. Now you can see that the random question I asked in Example 12.1.2 was not random after all!

That example calculated the product $\lambda_2 T$. Here λ_2 corresponds to an energy of 2.224 MeV. Thus, using Planck's quantization formula $E = hc/\lambda$, I get

$$\lambda_2 = \frac{hc}{E} = \frac{(6.63 \times 10^{-34})(3 \times 10^8)}{2224000(1.6 \times 10^{-19})} = 5.59 \times 10^{-13} \text{ m},$$

and then Equation (12.16) gives

$$T = \frac{5.33 \times 10^{-4}}{5.59 \times 10^{-13}} = 9.5 \times 10^8.$$

This is an order of magnitude smaller than the first estimate and places the deuteron formation in the fifth epoch.

---

[9] This is about a third the energy needed to pull a single nucleon out of ^{4}He.

Because ^{4}He is so tightly bound, it is energetically a more favorable final product than other multi-nucleon elements, which are also formed, though in much smaller quantities. For example, some deuterons may survive; or a proton can fuse with a deuteron to form ^{3}He; or a D can fuse with a ^{4}He to form a lithium. Although these other elements do form at the temperature of 950 million Kelvin, most of them disintegrate immediately, because of their small binding energies, leaving only a minute trace behind.

The formation of ^{4}He via the deuteron bottleneck, as predicted by the laws governing nuclear interactions, points to

Note 12.6.3. *Two important cosmological predictions:*
(a) A universal radiation must exist to prevent the formation of deuterons at earlier times.
(b) The ratio of the hydrogen mass abundance to helium mass abundance in the universe should be 74% to 26%.

If radiation did not exist, we have to look at other dissociation mechanisms for deuteron. One such mechanism is $e^- + D \to 2n + \nu$. To calculate the minimum electron energy for this process, I go to the CM frame. Since the total momentum is zero, Equation (8.29) becomes

$$\mathcal{E}_e + \mathcal{E}_D = \mathcal{E}_{n_1} + \mathcal{E}_{n_2} + \mathcal{E}_\nu.$$

The deuteron and neutrons are much heavier than the other two particles. Thus, I can set $\mathcal{E}_D \approx m_D$, $\mathcal{E}_{n_1} \approx m_n \approx \mathcal{E}_{n_2}$. Then, the equation above yields

$$\mathcal{E}_e + m_D = 2m_n + \mathcal{E}_\nu, \quad \text{or} \quad \mathcal{E}_e = 2m_n - m_D + \mathcal{E}_\nu.$$

Since I'm looking for the minimum $\mathcal{E}_e$, I let $\mathcal{E}_\nu \to 0$ and get

$$\mathcal{E}_e = 2m_n - m_D = 2 \times 939.565 \text{ MeV} - 1875.613 \text{ MeV} = 3.517 \text{ MeV}.$$

Radiation is a necessary ingredient of the universe.

This is 7 times electron mass, so I won't be too far off if I assume that the electron is relativistic. Problem 12.13 shows that this energy turns out to be insufficient, but a better estimate (also discussed in that problem) would lead to an abundance of 35% helium and 65% protons, in complete violation of observation! Radiation is a necessary ingredient of the universe.

The Last Epoch 3000 K $< T <$ 10^8 K

Approximately 5 hours have passed since the big bang. Ignoring the neutrinos, which have their separate existence, the ^{4}He nuclei, protons, and electrons are immersed in a sea of photons. For every nucleon, there are 1.6 billion photons, and this ratio persists for the rest of the life of the universe. Equations (12.36), (12.37), and (12.47) show that $\rho_\gamma = 0.838$ kg/m^3, $\rho_\nu = 0.57$ kg/m^3, and $\rho_m = 0.00013$ kg/m^3. Thus, the universe is still dominated by radiation and neutrinos, but matter has gained some ground in the competition for dominance: while radiation was 63,500 times more dense than matter 5 hours earlier (verify this!), it is only 6350 more dense now.

Not much happens for most of this epoch. The universe keeps expanding, the wavelength of the photons and neutrinos keep stretching, and because of the interaction between the photons and the plasma of the charged particles of the universe (the positive protons and helium nuclei and the negative electrons), electromagnetic radiation is trapped inside this plasma. Almost 66,000 years after the big bang, radiation surrenders to matter and the latter becomes the dominant contributor to the density. And for another 300 millennia nothing of interest occurs.

At the end of the this epoch, about 380,000 years after the big bang, the photons reach a temperature that is no longer sufficient to ionize the helium and hydrogen atoms that kept being formed all along. The ionization energy for hydrogen is 13.6 eV. If photons have less energy than this, they will not be able to dissociate the hydrogen atoms formed. If you use 13.6 eV as the average energy of photons, you'll find a temperature of 58,570 K. But that is an overestimate because there are 1.6×10^9 photons for every proton and many of them have much higher energies than we need. The right question to ask is, "At what temperature is the number of photons having an energy of 13.6 eV or higher $1/1.6 \times 10^9$ of the total number?" This is another illustration of why the random question asked in Example 12.1.2 was not so random!

The wavelength corresponding to 13.6 eV is (using the Planck formula)

$$\lambda_2 = \frac{hc}{E} = \frac{(6.63 \times 10^{-34})(3 \times 10^8)}{13.6(1.6 \times 10^{-19})} = 9.14 \times 10^{-8} \text{ m}.$$

Therefore, from (12.16), I get

$$T = \frac{5.33 \times 10^{-4}}{9.14 \times 10^{-8}} = 5832 \text{ K} \tag{12.52}$$

corresponding to about 170,000 years after the big bang. Detailed calculations beyond the scope of this book yield 380,000 years as the age of the universe when atoms were formed. The atoms could no longer interact with the photons, and the latter decoupled from matter, roaming the universe unnoticed until 1964, when they were picked up by the horn-shaped antennas of two radio astronomers on Earth. The helium and hydrogen gas mixture was now ready to form galaxies, stars, planets, and ... us.

12.7 PROBLEMS

12.1. Derive (12.6) from (12.4), and (12.9) from (12.7).

12.2. Consider the reaction $\gamma + \gamma \longrightarrow p + \bar{p}$, which can occur if the initial photons are extremely energetic. Here p is a generic particle and $\bar{p}$ its antiparticle.

(a) Use Equation (8.24) to show that

$$E_1 E_2 = \frac{m_p^2}{\sin^2(\theta/2)}$$

where m_p is the mass of p (or $\bar{p}$), E_1 and E_2 are the energies of the photons, and θ is the angle between the direction of motion of the initial photons.

(b) What value of θ minimizes $E_1 E_2$? If $E_1 > E_2$, what is the total momentum of the pair on the right-hand side (in terms of E_1 and m_p) for the θ that minimizes $E_1 E_2$?

(c) Using the θ in (b), show that the value of E_1 that minimizes the total energy $E_1 + E_2$ is $E_1 = m_p$. What is the total momentum of the pair on the right-hand side now?

12.3. In Example 12.1.3, I calculated the surface temperature of the Sun assuming that it was (approximately) a black body radiator. Using the result of that example and Equation (12.17),

(a) find the brightness of the Sun.

(b) How many 100-watt light bulbs do you have to put on a square meter to give this brightness?

(c) The radius of the Sun is 700,000 km. What is the power output of the Sun?

(d) If the mass of the Sun is the source of this energy, how many kilograms of its mass does the Sun lose every second?

(e) The mass of the Sun is 2×10^{30} kg. Assuming that its mass depletion rate is proportional to its mass, find the present proportionality constant.

(f) Assuming that the constant in (e) does not change with time, find $m(t)$, the Sun's mass as a function of time.

(g) How many years do you have to wait before the Sun loses half of its mass (and is no longer a shining star)? This is a huge overestimate! The Sun dies far sooner than this due to nuclear processes that give off colossal explosive (and implosive) energies.

12.4. Using the present temperature of the universe and Equations (12.27) and (12.36), calculate Ω_γ and compare it with the value given in (12.29).

12.5. In this problem you are asked to compare the present number of photons and baryons in the universe.

(a) Using the present temperature of the universe, calculate n_γ, the present photon number density.

(b) From (12.27) and the value of Ω_b in (12.29), calculate the present ρ_b.

(c) Ignore the negligible mass difference between a proton and a neutron and assume that only protons contribute to ρ_b. Moreover, since protons are not moving fast, you can assume that their energy is just $m_p c^2$. With $m_p = 1.67 \times 10^{-27}$ kg, find n_b, the present baryon number density.

(d) How many photons are there in the universe for each baryon?

12.6. Decide which particles of Table 12.1 were present at the beginning of the second epoch.

(a) Calculate α assuming that whatever particle you have is relativistic.
(b) How old was the universe at the beginning of the second epoch?

12.7. In this problem, you'll calculate the density of the universe at the beginning of the second epoch.

(a) Find α and, from that, the density of the universe.
(b) The Earth has a mass of 5.97×10^{24} kg. What would the radius of the Earth be if it were composed of such a dense material?
(c) At the confinement temperature, the density of the universe dropped to 6.5×10^{17} kg/m^3. What would the radius of the Earth be if it were composed of this material?

12.8. In this problem you'll calculate the age of the universe when quarks were confined at a temperature of 2×10^{12} K.

(a) Look at Table 12.1 and decide which particles were present just before that temperature was reached. Calculate α.
(b) Now use (12.49) to calculate the age of the universe.

12.9. Assume that at a temperature of 2×10^{12} K, quarks and gluons are bagged into hadrons, the lightest of which are pions with spin zero (therefore $\alpha = 0.5$, just like the Higgs boson). There are three kinds of pion: neutral pions π^0 (with no antiparticle) with a mass of 135 MeV and two charged pions (one being the antiparticle of the other) $\pi^{\pm}$ with a mass of 139.57 MeV. The next lightest hadron has a mass of approximately 500 MeV.

(a) What is the threshold temperature for π^0? For $\pi^{\pm}$? For the next lightest hadron?
(b) Determine the particle content of the universe at the beginning of the third epoch and from that, α.
(c) Now use (12.49) to calculate the age of the universe at the beginning of the third epoch.
(d) The mean life of π^0 is 8.5×10^{-17} s, and that of $\pi^{\pm}$ is 2.6×10^{-8} s. What fraction of charged pions decay between their production at confinement time and the beginning of the third epoch? What fraction of neutral pions?
(e) In light of (d), do you have to reconsider (b) and (c)? If so, recalculate them and find the new value for the age of the universe at the beginning of the third epoch.

12.10. Right after μ^+-μ^- annihilation, the temperature is 10^{11} K.

(a) Which particles are present then? Ignore the baryon "contamination." What is the value of α?
(b) How old is the universe?

(c) What is the density of the universe assuming that it is all ρ_{rel}? Do you have to worry about the contribution from matter density? Use (12.47) to find out!

(d) Assume that a grain of sand is a cube of side one millimeter. What would the mass of this grain be if it had a density you found in (c)?

12.11. Derive Equation (12.51) from the equations preceding it.

12.12. Things happen very quickly when the universe is very young. To see how quickly, suppose two points of the universe are one millimeter apart 10^{-30} second after the big bang.

(a) Which particles of Table 12.1 are present at this time?[10] Use (12.48) to find $a_\alpha(t)$.

(b) One second after the big bang, the fourth epoch starts. What is α now? What is $a_\alpha(t)$?

(c) How far apart are the two points now? Recall that the ratio of distances is the same as the ratio of the scale factors.

(d) Compare the distance in (c) with the Earth-Sun distance. From 1 mm to this distance in just one second! Now you can appreciate why we call it a "bang," a "BIG" bang!

12.13. In this problem, you are asked to estimate the helium-proton abundance if $e^- + D \to 2n + \nu$ were responsible for deuteron dissociation. I have already calculated the minimum energy for the electron. It is 3.517 MeV.

(a) What is the wavelength associated with this energy?

(b) Assuming that electrons (and positrons) outnumber protons and neutrons 1.6 billion to 1, use Example 12.1.2 to show that the threshold temperature for deuteron formation is 1.5×10^9 K. However, there is a problem here. This temperature is less than the e^+-e^- annihilation temperature. At 1.5×10^9 K, the electrons do not outnumber baryons 1.6 billion to 1. This means that 3.517 MeV is not sufficient for deuteron production. Let's assume that the minimum energy corresponds to the e^+-e^- annihilation temperature of 2.2×10^9 K.

(c) If the content of the universe is electrons, positrons, photons, and neutrinos, what is α? Use (12.49) to estimate the age of the universe when deuterons start to form at 2.2×10^9 K.

(d) Show that from the end of the third epoch, about 2.2% of the neutrons decay into protons. Even if the other contributions to neutron depletion remain at 28% (the actual number is smaller than this because 28%

---

[10] Hint: How old is the universe at the beginning of the second epoch?

corresponds to the entire fifth epoch), the total depletion rate is 30.2%. Now show that the helium-proton abundance is 34.9% and 65.1%, respectively.

12.14. Calculate the photon and neutrino mass densities at the beginning of the fifth epoch and compare them with matter density. What was the ratio ρ_γ / ρ_m then? Do the same for the beginning of the last epoch.

APPENDIX A

Maxwell's Equations

After Örsted's discovery of the production of magnetic fields from moving electric charges, an intense search was undertaken by many physicists such as Ampère and Faraday to find a connection between electric and magnetic phenomena. By the mid-1800s, a fairly good understanding of electromagnetism was attained which, in the contemporary language of vectors is translated in the following four equations:

The four equations that Maxwell inherited in integral form.

$$(1)\ \oiint_S \mathbf{E}\cdot d\mathbf{a} = \frac{Q}{\epsilon_0} \qquad (2)\ \oiint_S \mathbf{B}\cdot d\mathbf{a} = 0$$
$$(3)\ \oint_C \mathbf{E}\cdot d\mathbf{r} = -\frac{d\phi_m}{dt} \qquad (4)\ \oint_C \mathbf{B}\cdot d\mathbf{r} = \mu_0 I. \tag{A.1}$$

These equations should be familiar from your background in introductory calculus-based physics courses. In a nutshell, the first integral, Gauss's law (or Coulomb's law in disguise), states that the electric flux through the closed surface S is essentially the total charge Q in the volume surrounded by S. The second integral says that the corresponding flux for a magnetic field is zero. The fact that this holds for an arbitrary surface implies that there are no magnetic charges. The third equation, **Faraday's law**, connects the electric field to the rate of change of magnetic flux ϕ_m. Finally, the last equation, **Ampère's law**, states that the source of the magnetic field is the electric current I. The constant ϵ_0 and μ_0 arise from a particular set of units used for charges and currents.

A.1 THE BRIEFEST INTRODUCTION TO VECTOR ANALYSIS

To get to Maxwell's equations, the integral equations of (A.1) need to be transformed into differential equations. This can be accomplished by using the two

Special Relativity. DOI:10.1016/B978-0-12-810411-8.00020-1

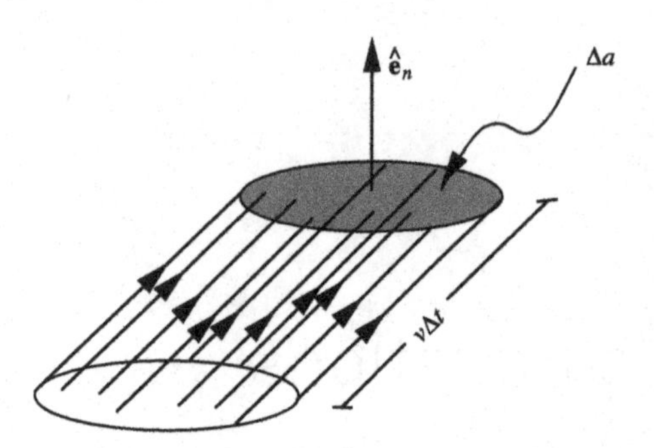

FIGURE A.1 The flux through a small area is just the amount present in the volume divided by time. The volume is $(v\Delta t)(\Delta a)\cos\theta$ where θ is the angle between velocity and the normal to the area.

key theorems of vector analysis, the **Divergence Theorem** and the **Stokes' Theorem**, which are stated without proof here.[1]

Divergence Theorem

Theorem A.1.1. (***Divergence Theorem***). *The surface integral (flux) of any vector field* **A** *through a closed surface S bounding a volume V is equal to the volume integral of the divergence (or flux density) of* **A**:

$$\oiint_S \mathbf{A}\cdot d\mathbf{a} = \iiint_V \nabla\cdot\mathbf{A}\,dV. \tag{A.2}$$

By convention, the direction of the vector $d\mathbf{a}$ is outward from the volume. The **divergence** $\nabla\cdot\mathbf{A}$ of the vector field **A** is defined by

Divergence or flux density.

$$\nabla\cdot\mathbf{A} = \frac{\partial A_x}{\partial x} + \frac{\partial A_y}{\partial y} + \frac{\partial A_z}{\partial z} \tag{A.3}$$

and is the outward flux of **A** through a small closed volume ΔV divided by ΔV in the limit that ΔV goes to zero. That is why $\nabla\cdot\mathbf{A}$ is also called **flux density**.

A useful application of the divergence theorem is to conserved quantities in motion. Suppose a physical quantity (conserved or not) has a density ρ and velocity **v** at a point on an infinitesimal area Δa. Then Figure A.1 shows that the amount of that physical quantity crossing Δa per unit time is simply the amount of the quantity in the volume ΔV shown divided by time, because that amount (and only that amount) crosses the area:

$$\frac{\rho\Delta V}{\Delta t} = \frac{\rho(\mathbf{v}\Delta t)\cdot\hat{\mathbf{e}}_n\Delta a}{\Delta t} = \rho\mathbf{v}\cdot\hat{\mathbf{e}}_n\Delta a \equiv \rho\mathbf{v}\cdot\Delta\mathbf{a}.$$

By definition, this is the flux $\Delta\phi$ of the physical quantity through $\Delta\mathbf{a}$,

$$\Delta\phi = \mathbf{J}\cdot\Delta\mathbf{a}, \qquad \text{where} \quad \mathbf{J}\equiv\rho\mathbf{v}. \tag{A.4}$$

---

[1] Any book on mathematical physics or vector calculus proves these theorems. The book with which I'm most familiar is [Hassani 08], where the divergence theorem is treated on pp. 371–376 and Stokes' theorem on pp. 391–396.

This equation is quite general, even though the path from the bottom to the top area in Figure A.1 is straight. This is because when Δt is small enough, the distances covered by the particles of the quantity are straight line segments. **J** is called the **current density** of the moving physical quantity. For example, if ρ is the electric charge density, then **J** is the familiar electric current density and $\Delta\phi$ is simply the current through $\Delta\mathbf{a}$.

Current density is $\rho\mathbf{v}$.

Now suppose that the physical quantity is conserved. Consider the amount of that physical quantity in a volume V with its bounding surface S. If $Q(t)$ represents this amount at time t, then the rate of depletion of $Q(t)$ must equal the *outward* flux through S:

$$-\frac{dQ}{dt} = \oint\!\!\oint_S \mathbf{J} \cdot d\mathbf{a}. \tag{A.5}$$

Integral form of the continuity equation.

This is the mathematical statement of the conservation of Q, sometimes called the **continuity equation**. The minus sign ensures that *positive* flux gives rise to a *depletion*, and vice versa.

The local or differential form of the continuity equation can be obtained as follows: Write the LHS of Equation (A.5) as[2]

$$-\frac{dQ}{dt} = -\frac{d}{dt}\iiint_V \rho(\mathbf{r},t)\,dV(\mathbf{r}) = -\iiint_V \frac{\partial\rho}{\partial t}(\mathbf{r},t)\,dV(\mathbf{r}),$$

while the RHS, with the help of the divergence theorem, becomes

$$\oint\!\!\oint_S \mathbf{J}\cdot d\mathbf{a} = \iiint_V \nabla\cdot\mathbf{J}\,dV.$$

Together they give

$$-\iiint_V \frac{\partial\rho}{\partial t}\,dV = \iiint_V \nabla\cdot\mathbf{J}\,dV$$

or

$$\iiint_V \left\{\frac{\partial\rho}{\partial t} + \nabla\cdot\mathbf{J}\right\} dV = 0.$$

This relation is true for all volumes V. In particular, you can make the volume as small as you please. Then, the integral will be approximately the integrand times the volume. Since the volume is nonzero (but small), the only way that

[2] Note how the total derivative outside the integral becomes a partial derivative when it moves inside. This is because the result of the integration is only a function of time, while the integrand is, in general, a function of position and time.

the product can be zero is for the integrand to vanish:

Differential form of the continuity equation.

Note A.1.2. *The* ***differential form*** *of the continuity equation is*

$$\frac{\partial \rho}{\partial t} + \nabla \cdot \mathbf{J} = 0, \qquad where \quad \mathbf{J} = \rho \mathbf{v}. \tag{A.6}$$

The second important theorem of vector analysis relates the line integral of a vector field to a certain surface integral:

Stokes' theorem

Theorem A.1.3. (***Stokes' Theorem***). *The line integral of a vector field* **A** *around a closed path* C *is equal to the surface integral of the curl of* **A** *on* any surface *whose only edge is* C. *In mathematical symbols,*

$$\oint_C \mathbf{A} \cdot d\mathbf{r} = \iint_S (\nabla \times \mathbf{A}) \cdot d\mathbf{a}. \tag{A.7}$$

The direction of the normal to the infinitesimal area $d\mathbf{a}$ *of the surface* S *is related to the direction of integration around* C *by the right-hand rule.*

Curl of a vector.

The quantity $\nabla \times \mathbf{A}$, called the **curl** of **A**, is given by

$$\begin{aligned} \nabla \times \mathbf{A} &\equiv \det \begin{pmatrix} \hat{\mathbf{e}}_x & \hat{\mathbf{e}}_y & \hat{\mathbf{e}}_z \\ \frac{\partial}{\partial x} & \frac{\partial}{\partial y} & \frac{\partial}{\partial z} \\ A_x & A_y & A_z \end{pmatrix} \\ &\equiv \left(\frac{\partial A_z}{\partial y} - \frac{\partial A_y}{\partial z}\right)\hat{\mathbf{e}}_x + \left(\frac{\partial A_x}{\partial z} - \frac{\partial A_z}{\partial x}\right)\hat{\mathbf{e}}_y + \left(\frac{\partial A_z}{\partial x} - \frac{\partial A_x}{\partial z}\right)\hat{\mathbf{e}}_z, \end{aligned} \tag{A.8}$$

where $\hat{\mathbf{e}}_x$, $\hat{\mathbf{e}}_y$, and $\hat{\mathbf{e}}_z$ are the unit vectors along the three Cartesian axes.

The vector derivative operator

$$\nabla \equiv \hat{\mathbf{e}}_x \frac{\partial}{\partial x} + \hat{\mathbf{e}}_y \frac{\partial}{\partial y} + \hat{\mathbf{e}}_z \frac{\partial}{\partial z}$$

introduced in both divergence and curl, whose action on a function f yields the gradient ∇f of that function, has some important properties, which are summarized below, and which you should verify:

$$\begin{aligned} \nabla \times (\nabla f) &= 0, \\ \nabla \cdot (\nabla \times \mathbf{A}) &= 0, \\ \nabla \cdot (\nabla f) \equiv \nabla^2 f &= \frac{\partial^2 f}{\partial x^2} + \frac{\partial^2 f}{\partial y^2} + \frac{\partial^2 f}{\partial z^2}, \\ \nabla \times (\nabla \times \mathbf{A}) &= \nabla(\nabla \cdot \mathbf{A}) - \nabla^2 \mathbf{A}. \end{aligned} \tag{A.9}$$

A.2 MAXWELL'S CONTRIBUTION

One of the great contributions of Maxwell was to cast Equations (A.1) in differential form. The differential form of the equations is important because it places particular emphasis on the fields which are the primary objects. The differential form of the first equation follows immediately from the divergence theorem:

$$\text{LHS} = \iiint\limits_V \nabla \cdot \mathbf{E}\, dV, \qquad \text{RHS} = \frac{1}{\epsilon_0} \iiint\limits_V \rho\, dV.$$

Equating the two sides and noting that the equality holds for arbitrary volume, you obtain $\nabla \cdot \mathbf{E} = \rho/\epsilon_0$. Following the same procedure, the second equation yields $\nabla \cdot \mathbf{B} = 0$.

For the third equation of (A.1), Stokes' theorem turns the left-hand side of that equation into

$$\text{LHS} = \iint\limits_S (\nabla \times \mathbf{E}) \cdot d\mathbf{a},$$

while the right-hand side becomes

$$-\frac{d\phi_m}{dt} = -\frac{d}{dt} \iint\limits_S \mathbf{B} \cdot d\mathbf{a} = \iint\limits_S \left(-\frac{\partial \mathbf{B}}{\partial t}\right) \cdot d\mathbf{a},$$

if we assume that the change in the flux comes about solely due to a change in the magnetic field. This makes it possible to push the time differentiation inside the integral, upon which it becomes a partial derivative because $\mathbf{B}$ is a function of position as well. Since the last two equations hold for arbitrary S, the integrands must be equal. The derivation of the last equation is very similar to that of the third, and you should prove it yourself, remembering that the electric current is the surface integral of the current density. Putting all of these together, you get the differential form of Equation (A.1):

The differential form of the four equations that Maxwell inherited.

$$\text{(1) } \nabla \cdot \mathbf{E} = \frac{\rho}{\epsilon_0} \qquad \text{(2) } \nabla \cdot \mathbf{B} = 0$$
$$\text{(3) } \nabla \times \mathbf{E} = -\frac{\partial \mathbf{B}}{\partial t} \qquad \text{(4) } \nabla \times \mathbf{B} = \mu_0 \mathbf{J}. \tag{A.10}$$

Maxwell discovers the inconsistency of Equation (A.10) with the conservation of electric charge, and modifies the last equation to resolve the inconsistency.

Maxwell inherited these equations and started pondering about them in the late 1850s and early 1860s. He noticed that while the second and third are consistent with other aspects of electromagnetism, the other two equations lead to a contradiction. Let us retrace his argument. By the second equation in (A.9), the divergence of the left-hand side of the last equation of (A.10) vanishes. Therefore, if you take the divergence of both sides, you get $\nabla \cdot \mathbf{J} = 0$. This contradicts the differential form of the continuity equation (A.6) for electric charges. Because of the firm establishment of the charge conservation,

Maxwell decided to try altering the fourth equation to make it compatible with charge conservation.

Add a vector field **V** (to be determined) on the right-hand side of the fourth equation:

$\nabla \times \mathbf{B} = \mu_0 \mathbf{J} + \mathbf{V}.$

Now take the divergence of both sides to obtain

$$0 = \mu_0 \nabla \cdot \mathbf{J} + \nabla \cdot \mathbf{V} \quad \text{or} \quad 0 = -\mu_0 \frac{\partial \rho}{\partial t} + \nabla \cdot \mathbf{V}.$$

Substitute from the first equation for ρ:

$$0 = -\mu_0 \frac{\partial(\epsilon_0 \nabla \cdot \mathbf{E})}{\partial t} + \nabla \cdot \mathbf{V} \quad \text{or} \quad \nabla \cdot \mathbf{V} = \nabla \cdot \left(\mu_0 \epsilon_0 \frac{\partial \mathbf{E}}{\partial t}\right).$$

The last equation follows from the fact that ∇, consisting of derivatives with respect to spatial coordinates, commutes with time derivative. The simplest choice for **V** is $\mu_0 \epsilon_0 \partial \mathbf{E}/\partial t$. With this modification of the last equation, the four equations in (A.10) become[3]

The four Maxwell equations.

$$(1)\ \nabla \cdot \mathbf{E} = \frac{\rho}{\epsilon_0} \qquad (2)\ \nabla \cdot \mathbf{B} = 0$$

$$(3)\ \nabla \times \mathbf{E} = -\frac{\partial \mathbf{B}}{\partial t} \qquad (4)\ \nabla \times \mathbf{B} = \mu_0 \mathbf{J} + \mu_0 \epsilon_0 \frac{\partial \mathbf{E}}{\partial t}. \tag{A.11}$$

Mathematics and the force of logic and human reasoning unravel one of the greatest secrets of Nature!

One of the greatest moments in the history of physics and mathematics occurred when Maxwell, prompted solely by the forces of logic and pure deduction, introduced the second term in the last equation. Such moments were rare prior to Maxwell. Theories and laws were empirical (or inductive); they were introduced to fit the data and summarize, more or less directly, the numerous observations made. Maxwell broke this tradition and set the stage for deductive reasoning which, after a great deal of struggle to abandon the inductive tradition, became the norm for modern physics.

Today, we aptly call all four equations in (A.11) **Maxwell's equations**, although his contribution to those equations was a "mere" introduction of the second term on the right-hand side of the last equation. However, no other "small" contribution has ever affected humankind so enormously. This very "small" contribution was responsible for Maxwell's prediction of the electromagnetic waves which were subsequently produced in the laboratory in 1887—only eight years after Maxwell's premature death—and put to technological use in 1901 in the form of the first radio. Today, Maxwell's equations are at the heart of every electronic device. Without them, our entire civilization, as we know it, would be nonexistent.

---

[3] When dielectric and magnetic materials are present, the first and last equations change. Instead of **E** and **B** we have to use the displacement field **D** and magnetizing field **H**, in which case the charge and current densities on the right-hand side become the free charge and current densities ρ_f and $\mathbf{J}_f$.

A.3 ELECTROMAGNETIC WAVES IN VACUUM

Practical applications of Maxwell's equations aside, they have a much more important theoretical implication. Consider the special case of these equations in vacuum, where there are no charges or currents. Is it possible to have a solution of these equations there? First write them down for this special case:

$$(1)\ \nabla\cdot\mathbf{E}=0 \qquad (2)\ \nabla\cdot\mathbf{B}=0$$

$$(3)\ \nabla\times\mathbf{E}=-\frac{\partial\mathbf{B}}{\partial t} \qquad (4)\ \nabla\times\mathbf{B}=\mu_0\epsilon_0\frac{\partial\mathbf{E}}{\partial t}. \tag{A.12}$$

The first two equations involve only one field, while the last two equations couple the electric and magnetic fields. To decouple them, note that if you take the curl of the third equation, you obtain the (time derivative of the) curl of **B** on the right-hand side, which is essentially the time derivative of **E** by the fourth equation. So, take the curl of the third equation and use the last equation in (A.9). You'll get

$$\text{LHS of (3)}=\nabla\times(\nabla\times\mathbf{E})=\nabla\underbrace{(\nabla\cdot\mathbf{E})}_{=0\text{ by (1)}}-\nabla^2\mathbf{E}=-\nabla^2\mathbf{E},$$

and

$$\text{RHS}=-\nabla\times\left(\frac{\partial\mathbf{B}}{\partial t}\right)=-\frac{\partial}{\partial t}(\nabla\times\mathbf{B})=-\frac{\partial}{\partial t}\left(\mu_0\epsilon_0\frac{\partial\mathbf{E}}{\partial t}\right).$$

Now set the two sides equal and obtain

$$\nabla^2\mathbf{E}=\mu_0\epsilon_0\frac{\partial^2\mathbf{E}}{\partial t^2}. \tag{A.13}$$

This is a three-dimensional wave equation.[4] Recall that the inverse of the coefficient of the second time derivative is the square of the speed of propagation of the wave. It follows that

Electromagnetic waves propagate at the speed of light.

$$v=\frac{1}{\sqrt{\mu_0\epsilon_0}}=\frac{1}{\sqrt{(4\pi\times10^{-7})(8.854\times10^{-12})}}=2.998\times10^8\ \text{m/s},$$

i.e., that the electric field propagates in empty space with the speed of light c. You should check that the magnetic field also satisfies the same wave equation, and that it too propagates with the same speed. In fact, it can be shown that the so-called plane wave solutions of Maxwell's equations consist of an electric and a magnetic component which are coupled to one another and, therefore do not propagate independently (see Problem A.3).

The innocent look of (A.12) or (A.13) is deceptive. The latter was derived without regard to the origin of the electric field. In particular, it is independent of its source. No matter how the source of the electromagnetic wave is moving, the wave that it produces always travels with the speed of light c!

---

[4] You may be familiar with the one-dimensional wave equation in which only one second partial derivative with respect to a single space coordinate appears.

Note A.3.1. *Electromagnetic waves travel with the speed of light in vacuum regardless of the motion of their source.*

A.4 ELECTROMAGNETIC WAVES IN A CUBIC CAVITY

When EM waves are confined inside a box, called a **cavity** in this context, they form standing waves. The study of EM waves in a cavity is significant in two respects: it is a neat mathematical exercise and its use led to the discovery of quantum mechanics.

Place one of the corners of a *conducting* cube of side L at the origin and align all its faces parallel to the planes of a Cartesian coordinate system. The cube is empty except for the presence of EM waves. Each component of **E** of Equation (A.13) satisfies the wave equation. So,

$$\frac{\partial^2 E_x}{\partial x^2}+\frac{\partial^2 E_x}{\partial y^2}+\frac{\partial^2 E_x}{\partial z^2}=\frac{1}{c^2}\frac{\partial^2 E_x}{\partial t^2} \tag{A.14}$$

with similar equations for E_y and E_z. To solve this equation, separate the variables, i.e., write $E_x(x, y, z, t)$ as a product of four functions each depending on a single variable, $E_x(x, y, z, t) = X_1(x)Y_1(y)Z_1(z)T_1(t)$, substitute it in the last equation, and divide the result by $X_1(x)Y_1(y)Z_1(z)T_1(t)$ to obtain

$$\frac{1}{X_1}\frac{d^2X_1}{dx^2}+\frac{1}{Y_1}\frac{d^2Y_1}{dy^2}+\frac{1}{Z_1}\frac{d^2Z_1}{dz^2}=\frac{1}{c^2T_1}\frac{d^2T_1}{dt^2}. \tag{A.15}$$

I have used the subscript 1 to refer to the "first component" of **E**. Each term of the last equation is dependent on a different independent variable. The only way that the equality can hold is for each term to be a constant:

$$\frac{1}{X_1}\frac{d^2X_1}{dx^2}=k_{1x}^2,\quad \frac{1}{Y_1}\frac{d^2Y_1}{dy^2}=k_{1y}^2\quad \frac{1}{Z_1}\frac{d^2Z_1}{dz^2}=k_{1z}^2,\quad \frac{1}{T_1}\frac{d^2T_1}{dt^2}=\omega_1^2 \tag{A.16}$$

with

$$\frac{\omega_1^2}{c^2}=k_{1x}^2+k_{1y}^2+k_{1z}^2. \tag{A.17}$$

The solutions of all these ordinary differential equations are exponentials. For example, the most general solution of the Y_1 equation is

$$Y_1(y)=A_1e^{k_{1y}y}+B_1e^{-k_{1y}y}.$$

As you can see there are a lot of unknown constants to be determined. Just in this equation, there are three: A_1, B_1, and k_{1y}. It turns out that Maxwell's equations and boundary conditions (which are also derived from Maxwell's equations) determine all the unknown constants. Let me first apply the boundary conditions.

The electric field has the property that its tangential component is continuous across a boundary (see Problem A.4). Since the field is zero inside a conductor, the components tangential to the sides of the cavity must vanish. Now note that E_x is tangent to the planes $y=0$ and $y=L$. Therefore, $Y_1(0)=0=Y_1(L)$. The first condition gives $B_1=-A_1$ and $Y_1(y)=A_1(e^{k_{1y}y}-e^{-k_{1y}y})$. The second condition now yields

$$Y_1(L)=A_1(e^{k_{1y}L}-e^{-k_{1y}L})=0 \quad \text{or} \quad A_1(e^{2k_{1y}L}-1)=0.$$

I don't want to set $A_1=0$, because then $Y_1(y)=0$ for all y, making $E_x(x,y,z,t)=0$. So, I have to look for nontrivial solutions for k_{1y} in $e^{2k_{1y}L}=1$. If you stick to *real* values for k_{1y}, the only solution would be $k_{1y}=0$, again making $Y_1(y)$ and $E_x(x,y,z,t)$ zero! However, if I set $k_{1y}L=in_{1y}\pi$, where n_{1y} is a positive integer, I have a nontrivial solution. Therefore, I can now write

$$Y_1(y)=A_1\left(e^{in_{1y}\pi y/L}-e^{-in_{1y}\pi y/L}\right)=2iA_1\sin\left(\frac{n_{1y}\pi y}{L}\right), \tag{A.18}$$

with a similar expression for $Z_1(z)$. I can't do the same thing with $X_1(x)$, because E_x is perpendicular to the planes $x=0$ and $x=L$. So, the most general solution for E_x is

$$\begin{aligned}E_x(x,y,z,t)=&\left(A_1e^{k_{1x}x}+B_1e^{-k_{1x}x}\right)\sin\left(\frac{n_{1y}\pi y}{L}\right)\sin\left(\frac{n_{1z}\pi z}{L}\right)\\&\times\left(C_1e^{\omega_1 t}+D_1e^{-\omega_1 t}\right)\end{aligned}$$

where I have redefined A_1 and B_1 and absorbed all constants in them. Similarly, I can obtain

$$\begin{aligned}E_y(x,y,z,t)=&\sin\left(\frac{n_{2x}\pi x}{L}\right)\left(A_2e^{k_{2y}y}+B_2e^{-k_{2y}y}\right)\sin\left(\frac{n_{2z}\pi z}{L}\right)\\&\times\left(C_2e^{\omega_2 t}+D_2e^{-\omega_2 t}\right)\\E_z(x,y,z,t)=&\sin\left(\frac{n_{3x}\pi x}{L}\right)\sin\left(\frac{n_{3y}\pi y}{L}\right)\left(A_3e^{k_{3z}z}+B_3e^{-k_{3z}z}\right)\\&\times\left(C_3e^{\omega_3 t}+D_3e^{-\omega_3 t}\right).\end{aligned} \tag{A.19}$$

There seems to be an overwhelming number of unknowns to be determined, and I have used up all the boundary conditions. However, there is one Maxwell equation that can amazingly determine all these unknowns: the first equation (in empty cavity) $\nabla\cdot\mathbf{E}=0$.

Consider the three terms of $\nabla\cdot\mathbf{E}$, each written separately

$$\begin{aligned}\frac{\partial E_x}{\partial x}&=k_{1x}\left(A_1e^{k_{1x}x}-B_1e^{-k_{1x}x}\right)\sin\left(\frac{n_{1y}\pi y}{L}\right)\sin\left(\frac{n_{1z}\pi z}{L}\right)\left(C_1e^{\omega_1 t}+D_1e^{-\omega_1 t}\right)\\\frac{\partial E_y}{\partial y}&=k_{2y}\sin\left(\frac{n_{2x}\pi x}{L}\right)\left(A_2e^{k_{2y}y}-B_2e^{-k_{2y}y}\right)\sin\left(\frac{n_{2z}\pi z}{L}\right)\left(C_2e^{\omega_2 t}+D_2e^{-\omega_2 t}\right)\\\frac{\partial E_z}{\partial z}&=k_{3z}\sin\left(\frac{n_{3x}\pi x}{L}\right)\sin\left(\frac{n_{3y}\pi y}{L}\right)\left(A_3e^{k_{3z}z}-B_3e^{-k_{3z}z}\right)\left(C_3e^{\omega_3 t}+D_3e^{-\omega_3 t}\right).\end{aligned} \tag{A.20}$$

When you add these and set the sum equal to zero and demand that the result be zero for all t, x, y, and z, then the only way that can happen is for all functions to be the same. For example, the equality of the time functions gives $\omega_1 = \omega_2 = \omega_3 \equiv \omega$, $C_1 = C_2 = C_3 \equiv C$, and $D_1 = D_2 = D_3 \equiv D$. The equality of the x functions gives $A_1 = B_1$, and $k_{1x} = in_{2x}\pi/L = in_{3x}\pi/L$, with similar results for the y and z functions. All this means that there is no need for the labels 1, 2, and 3 for the k's. Furthermore, (A.17) yields

$$\frac{\omega^2}{c^2} = -\left(\frac{n_x\pi}{L}\right)^2 - \left(\frac{n_y\pi}{L}\right)^2 - \left(\frac{n_z\pi}{L}\right)^2,$$

showing that ω must be imaginary. So, I replace ω with $i\omega$ and write

$$\frac{\omega^2}{c^2} = \left(\frac{n_x\pi}{L}\right)^2 + \left(\frac{n_y\pi}{L}\right)^2 + \left(\frac{n_z\pi}{L}\right)^2. \tag{A.21}$$

If I further assume that $\mathbf{E} = 0$ at $t = 0$, then $D = -C$ and the time function becomes $2iC \sin\omega t$. Collecting all the constants into a single one, the components of the electric field can finally be written as

$$\begin{aligned}
E_x(x,y,z,t) &= E_{0x} \cos\left(\frac{n_x\pi x}{L}\right) \sin\left(\frac{n_y\pi y}{L}\right) \sin\left(\frac{n_z\pi z}{L}\right) \sin\omega t \\
E_y(x,y,z,t) &= E_{0y} \sin\left(\frac{n_x\pi x}{L}\right) \cos\left(\frac{n_y\pi y}{L}\right) \sin\left(\frac{n_z\pi z}{L}\right) \sin\omega t \\
E_z(x,y,z,t) &= E_{0z} \sin\left(\frac{n_x\pi x}{L}\right) \sin\left(\frac{n_y\pi y}{L}\right) \cos\left(\frac{n_z\pi z}{L}\right) \sin\omega t
\end{aligned} \tag{A.22}$$

where E_{0x}, E_{0y}, and E_{0z} are the (constant) amplitudes of each component. The divergence equation now reduces to

$$n_x E_{0x} + n_y E_{0y} + n_z E_{0z} = 0, \tag{A.23}$$

indicating that the amplitudes are linearly related, and only two of them are independent. These are the two **polarizations** of the electric field. You can define a vector $\mathbf{n} = \langle n_x, n_y, n_z \rangle$ with integer components, then (A.21) yields

$$\frac{\omega_n^2}{c^2} = \left(\frac{\pi}{L}\right)^2 |\mathbf{n}|^2 \quad \text{and} \quad \frac{\omega_n}{c} = \frac{\pi}{L}|\mathbf{n}| = \frac{\pi}{L}\sqrt{n_x^2 + n_y^2 + n_z^2}. \tag{A.24}$$

Each $\mathbf{n}$ describes a **mode** of cavity oscillation, and the corresponding angular frequency ω_n is labeled accordingly.

A.5 PROBLEMS

A.1. Derive the differential form of the fourth equation in (A.1).

A.2. Starting with Maxwell's equations, show that the magnetic field satisfies the same wave equation as the electric field. In particular, that it, too, propagates with the same speed.

A.3. Consider $\mathbf{E} = \mathbf{E}_0 e^{i(\omega t - \mathbf{k}\cdot\mathbf{r})}$ and $\mathbf{B} = \mathbf{B}_0 e^{i(\omega t - \mathbf{k}\cdot\mathbf{r})}$, where $i = \sqrt{-1}$, $\mathbf{E}_0$, $\mathbf{B}_0$, k, and ω are constants. The **E** and the **B** so defined represent *plane waves* moving in the direction of the vector **k**.

(a) Show that they satisfy Maxwell's equations in free space if:

$$(1)\ \mathbf{k}\cdot\mathbf{E}_0 = 0; \qquad (2)\ \mathbf{k}\cdot\mathbf{B}_0 = 0;$$

$$(3)\ \mathbf{k}\times\mathbf{E}_0 = \omega\mathbf{B}_0; \quad (4)\ \mathbf{k}\times\mathbf{B}_0 = -\frac{\omega}{c^2}\mathbf{E}_0.$$

(b) In particular, show that **k**, the propagation direction, and **E** and **B** form a mutually perpendicular set of vectors.

(c) By taking the cross product of **k** with an appropriate equation, show that $|\mathbf{k}| = \omega/c$.

A.4. Take a rectangular loop with two long sides on either side of a boundary surface. All sides are infinitesimal, but the short side is infinitesimal even compared to the long side. Apply the third Maxwell equation in integral form (A.1) to this loop. Since everything is infinitesimal, you can forget about the integrals. The magnetic flux is zero because the area is triply infinitesimal. The contribution from the two short sides to the left-hand side is zero. Now complete the proof of the equality of the tangential components of the electric field on the two sides of the boundary.

APPENDIX B

Derivation of 4D Lorentz Transformation

Lorentz transformations (LT) have some important properties which we should explore before finding their generalization to Equation (6.16). All these properties are consequences of their defining equation:

$$\widetilde{\Lambda}\eta\Lambda = \eta. \tag{B.1}$$

B.1 GENERAL PROPERTIES

The first property is that the unit matrix is a LT, because it satisfies (B.1) trivially.

The second property is that the product of two LTs is also a LT. Let Λ_1 and Λ_2 be two LTs. This means that

$$\widetilde{\Lambda}_1\eta\Lambda_1 = \eta \quad \text{and} \quad \widetilde{\Lambda}_2\eta\Lambda_2 = \eta.$$

Then

$$\widetilde{\Lambda_1\Lambda_2}\eta\Lambda_1\Lambda_2 = \widetilde{\Lambda}_2\underbrace{\widetilde{\Lambda}_1\eta\Lambda_1}_{=\eta}\Lambda_2 = \widetilde{\Lambda}_2\eta\Lambda_2 = \eta.$$

This shows that $\Lambda_1\Lambda_2$ is a Lorentz transformation.

The third property is that every LT has an inverse, which is also a LT. To show this, take the determinant of both sides of (B.1) to get

$$\left(\det\widetilde{\Lambda}\right)\left(\det\eta\right)\left(\det\Lambda\right) = \det\eta \;\Rightarrow\; -\left(\det\widetilde{\Lambda}\right)\left(\det\Lambda\right) = -1$$

or $(\det\Lambda)^2 = 1$, because a matrix and its transpose have the same determinant. Therefore, $\det\Lambda = \pm 1$, and since its determinant is not zero, Λ must have an inverse Λ^{-1}. Multiply Equation (B.1) by Λ^{-1} on the right and by $\widetilde{\Lambda^{-1}} = (\widetilde{\Lambda})^{-1}$ on the left to get

$$(\widetilde{\Lambda})^{-1}\widetilde{\Lambda}\eta\Lambda\Lambda^{-1} = \widetilde{\Lambda^{-1}}\eta\Lambda^{-1} \;\Rightarrow\; \eta = \widetilde{\Lambda^{-1}}\eta\Lambda^{-1},$$

showing that Λ^{-1} is a LT.[1] Problem B.1 asks you to prove that the inverse of the transpose of a matrix is the transpose of its inverse.

---

[1] These three properties turn the set of all Lorentz transformations into a **group**, the **Lorentz group**. But you don't need to know anything about group theory to follow the rest of this appendix.

Special Relativity. DOI:10.1016/B978-0-12-810411-8.00021-3

Lorentz transformations have an important subset. Consider the set of 4×4 matrices of the form

$$\Lambda \equiv \begin{pmatrix} 1 & 0 & 0 & 0 \\ 0 & a_{11} & a_{12} & a_{13} \\ 0 & a_{21} & a_{22} & a_{23} \\ 0 & a_{31} & a_{32} & a_{33} \end{pmatrix} \equiv \begin{pmatrix} 1 & \tilde{\mathsf{0}} \\ \mathsf{0} & \mathsf{A} \end{pmatrix},$$

where on the right-hand side the 4×4 matrix is written in block diagonal form in which $\mathsf{0}$ is a column vector and A is the 3×3 matrix whose elements are given on the left-hand side. Write the matrix η also in block diagonal form and note that

$$\begin{aligned} \tilde{\Lambda}\eta\Lambda &= \begin{pmatrix} 1 & \tilde{\mathsf{0}} \\ \mathsf{0} & \tilde{\mathsf{A}} \end{pmatrix} \begin{pmatrix} 1 & \tilde{\mathsf{0}} \\ \mathsf{0} & -1 \end{pmatrix} \begin{pmatrix} 1 & \tilde{\mathsf{0}} \\ \mathsf{0} & \mathsf{A} \end{pmatrix} \\ &= \begin{pmatrix} 1 & \tilde{\mathsf{0}} \\ \mathsf{0} & -\tilde{\mathsf{A}} \end{pmatrix} \begin{pmatrix} 1 & \tilde{\mathsf{0}} \\ \mathsf{0} & \mathsf{A} \end{pmatrix} = \begin{pmatrix} 1 & \tilde{\mathsf{0}} \\ \mathsf{0} & -\tilde{\mathsf{A}}\mathsf{A} \end{pmatrix}. \end{aligned}$$

The right-hand side becomes η if $\tilde{\mathsf{A}}\mathsf{A} = \mathsf{1}$. You may recall that a matrix satisfying $\tilde{\mathsf{A}}\mathsf{A} = \mathsf{1}$ is called **orthogonal**. Orthogonal 3×3 matrices are essentially rotations. Thus, three-dimensional rotations are special kinds of LTs.

B.2 DERIVATION OF THE LORENTZ BOOST

Now we have all the machinery we need to obtain the most general LT, i.e., one for which the relative velocity $\vec{\beta}$ is along an arbitrary direction. All we need to do is use the freedom in choosing the orientation of the axes of our coordinate system: find a rotation which takes our coordinate system CS_1 to another coordinate system CS_2 in which $\vec{\beta}$ has the desired direction. First note that a rotation of the coordinate system is completely equivalent to the rotation of vectors *in the opposite direction*. Figure B.1 illustrates that. So, going from CS_1 to CS_2 is equivalent to rotating the vectors by some rotation matrix Λ_{rot}.

Next let's consider Equation (6.13),

$$\begin{pmatrix} ct' \\ x' \\ y' \\ z' \end{pmatrix} = \underbrace{\begin{pmatrix} \gamma & \gamma\beta & 0 & 0 \\ \gamma\beta & \gamma & 0 & 0 \\ 0 & 0 & 1 & 0 \\ 0 & 0 & 0 & 1 \end{pmatrix}}_{\text{Call this matrix } \Lambda_0.} \begin{pmatrix} ct \\ x \\ y \\ z \end{pmatrix} \quad \text{or} \quad \mathsf{r}'_0 = \Lambda_0 \mathsf{r}_0, \tag{B.2}$$

where the zero subscript indicates the initial (unrotated) quantities. Now multiply both sides of (B.2) by Λ_{rot} to get

$$\Lambda_{rot}\mathsf{r}'_0 = \Lambda_{rot}\Lambda_0\mathsf{r}_0 \quad \text{or} \quad \Lambda_{rot}\mathsf{r}'_0 = \Lambda_{rot}\Lambda_0\Lambda^{-1}_{rot}\Lambda_{rot}\mathsf{r}_0.$$

Since $\mathsf{r} \equiv \Lambda_{rot}\mathsf{r}_0$ is the new $\mathbf{r}$ 4-vector and $\mathsf{r}' \equiv \Lambda_{rot}\mathsf{r}'_0$ is the new $\mathbf{r}'$ 4-vector, $\Lambda_{rot}\Lambda_0\Lambda^{-1}_{rot}$ must be the new Lorentz transformation Λ connecting $\mathbf{r}$ and $\mathbf{r}'$.

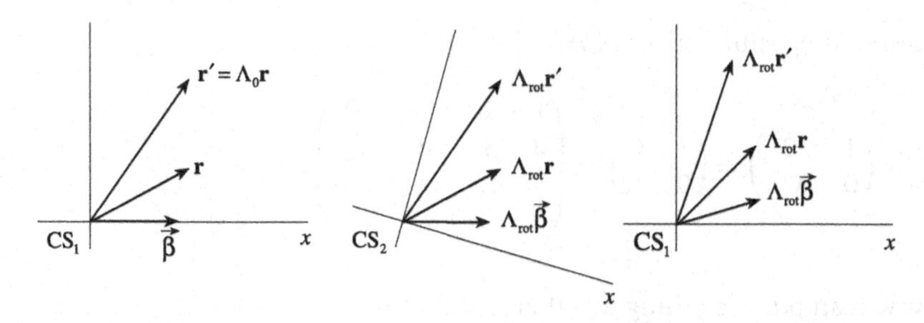

FIGURE B.1 If you rotate CS_1 you get CS_2. The (old) vectors have the same components in CS_2 as the rotated (in the inverse direction) vectors would have in CS_1.

The only thing left to do is to construct Λ, for which purpose we only need Λ_{rot}.

Let $\hat{\beta} = (\hat{\beta}_x, \hat{\beta}_y, \hat{\beta}_z)$ denote the unit 3-vector in the direction of $\vec{\beta}$. Our task, therefore, is to find a rotation which takes the three vector $(1, 0, 0)$ to $(\hat{\beta}_x, \hat{\beta}_y, \hat{\beta}_z)$. Let A, with elements a_{ij}, be the 3×3 matrix that does the job. Namely, let

$$\begin{pmatrix} \hat{\beta}_x \\ \hat{\beta}_y \\ \hat{\beta}_z \end{pmatrix} = \begin{pmatrix} a_{11} & a_{12} & a_{13} \\ a_{21} & a_{22} & a_{23} \\ a_{31} & a_{32} & a_{33} \end{pmatrix} \begin{pmatrix} 1 \\ 0 \\ 0 \end{pmatrix}.$$

Multiplying out the right-hand side shows that $a_{11} = \hat{\beta}_x$, $a_{21} = \hat{\beta}_y$, and $a_{31} = \hat{\beta}_z$. Therefore,

$$\mathsf{A} = \begin{pmatrix} \hat{\beta}_x & a_{12} & a_{13} \\ \hat{\beta}_y & a_{22} & a_{23} \\ \hat{\beta}_z & a_{32} & a_{33} \end{pmatrix}.$$

If this is to be a rotation, it should satisfy

$$\tilde{\mathsf{A}}\mathsf{A} = \mathsf{A}\tilde{\mathsf{A}} = 1.$$

You should convince yourself that if this holds and we consider the rows and columns of A as vectors, then each row (or each column) has a unit length and any two different rows (or different columns) are perpendicular to each other (see Problem B.2). The sought-after rotation has now been reduced to

$$\Lambda_{\text{rot}} = \begin{pmatrix} 1 & 0 & 0 & 0 \\ 0 & \hat{\beta}_x & a_{12} & a_{13} \\ 0 & \hat{\beta}_y & a_{22} & a_{23} \\ 0 & \hat{\beta}_z & a_{32} & a_{33} \end{pmatrix} \equiv \begin{pmatrix} 1 & \tilde{0} \\ 0 & \mathsf{A} \end{pmatrix}. \tag{B.3}$$

We also need the inverse of Λ_{rot}. Since Λ_{rot} is in block-diagonal form, its inverse is simply a block-diagonal matrix whose submatrices are inverses of the corresponding submatrices of Λ_{rot}:

$$\Lambda_{\text{rot}}^{-1} = \begin{pmatrix} 1 & \tilde{0} \\ 0 & \mathsf{A}^{-1} \end{pmatrix} = \begin{pmatrix} 1 & \tilde{0} \\ 0 & \tilde{\mathsf{A}} \end{pmatrix} = \begin{pmatrix} 1 & 0 & 0 & 0 \\ 0 & \hat{\beta}_x & \hat{\beta}_y & \hat{\beta}_z \\ 0 & a_{12} & a_{22} & a_{32} \\ 0 & a_{13} & a_{23} & a_{33} \end{pmatrix}. \tag{B.4}$$

Now we can put everything together and find Λ:

$$\begin{aligned} \Lambda &= \begin{pmatrix} 1 & 0 & 0 & 0 \\ 0 & \hat{\beta}_x & a_{12} & a_{13} \\ 0 & \hat{\beta}_y & a_{22} & a_{23} \\ 0 & \hat{\beta}_z & a_{32} & a_{33} \end{pmatrix} \begin{pmatrix} \gamma & \gamma\beta & 0 & 0 \\ \gamma\beta & \gamma & 0 & 0 \\ 0 & 0 & 1 & 0 \\ 0 & 0 & 0 & 1 \end{pmatrix} \begin{pmatrix} 1 & 0 & 0 & 0 \\ 0 & \hat{\beta}_x & \hat{\beta}_y & \hat{\beta}_z \\ 0 & a_{12} & a_{22} & a_{32} \\ 0 & a_{13} & a_{23} & a_{33} \end{pmatrix} \\ &= \begin{pmatrix} \gamma & \gamma\beta & 0 & 0 \\ \gamma\beta_x & \gamma\hat{\beta}_x & a_{12} & a_{13} \\ \gamma\beta_y & \gamma\hat{\beta}_y & a_{22} & a_{33} \\ \gamma\beta_z & \gamma\hat{\beta}_z & a_{32} & a_{33} \end{pmatrix} \begin{pmatrix} 1 & 0 & 0 & 0 \\ 0 & \hat{\beta}_x & \hat{\beta}_y & \hat{\beta}_z \\ 0 & a_{12} & a_{22} & a_{32} \\ 0 & a_{13} & a_{23} & a_{33} \end{pmatrix} \end{aligned}$$

because $\beta\hat{\beta}_x = \beta_x$ etc. You should do the rest of matrix multiplications and use Problem B.2 to come up with the final result:

The most general Lorentz boost.

$$\Lambda = \begin{pmatrix} \gamma & \gamma\beta_x & \gamma\beta_y & \gamma\beta_z \\ \gamma\beta_x & 1+\hat{\beta}_x^2(\gamma-1) & \hat{\beta}_x\hat{\beta}_y(\gamma-1) & \hat{\beta}_x\hat{\beta}_z(\gamma-1) \\ \gamma\beta_y & \hat{\beta}_x\hat{\beta}_y(\gamma-1) & 1+\hat{\beta}_y^2(\gamma-1) & \hat{\beta}_y\hat{\beta}_z(\gamma-1) \\ \gamma\beta_z & \hat{\beta}_x\hat{\beta}_z(\gamma-1) & \hat{\beta}_y\hat{\beta}_z(\gamma-1) & 1+\hat{\beta}_z^2(\gamma-1) \end{pmatrix}. \tag{B.5}$$

Equation (B.5) does not give the most general LT matrix, because it is fully described by the relative velocity $\vec{\beta}$ of two observers. As mentioned before the full set of Lorentz transformations include rotations as well. Equation (B.5) gives only a so-called Lorentz **boost**, which we sometimes write as $\Lambda(\vec{\beta})$. Lorentz boosts have the additional property that $\tilde{\Lambda} = \Lambda$, which is not shared by a general Lorentz transformation.

B.2.1 Transformation of Position and Time

Now that we have Λ, let's apply it to the new $\mathbf{r} \equiv (ct, x, y, z)$. Calling the result of this application $\mathbf{r}'$, and using matrices, you should be able to get

$$\begin{pmatrix} ct' \\ x' \\ y' \\ z' \end{pmatrix} = \begin{pmatrix} \gamma ct + \gamma\vec{\beta}\cdot\vec{r} \\ x + \beta_x\gamma ct + \hat{\beta}_x(\gamma-1)\hat{\beta}\cdot\vec{r} \\ y + \beta_y\gamma ct + \hat{\beta}_y(\gamma-1)\hat{\beta}\cdot\vec{r} \\ z + \beta_z\gamma ct + \hat{\beta}_z(\gamma-1)\hat{\beta}\cdot\vec{r} \end{pmatrix}, \tag{B.6}$$

where all components are *relative to the new coordinate system*. This shows that $ct' = \gamma(ct + \vec{\beta} \cdot \vec{r})$ and

$$\begin{pmatrix} x' \\ y' \\ z' \end{pmatrix} = \begin{pmatrix} x + \beta_x \gamma ct + \hat{\beta}_x(\gamma - 1)\hat{\beta} \cdot \vec{r} \\ y + \beta_y \gamma ct + \hat{\beta}_y(\gamma - 1)\hat{\beta} \cdot \vec{r} \\ z + \beta_z \gamma ct + \hat{\beta}_z(\gamma - 1)\hat{\beta} \cdot \vec{r} \end{pmatrix}$$

$$= \begin{pmatrix} x \\ y \\ z \end{pmatrix} + \gamma ct \begin{pmatrix} \beta_x \\ \beta_y \\ \beta_z \end{pmatrix} + (\gamma - 1)\hat{\beta} \cdot \vec{r} \begin{pmatrix} \hat{\beta}_x \\ \hat{\beta}_y \\ \hat{\beta}_z \end{pmatrix},$$

or in vector notation

$$\vec{r}' = \vec{r} + \gamma ct\vec{\beta} + \hat{\beta}(\gamma - 1)\hat{\beta} \cdot \vec{r} = \vec{r} + \gamma ct\vec{\beta} + \frac{1}{\beta^2}\vec{\beta}(\gamma - 1)\vec{\beta} \cdot \vec{r}.$$

With $\beta^2 = (\gamma^2 - 1)/\gamma^2$, the final result, which is stated in Equation (6.16), is trivially reached.

B.2.2 The 3 x 3 Submatrix

It is sometimes useful to separate the 3×3 matrix in Λ, which is obtained by eliminating the first row and column. So, first write Λ in block form:

$$\Lambda = \begin{pmatrix} \gamma & \gamma\tilde{\vec{\beta}} \\ \gamma\vec{\beta} & \overleftrightarrow{\lambda} \end{pmatrix}, \tag{B.7}$$

where $\overleftrightarrow{\lambda}$ is the 3×3 submatrix of Λ. Then, with the help of (B.5), write

$$\overleftrightarrow{\lambda} = \begin{pmatrix} 1 + \hat{\beta}_x^2(\gamma - 1) & \hat{\beta}_x\hat{\beta}_y(\gamma - 1) & \hat{\beta}_x\hat{\beta}_z(\gamma - 1) \\ \hat{\beta}_x\hat{\beta}_y(\gamma - 1) & 1 + \hat{\beta}_y^2(\gamma - 1) & \hat{\beta}_y\hat{\beta}_z(\gamma - 1) \\ \hat{\beta}_x\hat{\beta}_z(\gamma - 1) & \hat{\beta}_y\hat{\beta}_z(\gamma - 1) & 1 + \hat{\beta}_z^2 \end{pmatrix},$$

or

$$\overleftrightarrow{\lambda} = \begin{pmatrix} 1 & 0 & 0 \\ 0 & 1 & 0 \\ 0 & 0 & 1 \end{pmatrix} + (\gamma - 1) \begin{pmatrix} \hat{\beta}_x^2 & \hat{\beta}_x\hat{\beta}_y & \hat{\beta}_x\hat{\beta}_z \\ \hat{\beta}_x\hat{\beta}_y & \hat{\beta}_y^2 & \hat{\beta}_y\hat{\beta}_z \\ \hat{\beta}_x\hat{\beta}_z & \hat{\beta}_y\hat{\beta}_z & \hat{\beta}_z^2 \end{pmatrix}. \tag{B.8}$$

This can also be expressed in indexed form as

$$(\overleftrightarrow{\lambda})_{ij} = \delta_{ij} + (\gamma - 1)\hat{\beta}_i\hat{\beta}_j, \tag{B.9}$$

where δ_{ij} is the **Kronecker delta** defined by

$$\delta_{ij} = \begin{cases} 1 & \text{if } i = j \\ 0 & \text{if } i \neq j. \end{cases} \tag{B.10}$$

B.2.3 Inverse of a Boost

The inverse Lorentz boost is also important, because sometimes it is desired to go from the new RF to the old one. This transformation is accomplished by the inverse of the 4×4 matrix (B.5). Finding Λ^{-1} may appear daunting, but physics makes it trivial: the only difference between going from old to new and going from new to old is the direction of velocity. Therefore, changing $\vec{\beta}$ to $-\vec{\beta}$ should produce the inverse:

$$\Lambda^{-1} = \begin{pmatrix} \gamma & -\gamma\beta_x & -\gamma\beta_y & -\gamma\beta_z \\ -\gamma\beta_x & 1+\hat{\beta}_x^2(\gamma-1) & \hat{\beta}_x\hat{\beta}_y(\gamma-1) & \hat{\beta}_x\hat{\beta}_z(\gamma-1) \\ -\gamma\beta_y & \hat{\beta}_x\hat{\beta}_y(\gamma-1) & 1+\hat{\beta}_y^2(\gamma-1) & \hat{\beta}_y\hat{\beta}_z(\gamma-1) \\ -\gamma\beta_z & \hat{\beta}_x\hat{\beta}_z(\gamma-1) & \hat{\beta}_y\hat{\beta}_z(\gamma-1) & 1+\hat{\beta}_z^2(\gamma-1) \end{pmatrix}. \tag{B.11}$$

You should check that $\Lambda^{-1}\Lambda = \Lambda\Lambda^{-1} = 1$.

If you multiply both sides of Equation (B.1) on the right by $\Lambda^{-1}\eta$, you get

$$\tilde{\Lambda}\overbrace{\eta\underbrace{\Lambda\Lambda^{-1}}_{=1}\eta}^{=\eta^2=1} = \eta\Lambda^{-1}\eta \Rightarrow \tilde{\Lambda} = \eta\Lambda^{-1}\eta,$$

which in the case of Lorentz boosts (for which $\tilde{\Lambda} = \Lambda$) leads to

$$\Lambda(\vec{\beta}) = \eta\Lambda(-\vec{\beta})\eta \quad \text{or} \quad \Lambda(-\vec{\beta}) = \eta\Lambda(\vec{\beta})\eta. \tag{B.12}$$

These are useful relations which are used occasionally.

B.3 PROBLEMS

B.1. Take the transpose of $A^{-1}A = AA^{-1} = 1$ to show that the inverse of the transpose is the transpose of the inverse. You need both matrix products because an inverse must be a left inverse as well as a right inverse.

B.2. Show that $\tilde{A}A = A\tilde{A} = 1$ for a 3×3 matrix implies that each row (or column) of A has a unit length and the dot product of any two different rows (or different columns) vanishes.

B.3. By writing out the 4×4 matrices, verify that $\Lambda_{\text{rot}}^{-1}$ as given by Equation (B.4) is the inverse of Λ_{rot} as given by Equation (B.3).

B.4. Using relations such as $\hat{\beta}_x\hat{\beta}_y + a_{12}a_{22} + a_{13}a_{23} = 0$ and $a_{12}^2 + a_{13}^2 = 1 - (\hat{\beta}_x)^2$, obtained from Problem B.2, derive Equation (B.5).

B.5. Multiply the matrix of Equation (B.5) by the column 4-vector of $\mathbf{r}$ to obtain Equation (B.6).

B.6. Show directly that the matrices of (B.5) and (B.11) satisfy $\Lambda^{-1}\Lambda = \Lambda\Lambda^{-1} = 1$.

APPENDIX C

Relativistic Photography Formulas

In this appendix, I collect the derivation of all the formulas that are essential in understanding the photography of rapidly moving objects. My main goal is to map a point P in three dimensions onto a point in the two-dimensional plane of the photographic plate.

C.1 MAPPING FROM SPACE TO A PLANE

I focus the camera on a point Q—such as the center of a cube—in the vicinity of P and assume that the photographic plate is perpendicular to the line joining the pinhole of the camera to Q. Actually, instead of the photographic plate I'll choose a plane outside the camera that is parallel to the plate to avoid the confusion arising from the reflection through the pinhole and the consequent change in the sign of the coordinates of the image of P.

Using Figure C.1, you can show that the equation of the line joining Q to P_0 is

$$\mathbf{r}_1(t) = (1-t)\mathbf{r}_q + tb\hat{\mathbf{e}}_z \tag{C.1}$$

where $\mathbf{r}_q$ is the position vector of Q and $\hat{\mathbf{e}}_z$ is a unit vector in the z-direction. Points P' and Q' are the images of P and Q in the plane. Since Q' lies along the line of Equation (C.1), $\mathbf{r}_1(t_{q'}) = \mathbf{r}_{q'}$ for some $t = t_{q'}$, with $0 < t_{q'} < 1$. Let $d = \overline{Q'P_0}$ be the distance between the pinhole and the plane. Then $d = |\mathbf{r}_{q'} - b\hat{\mathbf{e}}_z|$, from which you can find $t_{q'}$ from (C.1):

$$t_{q'} = 1 - \frac{d}{|\mathbf{r}_q - b\hat{\mathbf{e}}_z|}. \tag{C.2}$$

I let $\mathbf{r}$ be the position vector of a point in the plane. Then the vector $\mathbf{r} - \mathbf{r}_1(t_{q'})$ lies in the plane and $\mathbf{r}_q - b\hat{\mathbf{e}}_z$ is perpendicular to the plane. Therefore, the equation of the plane is

$$[\mathbf{r} - \mathbf{r}_1(t_{q'})]\cdot(\mathbf{r}_q - b\hat{\mathbf{e}}_z) = 0. \tag{C.3}$$

The equation of the line connecting P to the pinhole of the camera is

$$\mathbf{r}_2(t) = (1-t)\mathbf{r}_p + tb\hat{\mathbf{e}}_z. \tag{C.4}$$

Special Relativity. DOI:10.1016/B978-0-12-810411-8.00022-5

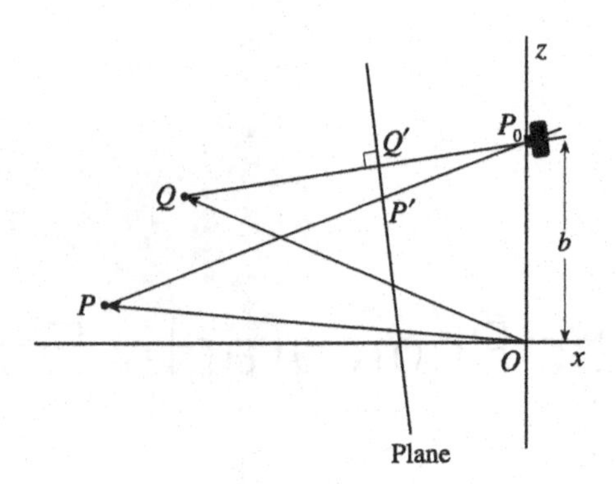

FIGURE C.1 The plane capturing the three-dimensional photons ought to be inside the camera. However, the reflection through the pinhole associated with that choice can cause some confusion.

Since P' is both on this line and in the plane, we must have $\mathbf{r}_{p'} = \mathbf{r}_2(t_{p'})$ and

$$[\mathbf{r}_2(t_{p'}) - \mathbf{r}_1(t_{q'})] \cdot (\mathbf{r}_q - b\hat{\mathbf{e}}_z) = 0 \tag{C.5}$$

for some $0 < t_{p'} < 1$. By substituting (C.1), (C.2), and (C.4) in (C.5), you should verify that

$$t_{p'} = 1 - \frac{d|\mathbf{r}_q - b\hat{\mathbf{e}}_z|}{(\mathbf{r}_p - b\hat{\mathbf{e}}_z)\cdot(\mathbf{r}_q - b\hat{\mathbf{e}}_z)}. \tag{C.6}$$

Remark C.1.1. Equation (C.6) puts a limitation on the points to be photographed, because its denominator can be zero. This happens when $\overline{PP_0}$ of Figure C.1 is perpendicular to $\overline{QP_0}$, i.e., when $\overline{PP_0}$ is parallel to the photographic plate. When this happens, no image is formed on the plate. What is interesting is that even if a stationary point P forms an image on the photographic plate, relativistic effects can render the denominator of Equation (C.6) zero when P is moving. In fact, if P is moving along OP toward the origin, then using Equation (6.16), and the fact that $ct = -|\vec{r}_P|$, you can (should) show that

$$\begin{aligned}\mathbf{r}'_p &= \mathbf{r}_p + \gamma\vec{\beta}\left(-|\mathbf{r}_p| + \frac{\gamma}{\gamma+1}\vec{\beta}\cdot\mathbf{r}_p\right)\\ &= \gamma(1+\beta)\mathbf{r}_p = \sqrt{\frac{1+\beta}{1-\beta}}\,\mathbf{r}_p\end{aligned}$$

because $-\vec{\beta}|\mathbf{r}_p| = \beta\mathbf{r}_p$. In the above equation, $\beta = |\vec{\beta}| > 0$. Thus, even though $\mathbf{r}_p - b\hat{\mathbf{e}}_z$ is not perpendicular to $\mathbf{r}_q - b\hat{\mathbf{e}}_z$ its extension, $\mathbf{r}'_p - b\hat{\mathbf{e}}_z$ (caused by its motion) could be, as Figure C.2 shows. ■

The components of the vector $\mathbf{s} \equiv \mathbf{r}_2(t_{p'}) - \mathbf{r}_1(t_{q'})$ are the coordinates of the image of P in the plane. I let you show that

$$\mathbf{s} = \frac{d|\mathbf{r}_q - b\hat{\mathbf{e}}_z|}{(\mathbf{r}_p - b\hat{\mathbf{e}}_z)\cdot(\mathbf{r}_q - b\hat{\mathbf{e}}_z)}(\mathbf{r}_p - b\hat{\mathbf{e}}_z) - \frac{d}{|\mathbf{r}_q - b\hat{\mathbf{e}}_z|}(\mathbf{r}_q - b\hat{\mathbf{e}}_z). \tag{C.7}$$

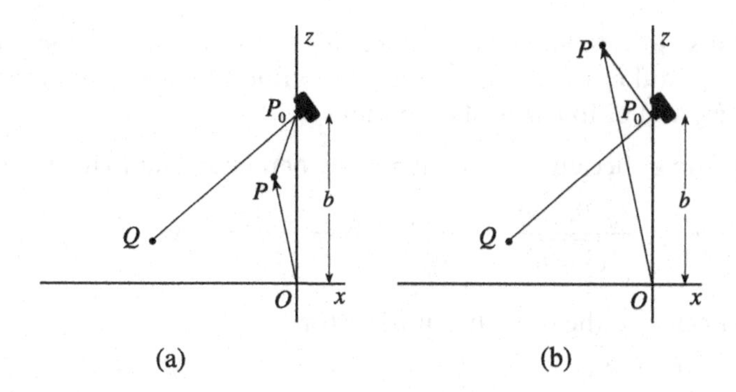

FIGURE C.2 (a) Although $\overline{P_0P}$ is not parallel to the photographic plate, (b) the extension of $\mathbf{r}_p$ by a factor of $\sqrt{(1+\beta)/(1-\beta)}$—as elaborated in Remark C.1.1—makes $\overline{P_0P}$ parallel to the plate when P is in motion.

The most interesting case is when Q is on the x-axis at $(x_q, 0, 0)$. Then, (C.2) and (C.6) give

$$t_{q'} = 1 - \frac{d}{\sqrt{x_q^2 + b^2}}$$

$$t_{p'} = 1 - \frac{d\sqrt{x_q^2 + b^2}}{x_q x - bz + b^2}, \tag{C.8}$$

and Equation (C.7) yields

$$s_x = \frac{dx\sqrt{x_q^2 + b^2}}{x_q x - bz + b^2} - \frac{dx_q}{\sqrt{x_q^2 + b^2}}$$

$$s_y = \frac{dy\sqrt{x_q^2 + b^2}}{x_q x - bz + b^2} \tag{C.9}$$

$$s_z = \frac{(z-b)d\sqrt{x_q^2 + b^2}}{x_q x - bz + b^2} + \frac{db}{\sqrt{x_q^2 + b^2}},$$

where (x, y, z) are the coordinates of P.

The vector $\mathbf{s}$ locates the image of the point P on the photographic plate. If I choose some (two-dimensional) coordinate system in that plane, I can designate the image of P by two numbers, its coordinates. Let $\hat{\mathbf{e}}_1$ and $\hat{\mathbf{e}}_2$ be two orthogonal unit vectors in the plane of the plate along the two desired axes, and let $\hat{\mathbf{e}}_3$ be perpendicular to the plane. It should be clear that $\mathbf{s} \cdot \hat{\mathbf{e}}_1$ and $\mathbf{s} \cdot \hat{\mathbf{e}}_2$ give the coordinates of the point P in the plane of the plate.[1] It should also

[1] The dot product of a vector with a unit vector is the projection of the vector along the unit vector.

be clear that $\mathbf{s} \cdot \hat{\mathbf{e}}_3 = 0$. Since I have the components of $\mathbf{s}$ along the three unit vectors $\hat{\mathbf{e}}_x$, $\hat{\mathbf{e}}_y$, and $\hat{\mathbf{e}}_z$ in Equation (C.9), I can find the new components once I know $\hat{\mathbf{e}}_1$, $\hat{\mathbf{e}}_2$, and $\hat{\mathbf{e}}_3$ in terms of $\hat{\mathbf{e}}_x$, $\hat{\mathbf{e}}_y$, and $\hat{\mathbf{e}}_z$.

I can easily find $\hat{\mathbf{e}}_3$ because $\mathbf{r}_q - b\hat{\mathbf{e}}_z$ is already perpendicular to the plane. Thus,

$$\hat{\mathbf{e}}_3 = \frac{\mathbf{r}_q - b\hat{\mathbf{e}}_z}{|\mathbf{r}_q - b\hat{\mathbf{e}}_z|} = \frac{\langle x_q, 0, -b\rangle}{\sqrt{x_q^2 + b^2}} = \frac{x_q}{\sqrt{x_q^2 + b^2}}\hat{\mathbf{e}}_x - \frac{b}{\sqrt{x_q^2 + b^2}}\hat{\mathbf{e}}_z.$$

From this, I can find the other two unit vectors. Let[2]

$$\hat{\mathbf{e}}_1 = \alpha_1\hat{\mathbf{e}}_x + \beta_1\hat{\mathbf{e}}_y + \gamma_1\hat{\mathbf{e}}_z$$
$$\hat{\mathbf{e}}_2 = \alpha_2\hat{\mathbf{e}}_x + \beta_2\hat{\mathbf{e}}_y + \gamma_2\hat{\mathbf{e}}_z.$$

Since these are perpendicular to $\hat{\mathbf{e}}_3$, you get

$$0 = \hat{\mathbf{e}}_3 \cdot \hat{\mathbf{e}}_1 = \frac{\alpha_1 x_q}{\sqrt{x_q^2 + b^2}} - \frac{\gamma_1 b}{\sqrt{x_q^2 + b^2}}$$
$$0 = \hat{\mathbf{e}}_3 \cdot \hat{\mathbf{e}}_2 = \frac{\alpha_2 x_q}{\sqrt{x_q^2 + b^2}} - \frac{\gamma_2 b}{\sqrt{x_q^2 + b^2}},$$

or $\gamma_1 = \alpha_1 x_q/b$ and $\gamma_2 = \alpha_2 x_q/b$. Furthermore, since $\hat{\mathbf{e}}_1 \cdot \hat{\mathbf{e}}_2 = 0$, you should also get

$$\alpha_1\alpha_2 + \beta_1\beta_2 + \gamma_1\gamma_2 = 0$$

or

$$\alpha_1\alpha_2\left(1 + \frac{x_q}{b}\right) + \beta_1\beta_2 = 0.$$

There are infinitely many solutions to this equation, corresponding to the infinitely many choices for the direction of the axes in the plane of the photographic plate. One convenient choice is $\beta_1 = 0 = \alpha_2$. Then, since $\hat{\mathbf{e}}_1$ and $\hat{\mathbf{e}}_2$ have unit length, $\beta_2 = 1$ and

$$1 = \alpha_1^2 + \gamma_1^2 = \alpha_1^2(1 + x_q^2/b^2) \quad \text{or} \quad \alpha_1 = \frac{b}{\sqrt{x_q^2 + b^2}}, \quad \gamma_1 = \frac{x_q}{\sqrt{x_q^2 + b^2}},$$

and therefore,

$$\hat{\mathbf{e}}_1 = \frac{b}{\sqrt{x_q^2 + b^2}}\hat{\mathbf{e}}_x + \frac{x_q}{\sqrt{x_q^2 + b^2}}\hat{\mathbf{e}}_z$$
$$\hat{\mathbf{e}}_2 = \hat{\mathbf{e}}_y$$
$$\hat{\mathbf{e}}_3 = \frac{x_q}{\sqrt{x_q^2 + b^2}}\hat{\mathbf{e}}_x - \frac{b}{\sqrt{x_q^2 + b^2}}\hat{\mathbf{e}}_z.$$

---

[2] I'm sorry, but I'm running out of symbols! I usually write components as Greek letters. But I have used all of the more common ones already. So, bear with me and don't confuse the letters in these equations with the ones I used in the text!

You can now calculate the new components of **s**:

$$
\begin{aligned}
s_1 &\equiv \mathbf{s}\cdot\hat{\mathbf{e}}_1 = \frac{bs_x + x_q s_z}{\sqrt{x_q^2+b^2}} = \frac{d(bx - bx_q + x_q z)}{x_q x - bz + b^2} \\
s_2 &\equiv \mathbf{s}\cdot\hat{\mathbf{e}}_2 = s_y = \frac{dy\sqrt{x_q^2+b^2}}{x_q x - bz + b^2} \\
s_3 &\equiv \mathbf{s}\cdot\hat{\mathbf{e}}_3 = \frac{x_q s_x - bs_z}{\sqrt{x_q^2+b^2}} = 0.
\end{aligned}
\tag{C.10}
$$

Divide all components by d and introduce the dimensionless coordinates

$$
\begin{aligned}
u &\equiv \frac{s_1}{d} = \frac{bx + x_q(z-b)}{x_q x - b(z-b)} \\
v &\equiv \frac{s_2}{d} = \frac{y\sqrt{x_q^2+b^2}}{x_q x - b(z-b)}
\end{aligned}
\tag{C.11}
$$

which map a point P with coordinates (x, y, z) in the reference frame O to a point (u, v) in a plane parallel to the photographic plate of Camera C.

To find the coordinates of the image of P in the plane parallel to the photographic plate of C', substitute the Lorentz-transformed coordinates (x', y', z') of P in Equation (C.11). It is convenient (though not necessary) to substitute the coordinates $(x'_q, 0, 0)$ of Q in (C.11) as well, so that both focus on the *same physical point.*

How do you find x'_q? Consider the event E_0 when the point $(x_q, 0, 0)$ in O passes the origin of O'. The time of E_0 according to O is given by $t_0 = |x_q|/\beta = -x_q/\beta$, because $x_q < 0$. The Lorentz transform of t_0 is

$$t'_0 = \gamma(t_0 + \beta x_q) = \gamma(t_0 - \beta^2 t_0) = \frac{t_0}{\gamma}.$$

This is the time at which Q reaches the origin of O' according to O'. Therefore, $x'_q \equiv -\beta t'_0 = x_q/\gamma$, and (C.11) becomes

$$
\begin{aligned}
u' &= \frac{bx' + x'_q(z-b)}{x'_q x' - b(z-b)} = \frac{b\gamma\left[x - \beta\sqrt{x^2+y^2+(z-b)^2}\right] + x_q(z-b)/\gamma}{x_q\left[x - \beta\sqrt{x^2+y^2+(z-b)^2}\right] - b(z-b)} \\
v' &= \frac{y\sqrt{x_q'^2+b^2}}{x'_q x' - b(z-b)} = \frac{y\sqrt{x_q^2/\gamma^2+b^2}}{x_q\left[x - \beta\sqrt{x^2+y^2+(z-b)^2}\right] - b(z-b)}.
\end{aligned}
\tag{C.12}
$$

There is another way of finding x'_q. This time place a firecracker at Q and let it explode at the same time that the two origins coincide. The explosion has spacetime coordinates $(x_q, 0)$ in O. Problem C.7 asks you to start with this event and go through some calculation to find x'_q.

You have seen two different ways of calculating x'_q, one above and the other in Problem C.7, which you should go through. In both cases x'_q turns out to be x_q/γ, i.e., the contracted length of $\overline{OQ}$. Is there, therefore, an easier way of finding the same thing using length contraction formula? Remember that for O' to measure a contracted length, the "events" at the two ends of the length have to occur at the same time *according to* O'. Camera C' takes the picture at $t' = 0$. Where is Q *at that time*? Place two firecrackers at O and Q. Let them explode simultaneously according to O', each one leaving a mark in the RF of O'. The location of Q is simply the distance measured between the mark at the origin and the mark left by Q. This is indeed $\overline{OQ}/\gamma$ as explained in Section 1.3.

When the camera aims at the origin, $u' \to x'$ and $v' \to y'$.

It is useful to note that

Note C.1.1. *When the camera aims at the origin (i.e., when $x'_q = 0 = x_q$), u' and v' are, respectively, proportional to x' and y' with the same proportionality constant.*

This can also be seen from Figure C.1, which shows that the uv-plane is parallel to the xy-plane in this particular case.

C.2 EQUATION OF THE CONTOUR

In this section we are going to find the equation of the contour of the region on the surface of a smooth solid object whose photons form the image of the object. The contour is found by drawing straight lines from the pinhole tangent to the surface.

Let the surface of the object be parametrized by the two variables ξ and η, so that

$$x = f(\xi, \eta), \quad y = g(\xi, \eta), \quad z = h(\xi, \eta) \tag{C.13}$$

gives the equation of the surface. Let P be a point on the surface of Equation (C.13) with parameters (ξ, η). Let P' be another point on the surface close to P with parameters $(\xi + \Delta\xi, \eta)$. The vector connecting P to P' has components

$$\Delta x = f(\xi + \Delta\xi, \eta) - f(\xi, \eta) \approx \frac{\partial f}{\partial \xi}\Delta\xi$$

$$\Delta y = g(\xi + \Delta\xi, \eta) - g(\xi, \eta) \approx \frac{\partial g}{\partial \xi}\Delta\xi$$

$$\Delta z = h(\xi + \Delta\xi, \eta) - h(\xi, \eta) \approx \frac{\partial h}{\partial \xi}\Delta\xi.$$

In the limit that $\Delta\xi \to 0$, this vector, i.e., the vector

$$\left\langle \frac{\partial f}{\partial \xi}, \frac{\partial g}{\partial \xi}, \frac{\partial h}{\partial \xi} \right\rangle \Delta\xi \equiv \left\langle \frac{\partial x}{\partial \xi}, \frac{\partial y}{\partial \xi}, \frac{\partial z}{\partial \xi} \right\rangle \Delta\xi,$$

and therefore the vector

$$\vec{\xi} \equiv \left\langle \frac{\partial x}{\partial \xi}, \frac{\partial y}{\partial \xi}, \frac{\partial z}{\partial \xi} \right\rangle, \tag{C.14}$$

is tangent to the surface obtained by varying ξ. Similarly,

$$\vec{\eta} \equiv \left\langle \frac{\partial x}{\partial \eta}, \frac{\partial y}{\partial \eta}, \frac{\partial z}{\partial \eta} \right\rangle \tag{C.15}$$

is tangent to the surface obtained by varying η. Therefore, $\vec{\xi} \times \vec{\eta}$ is normal to the tangent plane at (x, y, z). Let $\mathbf{r} = \langle x, y, z \rangle$ be the position vector of a point on the surface and $\mathbf{r}_0 = \langle 0, 0, b \rangle$ the position vector of the pinhole. Then the condition for the line along $\mathbf{r} - \mathbf{r}_0$ to be tangent to the surface is

$$(\mathbf{r} - \mathbf{r}_0) \cdot (\vec{\xi} \times \vec{\eta}) = 0 \tag{C.16}$$

or

$$x(\vec{\xi} \times \vec{\eta})_x + y(\vec{\xi} \times \vec{\eta})_y + (z - b)(\vec{\xi} \times \vec{\eta})_z = 0$$

or

$$(\vec{\xi} \times \vec{\eta})_x f(\xi, \eta) + (\vec{\xi} \times \vec{\eta})_y g(\xi, \eta) + (\vec{\xi} \times \vec{\eta})_z [h(\xi, \eta) - b] = 0. \tag{C.17}$$

This equation gives one of the parameters in terms of the other, which we can then insert in the parametric equation of the surface (C.13) to obtain the parametric equation of the curve bounding the region of the object exposed to the camera.

C.3 PROBLEMS

C.1. Verify that Equation (C.1) is a line passing through Q and the pinhole of the camera.

C.2. Derive Equation (C.2).

C.3. Substitute (C.1), (C.2), and (C.4) in (C.5) to show that

$$t_{p'} = 1 - \frac{d|\mathbf{r}_q - b\hat{\mathbf{e}}_z|}{(\mathbf{r}_p - b\hat{\mathbf{e}}_z) \cdot (\mathbf{r}_q - b\hat{\mathbf{e}}_z)}.$$

C.4. Derive Equation (C.7).

C.5. Derive Equations (C.8) and (C.9).

C.6. Derive Equation (C.12) from Equation (C.11).

C.7. In this problem you'll find x'_q in a different way.

(a) If the event of the explosion is at $(x_q, 0)$ in O, what is it in O'?

(b) What distance does Q travel in O' before C' takes the picture of the object?

(c) From (a) and (b) determine where Q is according to O' when C' takes a picture and show that $x'_q = x_q/\gamma$.

Bibliography

[Cameron 63] A.G.W. Cameron, Interstellar Communication, W. A. Benjamin, Inc., New York, 1963.

[Einstein 52] A. Einstein, The Principle of Relativity, Dover, New York, 1956, p. 48.

[Hassani 08] S. Hassani, Mathematical Methods for Students of Physics and Related Fields, 2nd ed., Springer, New York, 2008.

[Hickey 79] F.R. Hickey, Am. J. Phys. 47 (9) (1979) 711–714.

[Lampa 24] M. Lampa, Z. Phys. 72 (1924) 138–148.

[Okun 89] L. Okun, Phys. Today 42 (6) (1989) 31–36.

[Penrose 59] R. Penrose, Proc. Cambridge Philos. Soc. 55 (1959) 137–139.

[Scott 70] G.D. Scott, H.J. van Driel, Am. J. Phys. 38 (9) (1970) 971–9774.

[Terrell 59] J. Terrell, Phys. Rev. 116 (1959) 1041–1045.

[Weisskopf 60] V.F. Weisskopf, Phys. Today 13 (9) (1960) 24–27.

Index

B

C

D

E

T

Printed in the United States
By Bookmasters

Printed in the United States
By Bookmasters